International Conference on Building Materials, Products and Technologies (ICBMPT 2021)

IOP Conference Series: Materials Science and Engineering
Volume 1205

Telc, Czech Republic
29 September – 1 October 2021

Editors:

Martin Bohac
Martina Drdlova
Dana Kubatova
Theodor Stanek

ISBN: 978-1-7138-4664-2
ISSN: 1757-8981

Printed from e-media with permission by:

Curran Associates, Inc.
57 Morehouse Lane
Red Hook, NY 12571

Some format issues inherent in the e-media version may also appear in this print version.

Institute of Physics
Dirac House, Temple Back
Bristol BS1 6BE UK

Phone: 44 1 17 929 7481
Fax: 44 1 17 920 0979

techtracking@iop.org

Additional copies of this publication are available from:

Curran Associates, Inc.
57 Morehouse Lane
Red Hook, NY 12571 USA
Phone: 845-758-0400
Fax: 845-758-2633
Email: curran@proceedings.com
Web: www.proceedings.com

TABLE OF CONTENTS

Preface

All the world as well as our conference faced to the pandemic in 2021. Therefore, the 24th International Conference on Building Materials, Products and Technologies (ICBMPT 2021) had to be postponed from the usual date in June to the new date 29th September to 1st October 2021.
The participants appreciated our choice to organize the conference again in the beautiful town of Telč in the Czech Republic, which is popular with many tourists, because it belongs to the cultural heritage under protection of UNECSCO.
Despite the difficult situation, the organizers from the Research Institute for Building Materials collected 23 interesting papers among research workplaces in the Czech Republic and abroad. The conference brings not only interesting presentations, but also it creates a platform for scientific discussions and proposals of new mutual projects.
The main topics of ICBMPT 2021 were: Inorganic binders, Composite materials, Building materials based on secondary raw materials and Technology and test methods in building industry. Also, several keynote speakers were attracted to contribute to our conference. Besides the conference program, there was a possibility to take part in the specialized workshop Healthy buildings with many interesting topics, among others asbestos treatment and biodeteriogens in building environment.
We would like to thank all the participants for their valuable contribution to the conference. We would like to express our gratitude to the reviewers for their time, valuable recommendations and constructive evaluation, in order to guarantee high quality of all the papers. We are looking forward to the next conference in better times!

Scientific committee:

- doc. Ing. Vít Šmilauer, Ph.D., DSc., Faculty of Civil Engineering, Czech Technical University in Prague
- Ing. Martin Šperl, Ph.D., Institute of Theoretical and Applied Mechanics of the Czech Academy of Science
- doc. Ing. Tomáš Opravil, Ph.D., Faculty of Chemistry, Institute of Materials Science, Brno University of Technology
- doc. Ing. Jiří Zach, Ph.D., Faculty of Civil Engineering, Institute of Technology of Building Materials and Component, Brno University of Technology
- doc. Ing. František Šoukal, Ph.D., Faculty of Chemistry, Institute of Materials Science, Brno University of Technology
- prof. Dr. Ing. Martin Tchingnabé Palou, Faculty of Civil Engineering, Slovak university of technology in Bratislava
- prof. Ing. Zbynek Keršner, CSc., Faculty of Civil Engineering, Institute of Structural Mechanics, Brno University of Technology
- doc. Ing. Stanislav Rolc, CSc,, Military Research Institute, state enterprise
- prof. RNDr. Pavla Rovnaníková, CSc., Faculty of Civil Engineering, Brno University of Technology

Editors:

- Mgr. Martin Boháč, Ph.D.
- Ing. Martina Drdlová
- Ing. Dana Kubátová, Ph.D.
- RNDr. Theodor Staněk, Ph.D.

IOP Conf. Series: Materials Science and Engineering 1205 (2021) 011002 doi:10.1088/1757-899X/1205/1/011002

Peer review declaration

All papers published in this volume of *IOP Conference Series: Materials Science and Engineering* have been peer reviewed through processes administered by the Editors. Reviews were conducted by expert referees to the professional and scientific standards expected of a proceedings journal published by IOP Publishing.

- **Type of peer review:** Double-blind

- **Conference submission management system:**
  There are 4 editors, each for one topic area. According to the abstract, the editors will assess the suitability of the manuscript for the conference. The full manuscript is sent to two/three reviewers. According to the reviewers' reports and the editor's comments, the author makes corrections or additions.

- **Number of submissions received:** 27

- **Number of submissions sent for review:** 24

- **Number of submissions accepted:** 23

- **Acceptance Rate (Number of Submissions Accepted / Number of Submissions Received X 100):** 85

- **Average number of reviews per paper:** 2.3

- **Total number of reviewers involved:** 49

- **Any additional info on review process:**

  The novelty of the article and its impact on the field are the main aspects we consider. We evaluate whether the methodology used is appropriate. We also assess whether the exact technical specifications of the study materials and equipment are described, we consider the adequacy of the sample size and the quality of the figures. We assess whether the findings in the manuscript are aptly supplemented by the supplementary section and whether the authors have followed the submission guidelines. We consider whether the results or the proposed methodology have some potential broader applicability or relevance. We pay attention to the results and whether they have been discussed with other similar published studies.

- **Contact person for queries:**
  Name: Martin Nejedlik
  Affiliation: Reseach Institute for Building Materials, Brno, Czech Republic
  Email: nejedlik@vustah.cz

IOP Conf. Series: Materials Science and Engineering 1205 (2021) 012001 doi:10.1088/1757-899X/1205/1/012001

Effectiveness of common water-reducing admixtures in alkali-activated slag pastes with different types of activator

V Bilek Jr[1], **V Iliushchenko**[1], **P Hruby**[1] **and L Kalina**[1]

[1] Brno University of Technology, Faculty of Chemistry, Institute of Materials Science, Purkynova 464/118, 612 00 Brno, Czech Republic

Email: bilek@fch.vut.cz

Abstract. Rheology of alkali-activated slag (AAS) is a very complex issue, where the activator nature as well as its dose play an important role. Moreover, the use of water-reducing admixtures in these systems is an issue, as they often do not work properly. This could be attributed to the high pH as well as to the surface chemistry of AAS. Therefore, lignosulfonate-, polynaphthalene- and polycarboxylate-based superplasticizers were used to modify AAS pastes with sodium waterglass, hydroxide and carbonate activator. These pastes were tested using a rotational rheometer in an oscillatory shear mode of increasing shear strain to observe the evolution of viscoelastic moduli and to determine the oscillatory stress corresponding to the linear viscoelastic region limit ("yield point") and to the crossover point, where the storage modulus equals the loss modulus ("flow point"). In most cases, the used plasticizers did not improve the rheological properties; the only exception was the lignosulfonate one in sodium hydroxide-activated slag.

1. Introduction

Rheological parameters are the key characteristics of any binder system and strongly affect its further performance. They have to allow appropriate mixing, transport, casting, and compacting of the mixture. Simultaneously, the water content should be as low as possible to achieve higher mechanical properties, lower porosity and higher durability. For this purpose, plenty of water-reducing admixtures or superplasticizers (SPs) have been developed and their increasingly sophisticated development continues in an unprecedented way. Thanks to them, high-performance or even ultra-high-performance concretes can be designed and used.

However, the above-mentioned stands for the Portland cement-based concretes, while the use of conventional SPs and other organic admixtures in alkali-activated materials often fails. This is most often attributed to their decreased stability due to very high pH [1–3], but the surface chemistry and presence of dissolved species are also very important aspects [4, 5].

Usually, the rheological properties of alkali-activated slag (AAS) pastes are determined using rotational tests across a given range of shear rates and fitting the obtained data by rheological models for yield stress fluids [3, 6–8]. However, rheological investigations can also be carried out by oscillatory tests that, unlike rotational tests, enable the testing of the material in a non-destructed state close to the rest via sufficiently low amplitudes facilitating deformations within the so-called viscoelastic region (LVR).

Based on the above, comprehensive research on the relationships between surface chemistry, the composition of activating solution and the behaviour of organic additives in alkali-activated materials is the main topic of the implemented project GA20-26896S. The presented paper is only a small output of the corresponding research and combines its aims with oscillatory measurements that are only rarely used in AAS.

2. Experimental part

2.1. Materials and composition of the pastes

Granulated blast furnace slag ground to Blaine fineness of 400 m^2/kg was used in this study. The slag consisted of about 70% of an amorphous phase, 20% of akermanite, and lower amounts of calcite, quartz, and merwinite. Density of the slag was determined to be 2.88 g/cm^3 using the pycnometer method. The slag was activated by three different activating solutions, namely sodium hydroxide (SH), sodium waterglass (SWG) with silicate modulus adjusted to 1.5, and sodium carbonate (SC). The concentration of Na^+ in all activating solutions was 4.0 mol/dm^3. Their dose was adjusted to keep the volume fraction of slag in the resulting suspension constant at the value of 0.50 to ensure their comparativeness from the rheological point of view. In this case, the described activator dosing corresponds to 4.3% Na_2O with respect to the slag weight regardless of the activator type and water to slag ratio to 0.36, 0.33, and 0.34 for SWG-activated slag (SWGAS), SH-activated slag (SHAS) and SC-activated slag (SCAS), respectively.

These pastes were modified by three different types of commercially available SPs as follows: lignosulfonate-based (LS), polynaphthalene sulfonate-based (PNS), and polycarboxylate ether-based (PCE) one. Their doses were 0.3% of their dry matter to the slag weight regardless of the SP type. They were added to the activating solution and homogenized just prior to the paste's preparation. Volume of the SP was subtracted from the volume of activating solution to keep the total volume of liquid components constant and thus keep the slag volume fraction in the resulting paste of 0.50. The dilution of the activating solution with SP was taken into account for the design of the pastes' composition.

2.2. Rheological experiments

Slag was homogenized with liquid components in a beaker using a hand mixer. Total mixing time took three minutes and comprised of the following stages: During the first 60 s, a low mixing rate was set, after which the highest possible mixing rate was applied for 30 s. During the next 30 s, the paste was homogenized by a spatula and in the last 60 s again at the highest rate. After that, the paste was transferred into the rheometer (DHR-2, TA Instruments).

Crosshatched parallel plate geometry was used for the rheological investigations. The measuring gap of 1 mm was used. Each measurement started five minutes from the first contact of slag with a mixture of liquids. The first step was preshear at the shear rate of 5 s^{-1} for 10 s and after 10 s of rest, the amplitude sweep measurement at the frequency of 10 rad/s started. The strain was logarithmically increased from 10^{-4}% to 500%, 20 points per decade were measured, 2 cycles for each point. At least four measurements of each paste composition were carried out.

2.3. Evaluation of rheological experiments

Linear viscoelasticity region (LVR) was obtained from the storage modulus G' plotted against the oscillation stress as is shown in Figure 1. Usually, LVR is determined via the decrease of G' by 5 or 10% [9]. As the rheological parameters evolve in time, G' is not constant within the range of LVR, but increases during the measurement, which complicates the determination of LVR. Moreover, LVR of these suspensions is very short and therefore close to the limits of the instrument, which can lead to scattering of the results within the LVR. Therefore, several points at the top of the flat region of the observed "peak" were averaged and the last point higher than 95% of the averaged value was considered as the LVR limit. The corresponding oscillation stress was termed as the "yield point" in accordance with [9]. The values of other parameters, namely loss modulus G'', complex modulus G^*, loss factor ($\tan\delta$) as well as strain at the yield point, were obtained.

Besides the yield point, also the oscillation stress corresponding to the modulus crossover, where $G' = G''$ was determined and in accordance with [9] termed as the "flow point", i.e. the stress after which the rheological behaviour changes from rather viscoelastic solid-like to rather viscoelastic liquid-like.

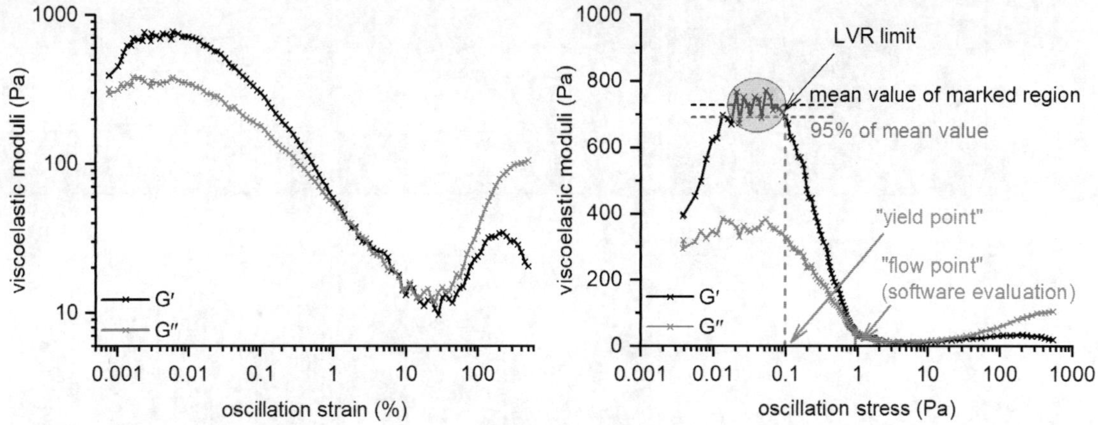

Figure 1. Typical outcome of the amplitude sweep measurement, where viscoelastic moduli are plotted against oscillation strain (left) and the procedure to evaluate important points on the same measurement with moduli plotted against oscillation stress (right).

3. Results and discussion

The average results for all tested combinations of activator and superplasticizer obtained using the procedure described in Section 2.3 are given in Figure 2 and Figure 3 for the yield point and flow point, respectively. The fact that $G' > G''$ at low or intermediate strains indicates that AAS pastes have cross-linked structure originating from attractive forces acting between the pastes' components.

First, it is worth comparing the rheological properties of AAS pastes with respect to the activator type and without the added plasticizer. Figure 2 shows that the rheology of SWGAS paste was noticeably different to that of SH and SC pastes. Its yield point was about 1.5 orders of magnitude lower and the viscoelastic moduli even about two times lower compared to the SHAS and SCAS pastes. Loss factor ($\tan\delta$) of SWG paste was determined to be 0.5, which is somewhat higher compared to the SH and SC pastes having a loss factor around 0.35 in both cases. The values of $\tan\delta$ below 1 indicate that all pastes behave rather as viscoelastic solids with quite narrow LVR, which was determined to be 0.005–0.010%.

If larger amplitudes are applied and therefore the oscillation stress increases, the bonds between particles in the paste start to break, which is manifested by a decrease in viscoelastic moduli and an increase in loss factor, which means that the storage modulus decreases faster than the loss modulus. One example of this situation is given in Figure 1. After reaching a certain strain (or stress, i.e. flow point), for which $G' = G''$, the paste is just in between the viscoelastic solid and viscoelastic liquid behaviour. The corresponding rheological parameters are given in Figure 3. Similarly to the yield point, the flow point is about two orders of magnitude lower for SWG compared to that of SHAS and SCAS, while G' and G'' by "only" one order of magnitude. Unlike the critical strains at the yield point, which were slightly higher for SWG and SC than for SH, the flow point of SWGAS paste was reached at much lower strains (1.5 orders of magnitude) compared to SHAS and SCAS pastes.

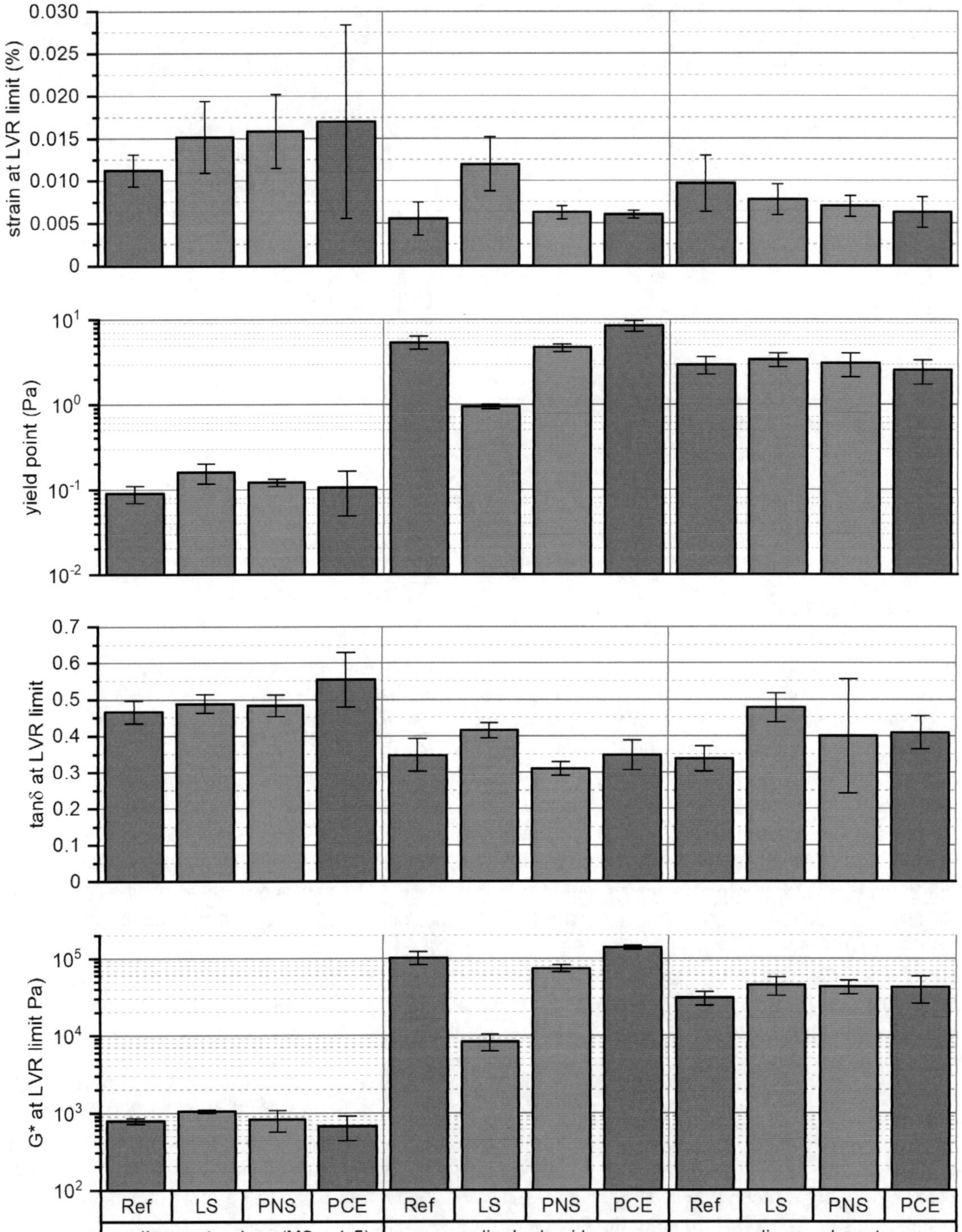

Figure 2. Viscoelastic parameters of alkali-activated slag pastes corresponding to the modulus crossover depending on the nature of the activator and superplasticizer type.

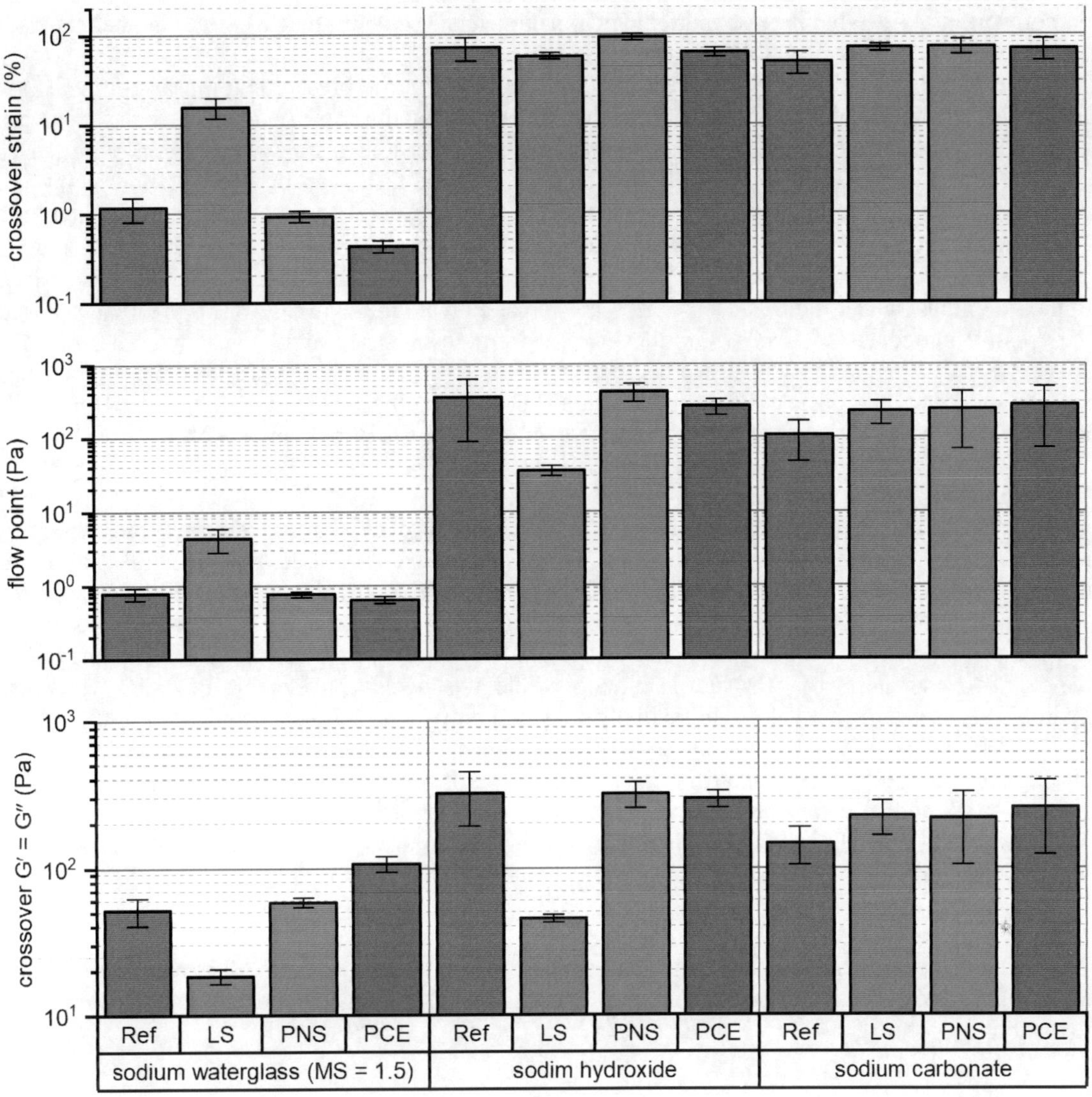

Figure 3. Viscoelastic parameters of alkali-activated slag pastes corresponding to the modulus crossover depending on the nature of the activator and superplasticizer type.

Lower yield point, flow point and viscoelastic moduli, as well as higher tanδ of SWG pastes compared to those activated by SH and SC are related to weaker structure and higher fluidity of SWG pastes. This correlates with the findings in the literature [5, 10, 11]. Beneficial effect of silicates is attributed to their adsorption onto the slag particles, which leads to lower values of zeta potential and increase in the magnitude of repulsive electrical double layer (EDL) forces [5]. The rheological behaviour of SCAS is between SWGAS and SHAS, but closer to SHAS, which suggests that carbonate anions are not able to increase repulsive EDL forces or increase the range of their action enough to noticeably reduce the mechanical properties of the SCAS or in other words increase its fluidity.

The nature of anionic group of activators, the ionic strength of the activating solution and related surface chemistry seems the key aspect affecting the effectiveness of superplasticizers in AAS. Figure 2 and Figure 3 show that PNS and PCE superplasticizers did not significantly affect the rheological behaviour of AAS pastes regardless of the activator nature. In most cases, the obtained values are equal to those of the reference paste without activator or even slightly higher, which indicates that these SPs do not work in AAS. The increase in rheological parameters in the presence of SPs could be related to their undesired interactions with AAS such as bridging forces [4].

The only SP, which showed the great effect on rheology in the present study, was LS in SHAS. In this case, the yield point was reduced more than five times and the corresponding viscoelastic moduli even by an order of magnitude. LS also slightly increased the loss factor, which again points to its effectiveness in SHAS, and a greatly beneficial effect can be also seen in terms of flow point and crossover modulus. The beneficial effect of LS in SHAS corresponds to the findings in the literature [11]. The reason LS did not reduce the yield point, the corresponding moduli and tanδ in SWG and SC pastes is likely that the negatively charged silicate or carbonate anions adsorbed on the surface of the slag grains hinder the LS adsorption. It is interesting that, unlike in SCAS, LS greatly decreased the crossover modulus of SWGAS, while increased crossover strain and flow point. These results suggest that despite the LS is not adsorbed onto the slag grains enough, the presence of its molecules in solution makes the paste's structure less rigid but more deformable prior to flow and therefore interacts with the dissolved species in solution. Nevertheless, this will be a subject of further study.

4. Conclusions

In this paper, the effect of common SPs on AAS rheology depending on the activator type was investigated by means of oscillatory amplitude sweep measurements that are unconventional for these materials so far. The results confirmed that AAS rheology is driven by the nature of both the activator and SP and that the workability improvement of AAS by SPs is a big issue. PNS- and PCE-based SPs showed incompatibility with slag activated by sodium hydroxide, waterglass as well as carbonate. For the last two activators, LS was inefficient too, while it showed considerable potential if sodium hydroxide is used as an activator. This is likely related mainly to the surface chemistry of AAS.

Besides, the paper briefly introduced a novel approach of expressing the activator dose based on the molar concentration of alkali in the activating solution and slag volume fraction in the whole suspension. Further research will be devoted to a wider establishment of this approach, to use oscillatory measurements for the characterisation of alkali-activated materials as well as to their interactions with organic substances.

Acknowledgement

This outcome was achieved with financial support within the project: GA20-26896S "Towards the organic admixtures tailoring for alkali-activated slag-based systems" founded by the Czech science foundation.

References

[1] Palacios M and Puertas F 2004 Stability of superplasticizer and shrinkage-reducing admixtures Stability of superplasticizer and shrinkage-reducing admixtures in high basic media *Mater. Construcc.* **54**

[2] Palacios M and Puertas F 2005 Effect of superplasticizer and shrinkage-reducing admixtures on alkali-activated slag pastes and mortars *Cem. Concr. Res.* **35** pp 1358–67

[3] Palacios M, Banfill P F G and Puertas F 2008 Rheology and Setting of Alkali-Activated Slag Pastes and Mortars: Effect of Organic Admixture *ACI Mater. J.* **105**

[4] Kashani A, Provis J L, Xu J, Kilcullen A R, Qiao G G and van Deventer J S J 2014 Effect of molecular architecture of polycarboxylate ethers on plasticizing performance in alkali-activated slag paste *J. Mater. Sci.* **49** pp 2761–72

[5] Kashani A, Provis J L, Qiao G G and van Deventer J S J 2014 The interrelationship between surface chemistry and rheology in alkali activated slag paste *Constr. Build. Mater.* **65** pp 583–91

[6] Mehdizadeh H and Najafi Kani E 2018 Rheology and apparent activation energy of alkali activated phosphorous slag *Constr. Build. Mater.* **171** pp 197–204

IOP Conf. Series: Materials Science and Engineering 1205 (2021) 012001 doi:10.1088/1757-899X/1205/1/012001

[7] Puertas F, Varga C and Alonso M M 2014 Rheology of alkali-activated slag pastes. Effect of the nature and concentration of the activating solution *Cem. Concr. Compos.* **53** pp 279–88
[8] Yang T, Zhu H, Zhang Z, Gao X, Zhang C and Wu Q 2018 Effect of fly ash microsphere on the rheology and microstructure of alkali-activated fly ash/slag pastes *Cem. Concr. Res.* **109** pp 198–207
[9] Mezger T G 2014 *Applied Rheology: With Joe Flow on Rheology Road* (Anton Paar GmbH)
[10] Aydın S and Baradan B 2014 Effect of activator type and content on properties of alkali-activated slag mortars *Compos. Part B-Eng.* **57** pp 166–72
[11] Shi C, Krivenko P V and Roy D 2006 *Alkali Activated Cements and Concretes* (CRC Press)

IOP Conf. Series: Materials Science and Engineering 1205 (2021) 012002 doi:10.1088/1757-899X/1205/1/012002

The use of limestone sludge for the geopolymer preparation

D Kubatova[1], I Khongova[1], M Krejci Kotlanova[1], A Zezulova[1] and M Bohac[1]

[1] Research Institute for Building Materials, Hněvkovského 30/65, 617 00 Brno, Czech Republic

Email: kubatova@vustah.cz

Abstract. The study investigates the effect of sludge waste from mining and washing of limestone on the properties of geopolymer binders based on metakaolin and blast furnace slag. The effect of adding two types of limestone sludge on workability, setting time, compressive strengths, durability and pore distribution was investigated. Limestone sludge was dosed in an amount of 0–30 %. Sodium water glass was used as an alkaline activator in the ratio of 5:4. A linear correlation between the flowability and limestone sludge content was observed, and very satisfactory results of compressive strength after 90 days were obtained for samples with sodium water glass (about 55 MPa). The positive effect of $CaCO_3$ on strength was confirmed. The increased content of clay components resulted in a reduction of compressive strength.

1. Introduction

The effect of the addition of waste sludge from limestone washing on the properties of geopolymers based on metakaolin and blast furnace slag was studied in this work. This waste is generated during the production of limestone, and it is landfilled at a sludge pond. The main component of this waste is calcite, so this raw material has no potential for use in alkali-activated materials at first sight. Geopolymers are a result of a reaction between aluminosilicate compounds and alkaline solutions. The main materials used in alkali-activated binders are kaolinitic clays, fly ash and blast furnace slag. During the reaction of the raw materials to the alkaline environment, dissolution occurs. After amorphisation and the formation of precursors, the geopolymer precipitates in an amorphous or semi-crystalline state [1]. There is an effort to reduce costs in the production of these materials. Therefore, alternative waste products are being sought, reducing the price and simultaneously maintaining or improving the final product properties.

In previous works, the beneficial effects of calcite on the properties of alkali-activated materials have been confirmed [2]. The addition of calcite and dolomite on the mechanical properties of metakaolin geopolymers is described in the study of Yip et al. (2008) [3], and improvement of strengths by the addition of 20% of calcite was confirmed. Qian (2015) [4] suggested that adding 10% to the metakaolin matrix resulted in a more compact structure. In Aboulayt´s work (2017) [5], the effect of the replacement of metakaolin by calcium carbonate was examined. Sodium and potassium hydroxides and alkaline solution were used as the alkaline-activator. The evolution of mechanical strengths vs the calcite contents was tested, and samples with calcite in the amount of 5–10% exhibited higher compressive strengths than the reference sample.

Cwirzen et al. (2014) [6] studied the effect of limestone on metakaolin-based geopolymers activated by sodium hydroxide solution. The study of Gao (2015) [7] deals with the influence of limestone powder (LS) on fresh mixture behaviour and mechanical properties of alkali-activated slag-fly ash blends in which the positive effect of limestone on workability and compressive strength was confirmed. Yuan

(2017) [8] studied the impact of LP on the reaction kinetics, workability, and compressive strength development of sodium carbonate activated slag. An optimal dosage of 10% of LP was determined to reaction kinetics and strength development.

This work aimed to verify the possibility of using dry waste limestone in a geopolymer mixture based on metakaolin and blast furnace slag. Two types of sludge with different compositions of 0–30% were used for the tests.

2. Experimental

2.1. Materials

Metakaolin (MK), blast furnace slag (BFS) and two types of limestone waste sludge (LS) were used as raw materials. MK Mefisto L05 was supplied by České lupkové závody, a.s. (Czech Republic), finely ground BFS originated from Třinec (Czech Republic). LS comes from the production of Vitošov limestone (Czech Republic). Waste sludge is generated from washing during limestone mining and processing. Sludge with washing water is drained to a sludge pond, where it sediments. The settled sludge is not used in any way, and its volume in the landfill is growing. Two types of samples were taken for the experiments, which differ in their composition. The samples were dried at 60°C and ground. Sodium water glass was used as the alkaline-activator (Ms = 1.65).

Chemical analysis of used materials is given in table 1. Characteristic particle diameters D_{10}, D_{50} and D_{90} are summarised in table 2.

Table 1. Chemical composition of materials used in mixtures (%).

	SiO_2	Al_2O_3	CaO	Na_2O	K_2O	MgO	Fe_2O_3	SO_3	TiO_2	LOI
LS1	19.60	3.11	41.50	0.09	0.46	0.26	1.40	0.01	0.24	33.25
LS2	20.32	9.04	34.37	0.18	1.2	0.67	3.16	0.02	0.46	30.23
BSF	53.21	42.94	0.20	0	0.78	0.28	0.92	1.56	0.05	2.10
MK	39.99	8.99	38.70	0.47	0.48	8.71	0.49	0.34	1.84	0.71

Table 2. Particle diameters of materials (μm).

	D_{10}	D_{50}	D_{90}	Mean diameter
LS1	1.43	14.74	54.11	22.07
LS2	2.26	17.14	49.29	22.13
BSF	1.66	10.73	28.63	13.38
MK	1.21	4.65	11.00	5.52

The XRD analysis of these raw materials was performed. 83% of the amorphous phase was identified in MK. MK also contained quartz (SiO_2), kaolinite ($Al_2Si_2O_5(OH)_4$), illite ($KAl_2(AlSi_3O_{10})(OH)_2$), anatase ($TiO_2$) and mullite ($Al_6Si_2O_{13}$). BFS contained calcite ($CaCO_3$), merwinite ($Ca_3MgSi_2O_8$) and melilite ($Ca_4Al_2MgSi_3O_{14}$). In addition, quartz (SiO_2) and a significant amount of the amorphous phase was present. The main component of LS1 was limestone ($CaCO_3$), quartz (SiO_2), albite ($NaAlSi_3O_8$), anorthite ($CaAl_2Si_2O_8$) illite ($KAl_2(AlSi_3O_{10})(OH)_2$) or muscovite ($KAl_3Si_3O_{12}H_2$). More clay components were identified in LS2. Kaolinite ($Al_2Si_2O_5(OH)_4$) and feldspar group minerals were identified in the second limestone sludge LS2. The detailed phase composition of limestone sludges is given in table 3. The specific surface area was evaluated according to the Blaine method 230 $m^2 \cdot kg^{-1}$ for LS1 and 436 $m^2 \cdot kg^{-1}$ for LS2.

Table 3. Mineralogical composition of limestone sludges.

%	LS1	LS2
Calcite	81.6	70.6
Quartz	13.5	7.2
Illite/muscovite	3.1	8.3
Kaolinite	-	4.8
Feldspar	1.8	9.1

2.2. Sample preparation
The weight ratio between MK and BFS was 1.25. MK and BFS were partially replaced by limestone sludge at 0–30%. Dry mixtures were homogenised and then mixed with sodium water glass in a mechanical mixer. Alkaline-solution had a molar SiO_2/Na_2O ratio of 1.65. The solids to solution mass ratio used was 1.25. Quartz sand was used for mortars preparation in a ratio of 1:1. Specimens 40 × 40 × 160 mm were stored at 20°C for 24 h in sealed moulds. After 24 h, hardened specimens were demoulded and stored in plastic bags under laboratory conditions (20°C) for 90 days.

2.3. Methods
The chemical composition of all raw materials was determined by XRF analysis (PANalytical Axios PW4400/30).

The particle size distribution was performed by the laser diffraction method on the CILAS 920L apparatus. The measuring range of the granulometry is 0.3–400 μm, and the measuring medium is isopropyl alcohol. The sample was sonicated for 60 seconds before measurement.

The composition of raw materials and final products was determined by X-ray powder diffraction. The X-ray diffraction analysis was conducted using Bruker D8 Advance apparatus with Cu anode ($\lambda K\alpha = 1.54184$ Å) and variable divergence slits at Θ-Θ Bragg-Brentano reflective geometry.

The extent of dissolution in an alkaline medium was studied for LS. A specified mass of LS was mixed with 2, 5 and 10 M NaOH solution at room temperature for 24 h using a magnetic stirrer. The samples were filtered (0.2 μm), and the clear liquid part was diluted by concentrated HCl. ICP-OES was used to analyse the concentration of Si, Al and Ca. The residue after filtration was analysed by X-ray powder diffraction.

The workability of the pastes was evaluated by the flow table test according to EN 1015-3 [9]. After mixing, the samples were applied into a conical ring with a diameter of 7 cm and a free flow without applying strokes was allowed.

The Vicat needle method was used to determine the initial and final setting time (EN 196-3) [10] of pastes.

The specimens (40 × 40 × 160 mm) were tested according to EN 196-1 [11], and compressive and flexural strength were evaluated at the ages 28, 56 and 90 days.

Total porosity of prepared material was determined by mercury porosimetry (MIP) performed on the apparatus Carlo Erba Strumentazione model 200, measuring a range of 3.75–10,000 nm. Various levels of pressure were applied to a sample immersed in mercury. The samples were tested after 90 days.

The freeze-thaw test was performed according to ČSN 72 2452 [12] on water-saturated specimens (40 × 40 × 160 mm), repeatedly subjected to freezing and thawing. The freezing cycle lasts four hours at -20°C. Thawing takes place in a water bath for 2 hours at 15°C. The flexural strength was determined on the specimens, and then the compressive strength was determined on the fractions of specimens. Comparisons were made with specimens aged 56 days after 25 cycles and 50 cycles. Two-month-old samples were used for the test.

3. Results and Discussion

The extent of dissolution in alkaline media was monitored in limestone samples. The content of Al, Si, Ca was determined in the filtrate after dissolving the initial materials in sodium hydroxide.

Limestone sludge is not an essential source of Si and Al due to its composition, but at the same time, it cannot be considered only as a filler that does not participate in the geopolymer structure [3]. The graphs (figure 1 a, b, c) show an increase in the ion content depending on the concentration of the sodium hydroxide solution [13]. LS1 decomposed more significantly only when it was exposed to 10 M NaOH. A more significant increase Ca, Al and Si are evident, especially in LS2. The concentration of Si in the filtrate may be higher due to the greater content in the material or due to the higher intrinsic extent of dissolution compared to Al [13]. As the X-ray diffraction results confirm (See figure 2), the kaolinite peaks disappeared, and the calcite peak decreased in samples treated with NaOH solution. Portlandite ($CaOH_2$) was identified in a decomposed sample with 2 M NaOH solution, and portlandite content increased with NaOH solution concentration. The LS2 sample disintegrated more significantly under the action of NaOH solution due to higher surface area and a larger ratio of clay and feldspar minerals.

The slump flow of the fresh alkali-activated pastes is presented in figure 1d. The data show that the addition of LS1 slightly increases workability. Slump flow increased from 250 mm to 290 mm when 30% of limestone was used due to relatively low water demand of limestone particles compared to slag and metakaolin [7]. That is consistent with the results of Yuan (2017) [8], who has observed improvement in the workability of pastes in BSF and PC-based materials after incorporating LS. On the contrary, the LS2 decreased workability from 250 mm to 141 mm. Probably due to the kaolinite content and its plate-like structure, which resulted in inter-particle friction [14].

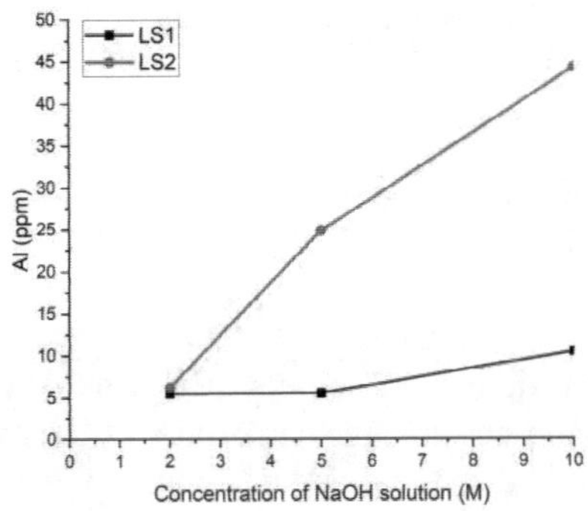

a)

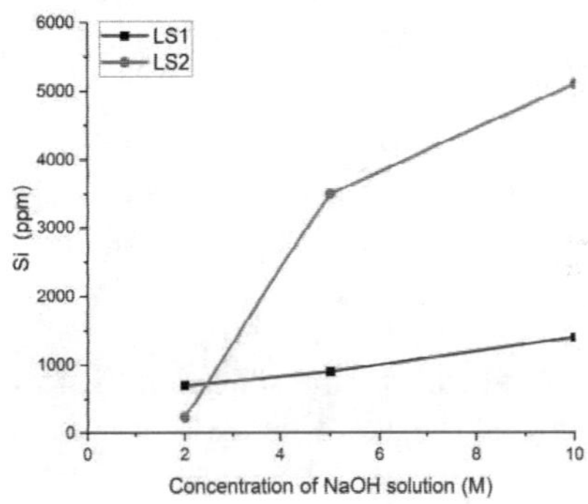

b)

Figure 1. The extent of dissolution of LS in NaOH solution: Al content in the filtrate (a), Si content in the filtrate (b),

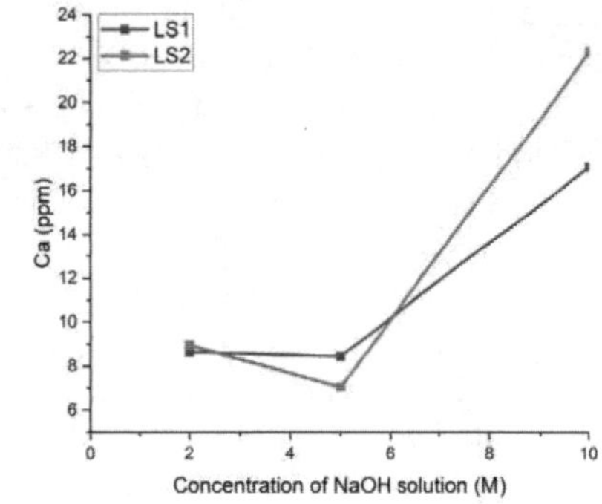

c)

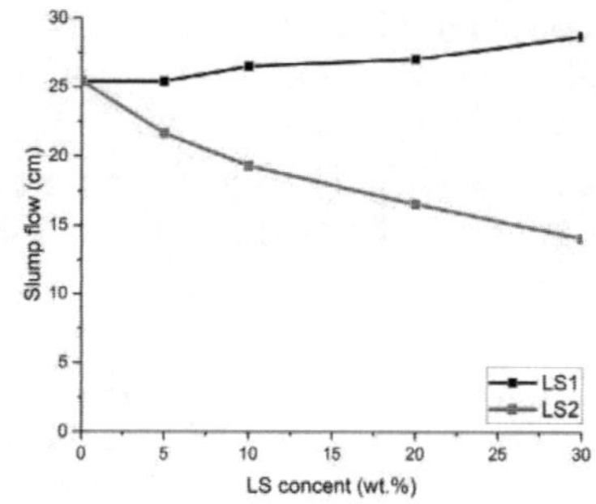

d)

Ca content in the filtrate (c), results of flow table test (d).

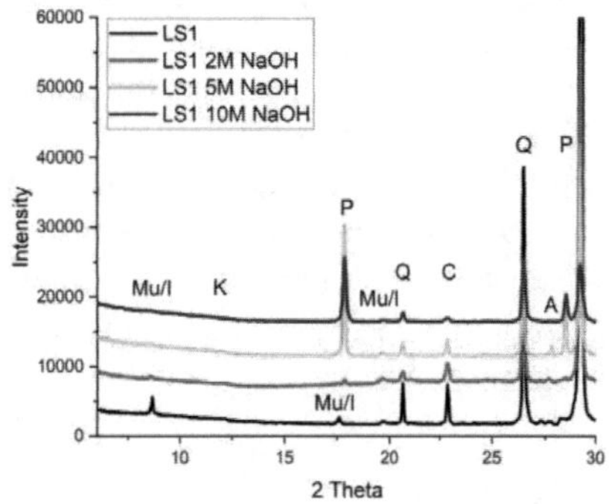

a)

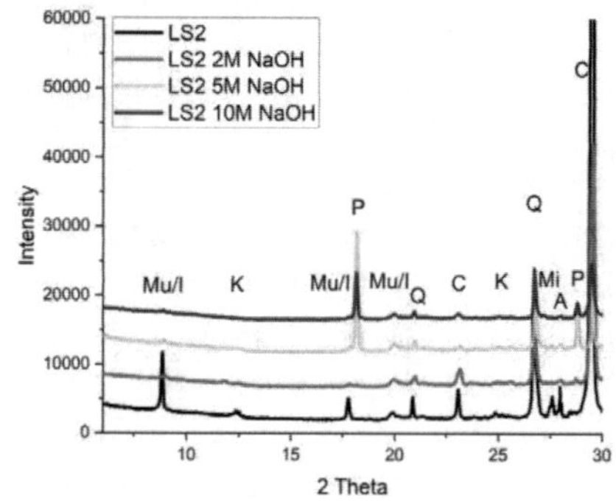

b)

Figure 2. XRD diffractograms of samples of LS after dissolution in NaOH solution a) LS1 b) LS2 (C-calcite, P-portlandite, Q-quartz, A-albite, Mi-microcline, K-kaolinite, Mu/I muscovite or illite).

The effect of LS on the setting of the mixture is shown in figure 3. The reference sample based on metakaolin and slag had an initial setting time of about 260 minutes, and the end of the setting occurred 10 minutes later. As the LS content in the mixture increased, the initial setting time increased up to 320 min (30LS1), and the final setting time was increased up to 360 minutes. Longer setting times were observed by using LS2. 5% LS2 prolonged the setting time by 60 minutes. The higher content of LS2 in the mixture had only a slight effect on the setting. The final setting time was observed up to 380 minutes in the sample 30LS2. The main reason for prolonging the setting time is the reduction of the slag content in the mixtures. Gao (2015) [7] studied geopolymer mixtures based on fly ash, blast furnace slag and calcium carbonate and found the effect of limestone on setting time as negligible. In another work [3], a small content of finely ground $CaCO_3$ in the metakaolin matrix increased the setting rate.

The setting time and polymerisation process of LS2 samples could be partly influenced by clay and feldspar minerals, which are present in LS2 in an amount of about 22%. Xu and Deventer (2000) [13] dealt with the geopolymerisation properties of various natural Al-Si minerals, including kaolinite and anorthite. They confirmed the complexity of the reaction mechanisms involved in dissolution, setting and formation of the gel structure, and it is impossible to predict their behaviour effectively.

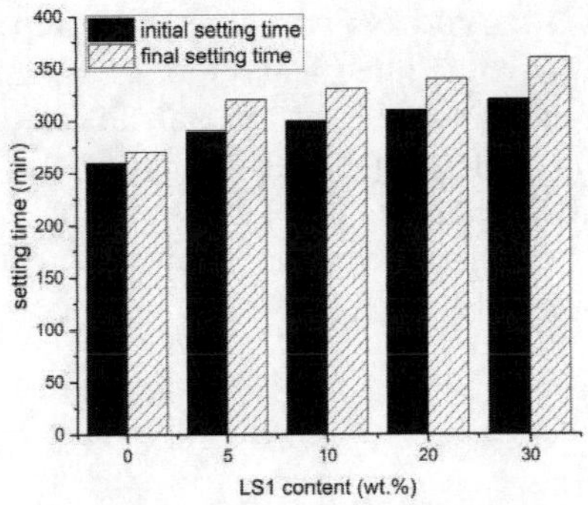

a)

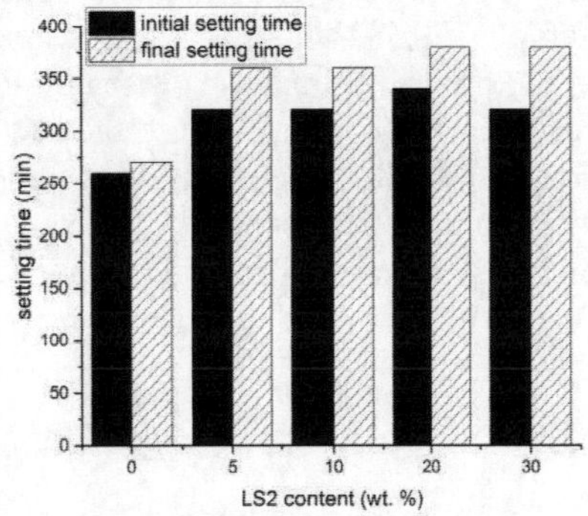

b)

Figure 3. Setting time depending on LS content in the matrix a) LS1 b) LS2.

The samples of the paste after 28, 56 and 90 days were ground to analyse phase composition by X-ray diffraction. The XRD pattern of all samples had a broad hump in the range of 26-29° (2 Θ).

Calcite ($CaCO_3$), quartz (SiO_2), akermanite ($Ca_2MgSi_2O_7$), merwinite ($Ca_3MgSi_2O_8$) and anatase (TiO_2) were identified in the reference sample after 28 days. No significant changes were seen on the XRD patterns after 90 days. Illite ($KAl_2(AlSi_3O_{10})(OH)_2$) or muscovite ($KAl_3Si_3O_{12}H_2$) were identified in all samples with LS. Kaolinite ($Al_2Si_2O_5(OH)_4$) peaks are not noticeable after 28 days already, which is consistent with other works [10]. Feldspar mineral, albite ($NaAlSi_3O_8$) and anorthite ($CaAl_2Si_2O_8$), which are components of the LS sludge, were identified in samples with more than 5% LS.

The replacement of calcite does not result in the formation of new phases. The calcite peaks increase in intensity with increasing LP content. The XRD results show that although calcite dissolves in an alkaline solution, it does not actively form the new crystalline phase. Yuan (2017) [8] reported that Ca^{2+} can increase the release of Al and Si ions from metakaolin, and then carboaluminates are formed in the metakaolin-based system. In the slag-based system, the effect of Ca has not been confirmed, and LP acts as a filler. Ca^{2+} ions released from slag can affect the dissolution of Ca^{2+} from LP. The samples were analysed even after 90 days and, no visible changes in structure were found.

Flexural strengths and compressive strengths were tested on mortar specimens after 28, 56 and 90 days. The reference sample with 0% of LS had 6.5 MPa flexural strength and 55 MPa compressive strength after 28 days. The flexural strength increased to 8.5 MPa and the compressive strength to 63 MPa, after 90 days.

Comparable results with the reference specimen were obtained with sample 5LS1 after 90 days. The higher LS1 content resulted in a slight decrease in flexural strength. The amount of LS2 in the mixture had no significant effect on the flexural strength, which in all samples exceeded 6 MPa after 28 days and over 7 MPa after 90 days.

The compressive strength decreased after 28 days with the increasing amount of LS1 in the matrix. In the case of sample 30LS1, it is up to 45 MPa and 30LS2 35 MPa. All samples with LS1 had a compressive strength of about 55 MPa after 90 days, about 12% less than the reference sample. It has been shown that calcite is an inert filler but dissolves in an alkaline environment. X-ray records of LS samples dissolved in NaOH solution identified portlandite peaks. In previous works [3], it was confirmed that the $CaCO_3$ content in the matrix increases compressive strength after 90 and 560 days in geopolymers where metakaolin was replaced by 20% calcite. A positive effect on compressive strength was observed for the alkali-activated blast furnace slag, which was replaced by 30% LS [15]. Yip (2008) [3] submits that Ca^{2+} cations play a limited structural role in the geopolymer gel formation. The concentration of these ions in the system remains low to form a separate hydrated gel with surface binding to unreacted mineral particles.

On the contrary, the presence of LS2 results in a lower compressive strength after 28 and 90 days. For the samples 20LS2 and 30LS2, a decrease in strengths after 90 days was observed compared to 56-day strengths. The compressive strength decreased by 25% to 39% after 90 days for LS2 samples.

Low strength can be attributed to the lower reactivity of clay and feldspar minerals in LS2, which needs a longer time for reaction and gel structure formation [13, 16, 17]. Due to the layered structure of kaolinite at almost zero charge on the surface, the penetration of alkalis through the structure is very slow, which results in the formation of a weak structure in most clay-based geopolymers. Kaolinite is not completely incorporated into the polymer structure [13, 14, 18].

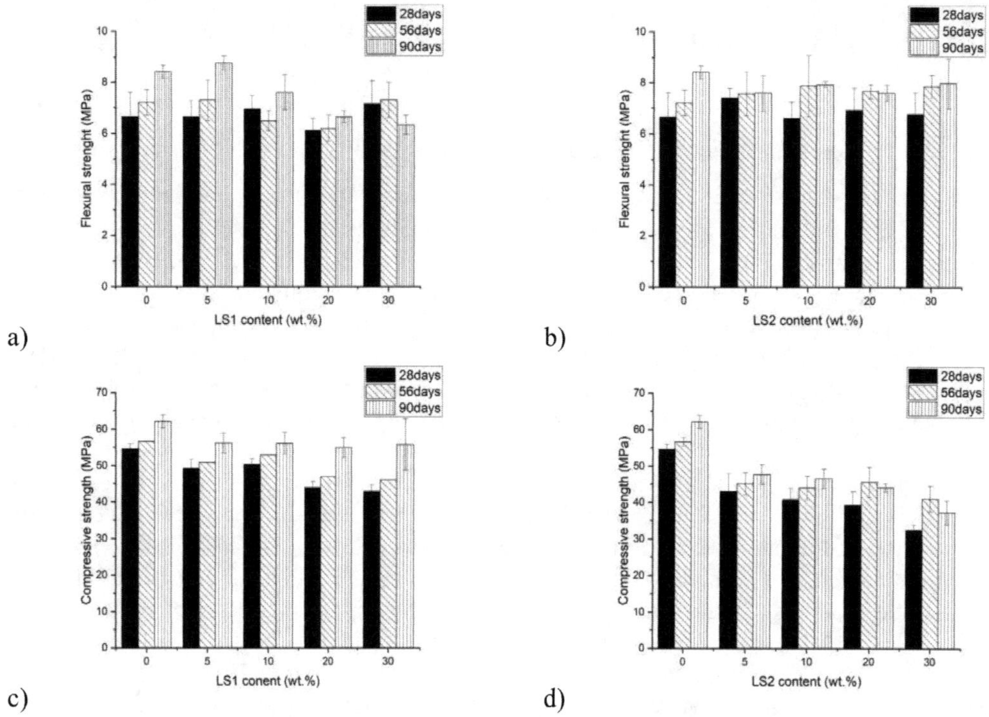

Figure 4. The extent of flexural strength (2a, 2b) and compressive strength (2c, 2d) after 28, 56 and 90 days.

After 90 days of curing under laboratory conditions, the pore structure of mortars was investigated by MIP (figure 5). Mesopores (2 < 30 nm) are the most represented, considered spaces between individual precipitates with free water [19]. The total porosity is slightly affected by the addition of LS. The reference sample has an overall porosity of 0.074 $cm^3 \cdot g^{-1}$, with LS1 and LS2 slightly increasing to values around 0.085 $cm^3 \cdot g^{-1}$ and 0.095 $cm^3 \cdot g^{-1}$, respectively.

The pore distribution was slightly changed by the addition of LS. The reference sample has a bimodal pore distribution: 5–7.5 nm small gel pores and transition pores 7.5–30 nm. Other samples except 30LS1 have a similar pore distribution. For 30LS1 the pore volume disappeared in the region 5–7.5 nm [20].

The pore distribution of LS2 samples was changed. The distribution curve has only one peak in the range of 3.75–25 nm, and the pore volume in this area increased. Changes in the pore distribution can be attributed to the particle distribution of the raw materials rather than the formation of new products that have not been confirmed by XRD. Evidenced by results shown in figures 5a and 5b, LS2 contrary to LS1, in combination with BFS and MK, was involved in forming a structure with smaller pores up to 10 nm.

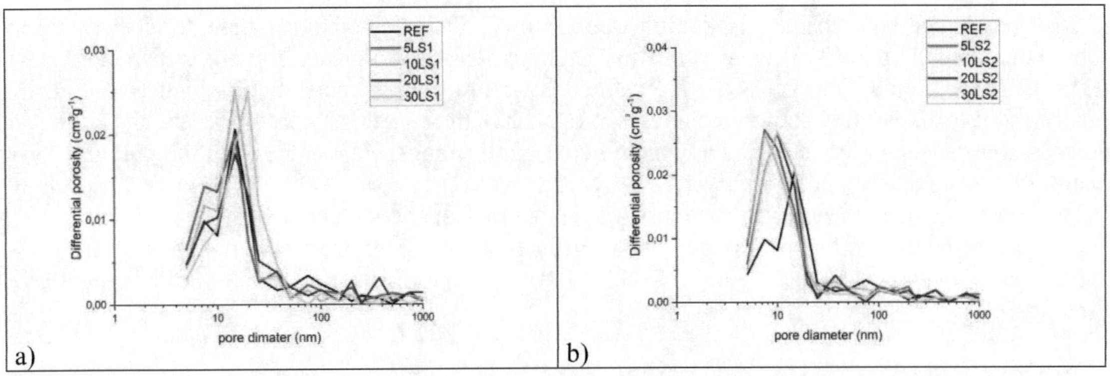

Figure 5. Pore diameter distribution of samples with LS.

The durability of specimens was tested by freez-thaw cycles (figure 6). Specimens were tested after 56 days of curing. The compressive and flexural strengths decreased with increasing LS content in the matrix. Compared to LS2, LS1 specimens showed worse freeze-thaw resistance. The strength characteristics after freeze-thaw cycles of 20LS1 and 30LS1 were not possible to be determined because of cracking and deterioration of samples. The compressive strength was reduced by 6% and 7%, respectively, in the reference sample. 5LS1 and 10LS1 had lower compressive strengths, 48 MPa and 45 MPa, respectively, but no significant degradation occurred during cycling, as in the reference sample. The flexural tensile strengths were comparable to the reference sample.

Degradation of samples with a higher content of limestone sludge does not occur if LS2 was used. This is probably due to the finer LS2 granulometry and different types of pores formed due to the higher content of clay and feldspar minerals.

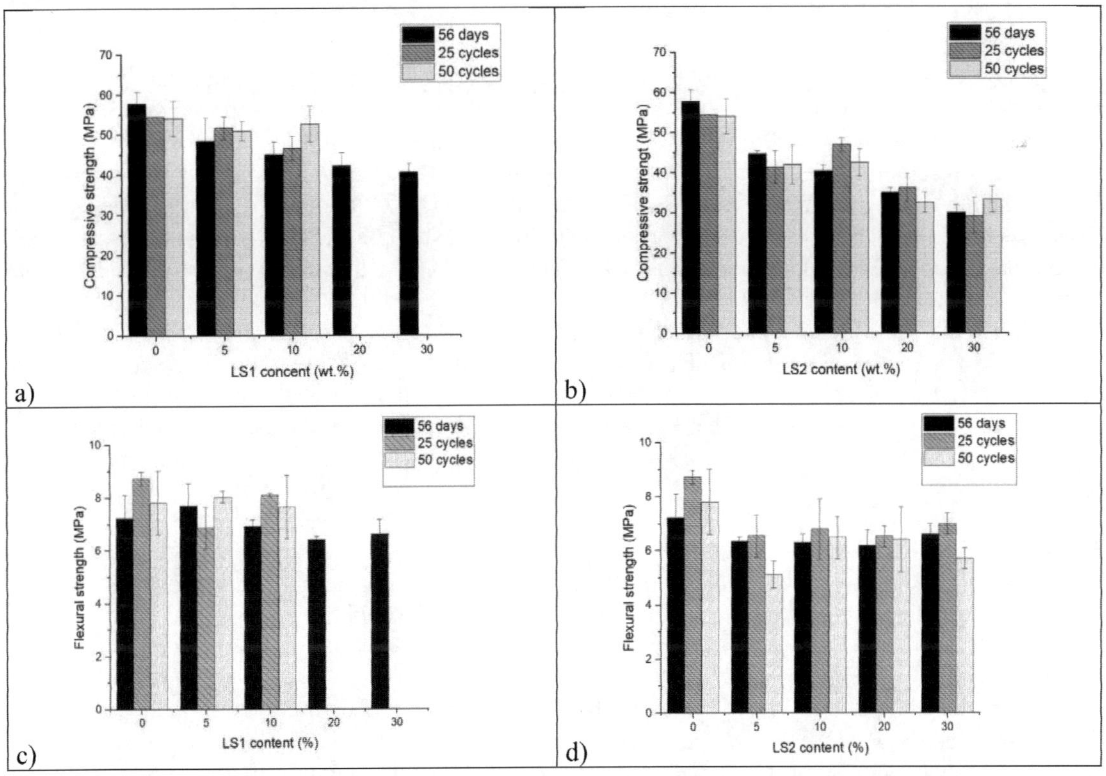

Figure 6. Compressive and flexural strength of samples after freezing-thawing test.

IOP Conf. Series: Materials Science and Engineering 1205 (2021) 012002 doi:10.1088/1757-899X/1205/1/012002

MIP performed pore characterisation on samples after 50 cycles of freeze-thaw tests. It has been shown that the total porosity decreased due to the effect of frost. An increase in pores diameter below 10 nm was monitored. The gel pores are refined by continual polymerisation, which is caused by immersing the samples in water. In previous work [21, 22], the beneficial effect of freeze-thaw tests on pore volume reduction was confirmed, which increased in strength. Figure 7 shows the pore diameter distribution of samples after 50 test cycles. A significant increase in gel pores was observed, especially in LS2 samples, due to the slower dissolution of clay and feldspar components.

The worse resistance is probably because a sufficiently strong structure was not formed, which is evidenced by the lower volume of gel pores (below 10 nm) in the structure of samples 20LS1 and 30LS2 (See figure 5).

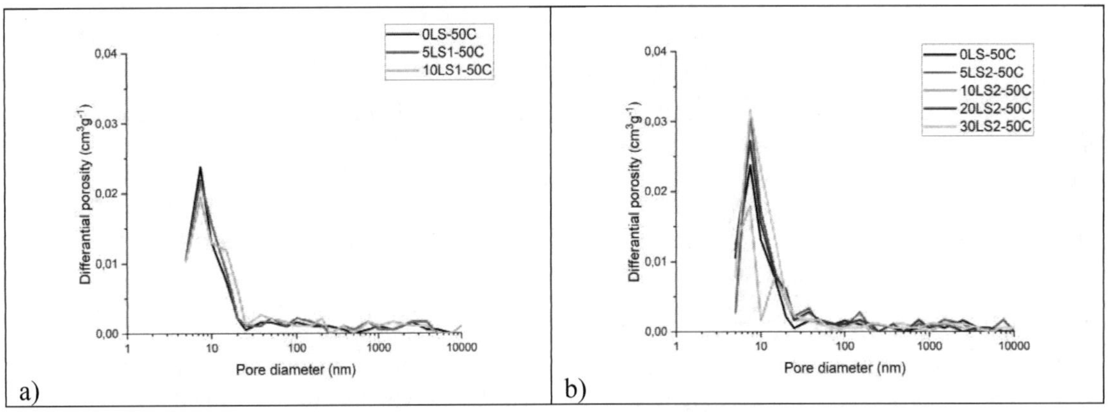

Figure 7. Pore diameter distribution of samples after 50 cycles of freeze-thaw test.

4. Conclusions

The possibility of using waste limestone sludge to prepare geopolmyer based on metakaolin slag was investigated in this work. The influence of LS on workability, setting time, compressive strength and flexural strength was studied. The samples were subjected to the freeze-thaw cycles. A positive effect of $CaCO_3$ on workability was found out while the higher content of clay and feldspar minerals in waste material had the opposite effect. The effect of waste sludge on the prolongation of setting time can also be assessed positively. The results of compressive and flexural strength and frost resistance tests suggest using a maximum of 10% limestone sludge with higher $CaCO_3$ content. The compressive strength was found to be around 55 MPa after 90 days.

A higher ratio of clay and feldspar components in waste sludge resulted in a strength decrease. The sufficient strengths of about 35 MPa were achieved using 30 % sludge in the matrix after 90 days of curing in laboratory conditions. After 50 cycles of freeze-thawing tests, there is no significant degradation of prepared samples. The maximum dosage that can be recommended is 20% in the matrix due to reduced workability. The incorporation of LS contributes to strength improvement and lowered costs.

Acknowledgements

This paper was elaborated with the institutional support for long-term development of research organisations by the Ministry of Industry and Trade of the Czech Republic.

References

[1] Merabtene M, Kacimi L and Clastres P 2019 Elaboration of geopolymer binders from poor kaolin and dam sludge waste *Heliyon* **5** e01938

[2] Moseson A J, Moseson D E and Barsoum M W 2012 High volume limestone alkali-activated cement developed by design of experiment *Cement and Concrete Composites* **34** pp 328–336

[3] Yip C K, Provis J L, Lukey G C and van Deventer J S J 2008 Carbonate mineral addition to

metakaolin-based geopolymers *Cement and Concrete Composites* **30** pp 979–985
[4] Qian J and Song M 2015 Study on influence of limestone powder on the fresh and hardened properties of early age metakaolin based geopolymer *RILEM Bookseries* **10** pp 253–259
[5] Aboulayt A, Riahi M, Ouazzani Touhami M, Hannache H, Gomina M and Moussa R 2017 Properties of metakaolin based geopolymer incorporating calcium carbonate *Advanced Powder Technology* **28** pp 2393–2401
[6] Cwirzen A, Provis J L, Penttala V and Habermehl-Cwirzen K 2014 The effect of limestone on sodium hydroxide-activated metakaolin-based geopolymers *Construction and Building Materials* **66** pp 53–6
[7] Gao X, Yu Q L and Brouwers H J H 2015 Properties of alkali activated slag-fly ash blends with limestone addition *Cement and Concrete Composites* **59** pp 119–128
[8] Yuan B, Yu Q L and Brouwers H J H 2017 Assessing the chemical involvement of limestone powder in sodium carbonate activated slag *Materials and Structures/Materiaux et Constructions* **50** 136
[9] EN 1015-3:1999 Methods of test for mortar for masonry - Part 3: Determination of consistence of fresh mortar (by flow table)
[10] EN 196-3:2016 Methods of testing cement - Part 3: Determination of setting times and soundness
[11] EN 196-1:2016 Methods of testing cement – Part1: Determination of strength
[12] ČSN 72 2452 (722452) Zkouška mrazuvzdornosti malty
[13] Xu H and van Deventer J S J 2000 The geopolymerisation of alumino-silicate minerals *International Journal of Mineral Processing* **59** pp 247–266
[14] Heah C Y, Kamarudin H, Mustafa Al Bakri A M, Bnhussain M, Khairul Nizar I, Ruzaidi C M and Liew Y M 2012 Study on solids-to-liquid and alkaline activator ratios on kaolin-based geopolymers *Construction and Building Materials* **35** pp 912–922
[15] Rakhimova N R, Rakhimov R Z, Naumkina N I, Khuzin A F and Osin Y N 2016 Influence of limestone content, fineness, and composition on the properties and microstructure of alkali-activated slag cement *Cement and Concrete Composites* **72** pp 268–274
[16] Komnitsas K, Zaharaki D and Perdikatsis V 2007 Geopolymerisation of low calcium ferronickel slags *Journal of Materials Science* **42** pp 3073–3082
[17] van Jaarsveld J G S, van Deventer J S J and Lukey G C 2002 The effect of composition and temperature on the properties of fly ash- and kaolinite-based geopolymers *Chemical Engineering Journal* **89** pp 63–73
[18] Liew Y M, Heah C Y, Mohd Mustafa A B and Kamarudin H 2016 Structure and properties of clay-based geopolymer cements: A review *Progress in Materials Science* **83** pp 595–629
[19] Zhang W, Yao X, Yang T and Zhang Z 2019 Effect of calcined dolomite addition on sodium carbonate-activated slag cements with different curing methods *Advances in Cement Research* **31** pp 370–381
[20] Zhao R, Yuan Y, Cheng Z, Wen T, Li J, Li F and Ma J L 2019 Freeze-thaw resistance of Class F fly ash-based geopolymer concrete *Construction and Building Materials* **222** pp 474–483
[21] Zhao M, Zhang G, Htet K W, Kwon M, Liu C, Xu Y and Tao M 2019 Freeze-thaw durability of red mud slurry-class F fly ash-based geopolymer: Effect of curing conditions *Construction and Building Materials* **215** pp 381–390
[22] Yan D, Xie L, Qian X, Ruan S and Zeng Q 2020 Compositional dependence of pore structure, strengthand freezing-thawing resistance of metakaolin-based geopolymers *Materials* **13** pp 1–16

Effect of mechanical activation on the properties of Portland cement

S Ravaszová[1] and K Dvořák[1]

[1] Brno University of Technology, Faculty of Civil Engineering, Veveri 331/95, 602 00 Brno, Czech Republic

Email: ravaszova.s@fce.vutbr.cz

Abstract. The paper deals with the mechanical activation of cement by grinding in a high-speed mill, and compares the parameters found with the commonly used grinding method in cement production, which uses grinding of cement by means of a ball mill. The aim is to verify the influence of the aging time of the ground material on the preservation of the mechanical activation effect and on the properties of the final product at different stages of hydration. It evaluates the physical-mechanical properties, the compressive and tensile strength after 1, 2, 7, 14 and 28 days, as well as the course of the hydration process. The evaluated results suggest the existence of mechanical activation, but this effect is very difficult to achieve and depends on many other factors.

1. Introduction

Mechanical activation is a complex process of changing the structure and thermodynamic state of solid particles. The principle of mechanical activation has been known for a long time. K. Tkacova and a team of authors [1] have been studying mechanical activation of materials for several years. In the publication by Sekulic et al [2], they consider mechanical activation as a phenomenon that can only be determined on a material of the same specific surface area and its resulting properties are evaluated. Z. Sekulic et al. ground Portland cement on different mills to approximately 400 m^2/kg and then evaluated the resulting strengths depending on the type of mill. They attributed the added strength value, in their case an increment of 10 MPa at 28-day strengths, to the principle of mechanical activation [2].

The relationship and difference between fine grinding and mechanical activation was addressed in V. V. Boldyrev with S. V. Pavlov and E. L. Goldberg [3]. He divides the grinding process into coarse grinding, fine grinding, and mechanical activation. However, he admits that it is not possible to use such a simple scale, because of the more complex physical background of the whole grinding process. Mechanical activation can also take place simultaneously with fine grinding.

To achieve mechanical activation, two conditions must be met. The first condition is the particle size, which must be below the brittle-refractive transition. The second condition is that sufficient grain loading due to stress must be achieved [4–6].

It has long been shown [7] that a higher specific surface area of cement brings these benefits. The rate of hydration is higher with higher specific surface area of the cement. Thus, the smaller the particle size of the cement, the faster the formation of hydration products. The initial strength increases with increasing specific surface area of the cement. The final strengths will also increase as the specific surface area of the cement increases. The development of hydration heat will increase with increasing specific surface area of the cement. Finer cement has a positive effect on the workability of mortars and concretes.

At the same time, there are disadvantages. The price increases disproportionately compared to the increase in specific surface area. The amount of water mixed into mortars and concretes increases and the risk of water separation increases. More finely ground cements are subject to greater shrinkage, with a higher risk of cracking due to shrinkage. The shelf life of finely ground cement is shorter because finer ground cement is more prone to lumping and is more reactive, hence more sensitive to air moisture [7–9].

The aim of the paper is to investigate the mechanical activation of cement using a high-speed mill, and to compare the parameters found with the commonly used grinding method in cement production. A ball mill will be used as a commonly used mill in cement manufacturing technology. The aim is also to verify the influence of the aging time of the ground material on the preservation of the mechanical activation effect and on the properties of the final product at different stages of hydration. It is important to evaluate the resulting physical and mechanical properties and to evaluate the course of the hydration process.

2. Materials and methods

2.1. Grinding in a disintegrator (DESI 11)

The clinker (Cement Hranice a.s.) with gypsum (Precheza a.s.) was first ground in a ball mill (Brio Hranice with an OM20f plastering drum). The mill settings were not altered throughout the mill operation and the speed was kept at a maximum of 65 rpm. To achieve a final specific surface area of cement of 400 m^2/kg was proceeded in short grinding cycles of 5 minutes each time with control of the specific surface area by ZEB PC-Blaine Star device.

Since the grinding process in the ball mill ends with the discharge phase, during which the cover of the grinding drum is replaced by a perforated plate and the discharge time is set to 10 to 15 minutes, it had to be considered that the cement is still partially ground during the discharge. By checking the specific surface at intervals of 5 minutes, the average increment of the specific surface over the last cycles was determined and, accordingly, the grinding was stopped just before the desired specific surface area was reached when the cement started to discharge.

After the cement was poured, homogenization was carried out in a laboratory screw homogenizer for 40 minutes. To ensure even better homogenisation, the process was interrupted every ten minutes, the cement was emptied, and the homogeniser cleaned. This was done three times, i.e., after 10, 20 and 30 minutes. After thorough homogenisation of the cement, the output specific surface area was again verified on PC-Blaine-Star automatic device.

This pre-prepared cement was subjected to grinding on a high-speed mill. It was a disintegrator DESI 11 with two counter-rotating rotors with a speed of 12000 rpm. The power of the motors was 2×4.5 kW, with a grinding speed of 1 kg per minute. Three sets of new rotors were available for grinding on the high-speed mill. To carry out the aging experiment, it was necessary to carry out the actual grinding on 3 consecutive days. Therefore, a new set of rotors was used for each grinding, more precisely CR rotors with cubic teeth developed by FFservis. The samples of dimensions $20 \times 20 \times 100$ mm was prepared from the ground cement in the disintegrator. The number of samples was designed to allow strength tests after 1, 2, 7, 14 and 28 days. To ensure the accuracy of the results, 3 samples then produced from each casting phase for each strength test. Prior to preparing the samples, a slurry test of normal consistency was performed to determine the optimum water coefficient.

2.2. Grinding in a ball mill (BRIO Hranice)

The grinding of cement in the ball mill was as close as possible to the grinding process in the disintegrator to ensure comparability of results. Again, the cement was pre-ground to a specific surface area of 400 m^2/kg. In this case, however, the preparation of the cement was partly different from that for the disintegrator. In the disintegrator preparation, the cement was prepared several days in advance. In the ball mill preparation, on the first day of grinding the cement was divided into three weights, one of which was ground directly to the required specific surface of 465 m^2/kg, and the other two weights were only prepared to a specific surface of 400 m^2/kg. These bales were then milled successively over the next two days to the required specific surface of 465 m^2/kg.

After finishing the last part of the cement, the same number of samples was formed as in the case of grinding in the disintegrator. Thus, a total of 45 samples with dimensions 20 × 20 × 100 mm was obtained from the ball mill grinding.

2.3. Preparation and mortification of the sample

The samples were made from cement putty, which was made from cement and water (keeping the same water coefficient to a slurry of normal consistency) and processed by hand with a whisk in a stainless-steel bowl.

The cementitious putty thus prepared was applied to the moulds and compacted in two layers using a vibrating table. The specimens were covered with plastic wrap and de-moulded after 24 hours. The first set of specimens was immediately tested for compressive strength. The remaining samples were stored at laboratory temperature immersed in water in a laboratory environment.

After the strength tests, the samples were always mortared for later investigation of the hydration process by DTA Thermogravimeter - Mettler Toledo DSC1 and X-Ray analysis by XRD PANalatical Empyrean, Cu - anode $\lambda = 1.540598$ for $K\alpha1$ radiation. The residue from the undercut strength test samples was hammered and ground in a laboratory planetary mill FRITCH Pulverisette 6 using isopropyl alcohol. The mill setting was the same each time, 500 rpm, for 105 seconds. The obtained mass was further filtered on a chemical filter assembly through a filter cloth and was washed three times with isopropyl alcohol. The resulting mortified samples were placed in a resealable bag and stored in a freezer. After all strength tests were completed, X-ray and DTA analyses were performed to map the hydration process of the mechanically activated cements.

3. Results and discussion

3.1. Grinding fineness

Table 1 contains the values of specific surface areas measured on the automatic Blain apparatus, which refer to the cement ground in the disintegrator and in the ball mill.

Table 1. Cement specific surface areas measured on PC-Blaine-Star automatic device.

	Specific surface area [m^2/kg]	
Aging of the sample	disintegrator DESI	ball mill Brio Hranice
OO*	449	452
O1*	464	484
O2*	465	462
ϕ [m^2/kg]	**459**	**466**

*marking of sample aging times (O: without aging; 1: one-day aging; 2: two-days aging)

When grinding in a ball mill, slight variations of the specific surface areas within the grinding batches are seen according to the aging. This is mainly because the values given are control measurements after homogenisation. Moreover, achieving a perfectly accurate specific surface area is not an easy task, if only because multiple measurements of the same sample show slight deviations in the order of units.

3.2. Physical-mechanical properties

The samples were tested for compressive strength. In the evaluation, the evolution of the strengths over time was monitored, as well as the comparison of grinding technologies. According to the assumptions of proving the effect of mechanical activation, the strengths of cement ground in a disintegrator should be higher for the same specific surface area. The mechanical activation effect could also be reflected in increased strengths of cement without aging compared to cement with one- or two-days aging.

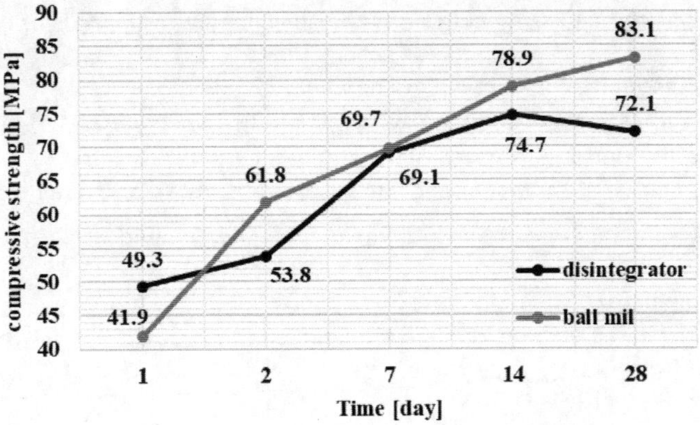

Figure 1. Comparison of compressive strengths of cements ground at different mills (without aging).

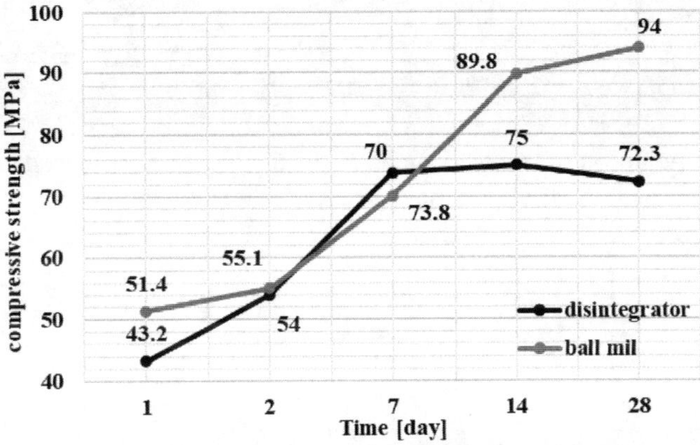

Figure 2. Comparison of compressive strengths of cements ground at different mills (one-day aging).

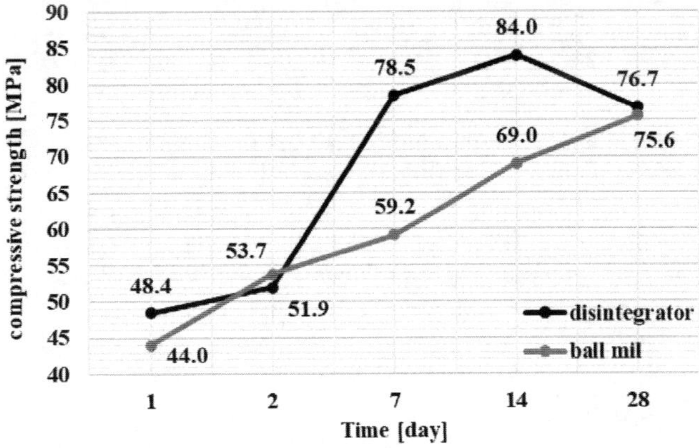

Figure 3. Comparison of compressive strengths of cements ground at different mills (two-days aging).

When comparing the strengths depending on the type of mill, the effect of grinding at different mills on the strengths of the cementitious putty was not demonstrated. For the cements without aging (figure

1) the values are comparable, the final strengths were higher for the cement ground on a ball mill. For the one-day aging (figure 2), the final strengths also came out higher for the cement ground in the ball mill. For the two-day aging (figure 3), the higher strengths were predominantly for cement from the disintegrator. Based on the results, it can be concluded that none of the grinding technologies seems to be more advantageous for achieving better physical-mechanical properties of cement.

3.3. Hydration process

The course of hydration was monitored mainly by X-ray diffraction analysis. DTA analysis was also performed to evaluate the portlandite increment. The evaluation of the X-ray analysis of the samples from the disintegrator can be seen in figure 4. Figure 5 then shows the evaluation of the samples from the ball mill.

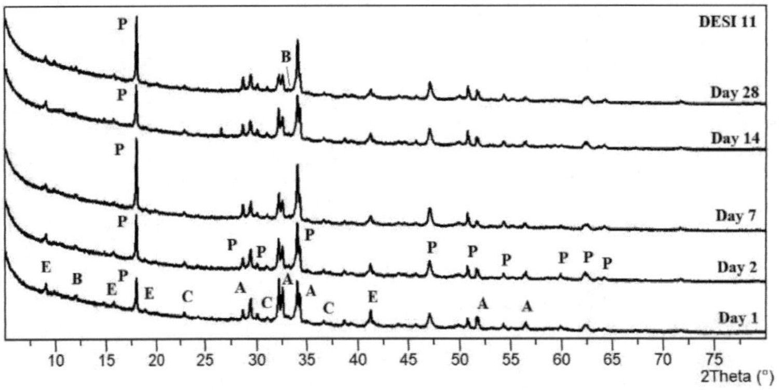

Figure 4. Hydration of cement ground in the disintegrator (P-portlandite; a-alite; B-brownmilerite; E-ettringite; C-cebollite $Ca_5Al_2Si_3O_{12}(OH)_4$).

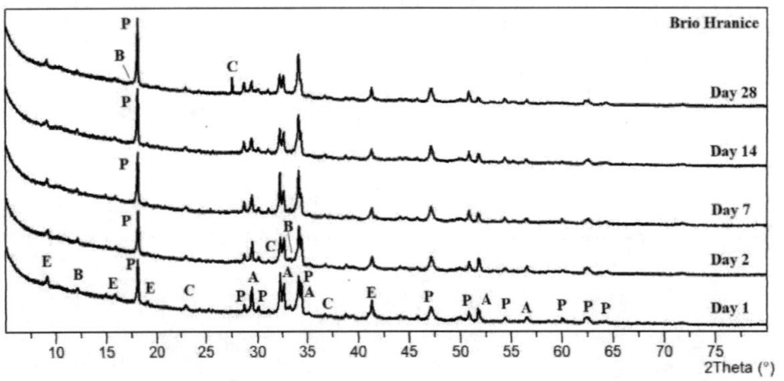

Figure 5. Hydration of cement ground in the ball mill (P-portlandite; a-alite; B-brownmilerite; E-ettringite; C-cebollite $Ca_5Al_2Si_3O_{12}(OH)_4$).

X-ray analysis confirmed that hydration was stable, so the lower strengths of cement ground in the disintegrator after 28 days were probably due to a defective set of samples or a factor other than the hydration process. From figures 4 and 5, the portlandite content increases with increasing strength, as best seen at 18 °2Θ. The intensity of alite, evident at one-day strengths, then decreases to disappear during hydration, like that of ettringite. Brownmillerite is only slightly visible, and the intensity does not change much during hydration.

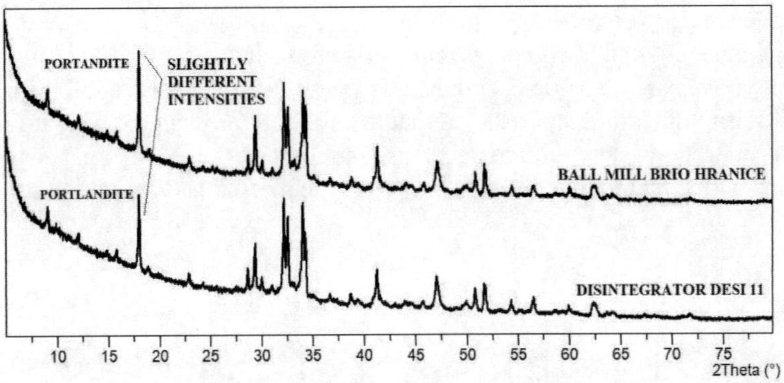

Figure 6. Comparison of grinding technologies on the X-ray curves.

If we place the X-ray analysis outputs of the cement from the ball mill and the disintegrator side by side (figure 6), the curves are almost identical, even though the actual experiments were carried out many months apart. For comparison, the curves of the cements from the one-day strengths from the cement without aging were selected. We can notice only a slight difference in the portlandite intensities, as indicated in the graph. Cement ground in a ball mill has slightly higher portlandite intensity than cement ground in a disintegrator, but this is the only difference. The other peaks of portlandite did not differ between the grinding technologies.

Furthermore, the portlandite content was evaluated from the measured DTA analyses as a representative of the output hydration products. The following table 2 summarizes the results in percentages.

Table 2. Portlandite content in samples with stopped hydration process from different mills.

		The portlandite content [%]	
days	aging	disintegrator	ball mill
	OO	7.45	5.17
1	O1	6.94	5.27
	O2	6.8	5.62
	OO	7.59	6.18
2	O1	7.74	6.39
	O2	7.36	6.23
	OO	8.06	6.89
7	O1	8.34	7.11
	O2	8.12	7.09
	OO	9.13	7.25
14	O1	9.08	7.52
	O2	8.98	7.23
	OO	9.64	8.67
28	O1	9.61	8.54
	O2	9.52	8.61

From table 2, the cementitious putty prepared from ball mill cement has a lower portlandite content than the cementitious putty prepared from cement milled in the disintegrator, as calculated from the DTA analysis.

3.4. Effect of the specific surface area

In the following figure 7, which compares cement before and after milling in the disintegrator, the clinker minerals are apparently visible, and there is a very slight decrease in all the intensities shown. Sources [8] and [9] from the introduction stated that the effect of mechanical activation is evident from the increase in the amorphous phase and the decrease in intensities on X-ray analysis. Visually, we are unable to determine with certainty from figure 7 whether the mechanical-activation effect has occurred.

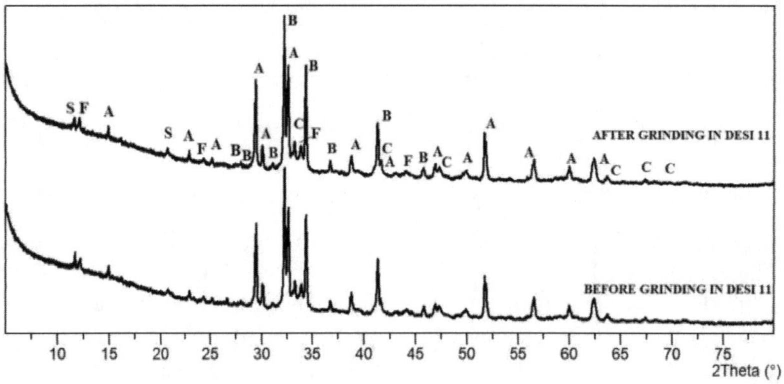

Figure 7. Result of X-Ray analysis of cement before and after grinding in disintegrator.

4. Conclusion

The cement ground in the disintegrator had an average specific surface area about 459 m^2/kg. For both compressive and flexural tensile strengths, the cement from the disintegrator showed the highest strengths after one day for the cement bodies without aging. However, in the further course of hydration this phenomenon disappeared.

The compressive strengths at 7, 14 and 28 days were compared with the grinding fineness of each aging, and it was found that as the hydration progressed, the strengths levelled off as a function of specific surface area, such that the finer the cement was, the higher the compressive strengths it exhibited, even with unintended small variations in specific surface area (approximately 10 m^2/kg). This phenomenon of differing initial strengths and subsequent comparison of strengths according to grinding fineness could be an effect of mechanical activation due to high-speed grinding in the disintegrator since it did not occur in the cement ground in the ball mill.

The highest compressive strength ever achieved was 94 MPa for cement ground in a ball mill with a one-day anging. The 28-day strengths of the cement from the disintegrator were unusually low and given the normal hydration history of the cement under investigation, as demonstrated by X-ray analysis, this set of samples was deemed to have failed. It can therefore be assumed that if the set had been successful, the strengths of the cement from the disintegrator after 28 days would have been like those of the ball mill.

The DTA analysis allowed the calculation of the portlandite content, with the highest portlandite content at the one-day strengths, which corresponded to the cement without annealing strengths. The portlandite content was higher in the cement from the disintegrator than in the cement from the ball mill throughout the hydration time.

The results evaluated suggest the existence of mechanical activation, but this effect is very difficult to achieve and certainly depends on many factors. Harnessing this effect in practice is a challenging but not easy task. It would also be important to assess the potential increased energy costs associated with milling. It should be considered whether the increase in utility is effective enough.

Due to the nature of the mechanical activation of cement, the effect of ageing on long-term strength or durability needs to be further investigated. Because of the short-term nature of the mechanical activation effect, its effect could be better exploited in activation mixing, for example in precast concrete plants.

Acknowledgments
We acknowledge the financial support to project "Effects of mechano-chemical activation on the process of formation, structure and stability of selected clinker minerals" (20-00676S).

References

[1] Tkacova K, Heegn H and Stevulova N 1993 Energy transfer and conversion during comminution and mechanical activation *International journal of mineral processing* **40** pp 17–31
[2] Sekulic Z, Popov S and Milosevic S 1998 Comminution and mechanical activation of Portland cement in different mill types *Ceramics* **42** pp 25–28
[3] Boldyrev V V, Pavlov S V and Goldberg E L 1996 Interrelation between fine grinding and mechanical activation *International Journal of Mineral Processing* **44-45** pp 181–185
[4] Kumar S, Kumar R, Bandopadhyay A, Alex T C, Kumar K R, Das S K and Mehrotra S P 2008 Mechanical activation of granulated blast furnace slag and its effect on the properties and structure of Portland slag cement *Cement and Concrete Composites* **30** pp 679–685
[5] Sekulic Z et al 1999 Mechanical activation of cement with addition of fly ash *Materials letters* **39** pp 115–121
[6] Blanco F, et al 2006 The effect of mechancally and chemically activated fly ashes on mortar properties *Fuel* **85** pp 2018–2026
[7] Vavrin F 1980 *Maltoviny* (Praha: SNTL-Nakladatelstvi technicke literatury) 103i
[8] Sobolev K et al 2016 The influence of mechanical activation by vibro-milling on the early-age hydration and strength development of cement *Cement and Concrete Composites* **71** pp 53–62
[9] Mucsi G, Racz A and Madai V 2013 Mechanical activation of cement in stirred media mill *Powder Technology* **235** pp 163–172

IOP Conf. Series: Materials Science and Engineering 1205 (2021) 012004 doi:10.1088/1757-899X/1205/1/012004

Shotcrete using ternary binder made from coal combustion products: from lab tests to an application

Vít Šmilauer[1*], Radoslav Sovják[1], Šárka Pešková[1], Rostislav Šulc[1], František Škvára[2], Martina Šídlová[2], Roman Snop[3], and Pavel Růžička[4]

[1]Czech Technical University in Prague, Faculty of Civil Engineering, Thákurova 7, 166 29 Prague 6, Czech Republic

[2]University of Chemistry and Technology Prague, Faculty of Chemical Technology, Institute of Glass and Ceramics, Technická 5, 166 28, Prague 6, Czech Republic

[3]ČEZ Energetické produkty, Komenského 534, 253 01 Hostivice, Czech Republic

[4]Hochtief CZ a.s., Plzeňská 16, 150 00 Prague 5, Czech Republic

E-mail: `vit.smilauer@fsv.cvut.cz`

Abstract. The paper presents a ternary binder development and its utilization in shotcrete. The binder is made from fluidized bed combustion (FBC) fly ash, siliceous fly ash, and $Ca(OH)_2$ addition, now available under the name Sorfix. XRD and TGA identified ettringite and C-S-H as two main hydration products. In addition, thermodynamic modeling verified robustness in terms of space-filling capabilities when varying input oxide composition.

Since alkali-free accelerators produce mostly ettringite in Portland-based systems, a fraction of Portland cement was advantageously replaced with the ternary binder, forming early ettringite as well. Extensive testing led to 45% replacement of Portland cement, following J2 curve for early strength gain used commonly in shotcrete tunnel linings. The shotcrete was successfully tested in a mock-up experiment in a 2 m^3 batch. Primary lining of a newly built metro D in Prague served for the full-scale application, utilizing over 1000 tons of Sorfix and saving over 700 tons of CO_2.

1. Introduction

Coal combustion products (CCP), such as fly ash or flue-gas desulfurization gypsum, are produced in high quantities around the world. CCP serve for various purposes, such as mineral admixture for concrete [1]. During 2010, 780 Mt of CCP were generated worldwide with an average utilization rate as 53% [2]. There are two useful fly ashes which emerge from coal burning, particularly:

- Siliceous fly ash.
- Fluidized bed combustion (FBC) fly ash. It emerges from co-combustion with limestone, binding SO_2 into anhydrous $CaSO_4$.

Hydration of FBC fly ash itself yields C-S-H, ettringite and gypsum as the main hydration products [3]. Volume expansion commonly occurring due to ettringite growth prevents simple addition to Portland-based binders. Ettringite is known from other systems widely used since 1970s in China; two common binders include calcium sulfoaluminate (CSA) cements and CSA-belite cements [4, pp. 341-361]. It has been documented that common hydrated products cover

IOP Conf. Series: Materials Science and Engineering 1205 (2021) 012004 doi:10.1088/1757-899X/1205/1/012004

ettringite, monosulphate, strätlingite, amorphous $Al(OH)_3$, and also portlandite and C-S-H with belite addition. As discussed further, the newly designed ternary binder Sorfix produces similar phases such as the CSA cement, which proved 40 year durability and excellent performance [4, pp. 341-361].

2. Materials

Design, development and properties of the ternary binder Sorfix have been previously summarized [5], together with years-lasting monitoring of volume changes and compressive strength evolution showing no detrimental strength loss or excessive shrinkage [6]. Let us recall the basic constituents of the ternary binder; FBC fly ash originates from fluidized bed combustion power plant Tisová (d_{50}=36.3 μm, Blaine 603 m^2/kg) while fly ash class F comes from Počerady (d_{50}=44.9 μm, Blaine 484 m^2/kg). Their chemical compositions are summarized in Table 1.

Table 1. Chemical composition of FBC and fly ashes (wt. %).

	SiO_2	Al_2O_3	Fe_2O_3	CaO	Na_2O	K_2O	TiO_2	SO_3
FBC ash	31.4	26.2	4.9	21.9	0.24	0.36	5.5	8.0
Fly ash	52.4	35.0	5.1	1.8	0.41	2.1	1.5	0.2

Binder composition was optimized for workability, for reasonable 28 day compressive strength. Further optimization focused on volumetric stability which was achieved in accelerated aging test (curing 28 days in 95°C water). Superplasticizer ensured a feasible workability on a specified water/binder ratio. Table 2 summarizes the optimal design of the paste, yielding 32 MPa average compressive strength at the age of 28 days. Bulk density of paste is 1710 kg/m^3 and the initial volume fraction of water yields 0.47 for this particular paste. Strength increase up to 50 MPa is further possible increasing superplasticizer dosage and lowering the water content.

Table 2. Paste composition for the ternary binder (wt. %).

FBC fly ash	Siliceous fly ash	$Ca(OH)_2$	Water	Superplasticizer
0.313	0.313	0.094	0.263	0.017

Shotcrete composition is summarized in Table 3, yielding concrete strength class of C25/30. In this particular mix design, ternary binder Sorfix makes 45% of the total binder mass. Wet mix from a concrete plant was transported to a construction site. An alkali-free accelerator was added in the nozzle as 7.5% mass of the total binder.

Table 3. Shotcrete mix design.

Component	Mass (kg/m^3)
CEM I 42.5 R	250
Ternary binder Sorfix	200
Water	180
Aggregate 0-2 mm	250
Aggregate 0-4 mm	760
Aggregate 4-8 mm	480
High-Range Water Reducer	4
Alkali-free accelerator	33.8 (7.5%)

3. Results and discussion

3.1. Volume phase evolution

Four overlapping techniques were used to deduce volumetric phase evolution; XRD, deconvolution of DTG, SEM/EDX mappings and porosimetry [5]. Seven phases were distinguished, namely non-reactive crystalline phases, calcium hydroxide, amorphous glass, ettringite, C-S-H globules, gel and capillary porosity, see Fig. 1.

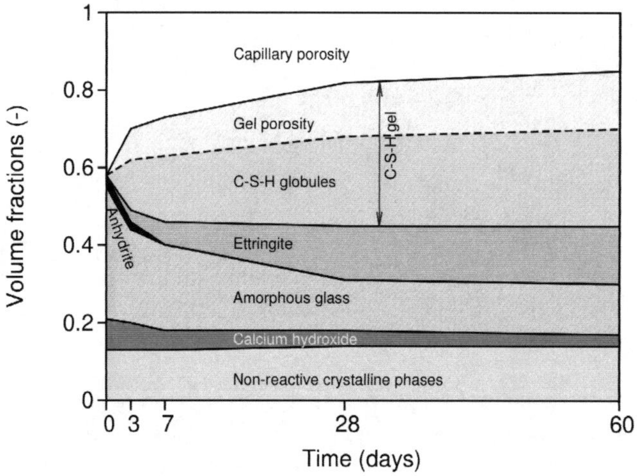

Figure 1. Phase evolution due to hydration of ternary binder Sorfix [5].

XRD analysis quantified the amount of calcium hydroxide, non-reactive crystalline phases, and amorphous content. DTG deconvolution yielded C-S-H gel amount and ettringite [5]. Volume fractions stemmed from mass fractions and known densities of particular phases [5]. The density of C-S-H gel is assumed as 2000 kg/m^3 for saturated LD C-S-H [7]. The total measured porosity is divided in to the capillary and gel porosity parts, where the gel porosity is calculated as 37% of the LD C-S-H gel (packing density 0.63) [8].

3.2. Thermodynamic modeling and binder robustness

Thermodynamic modeling contributes to the design and optimization of cementitious binders, yielding hydrated phase composition in solid solution [9]. The non-stoichiometric method, often called "Gibbs free energy minimization approach" assumes no chemical reactions' system and a stable equilibrium condition is reached on the minimization of the Gibbs free energy.

GEM-Selektor provides such a software package, containing a built-in thermodynamic database and Cemdata18 database for Portland cements, blended systems and alkali-activated materials [9]. It is possible to simulate phase assemblage in the Sorfix ternary binder and to explore sensitivity in terms of oxide composition.

The simulation took 100 g of the ternary binder interacting in water environment. First, it was necessary to assess dissolved components in the solution, which may form a hydrate assemblage. Fig. 1 provides an indication of dissolution from the components when hydration proceeds for 60 days. It transpires that 45% of fly ashes and 62% of $Ca(OH)_2$ had reacted by that time. Further estimations yielded that FBC fly ash reacted by 80% and siliceous fly ash by only 15%. The reactions have impact on SiO_2, CaO and SO_3 availability in the pore solution.

The ternary binder was simplified in the GEM-Selektor, leaving out Fe_2O_3 and TiO_2 contribution, increasing unreacted part by 2% vol. Further volume reduction occurred due to chemical shrinkage, estimated as 5% at 60 days of hydration.

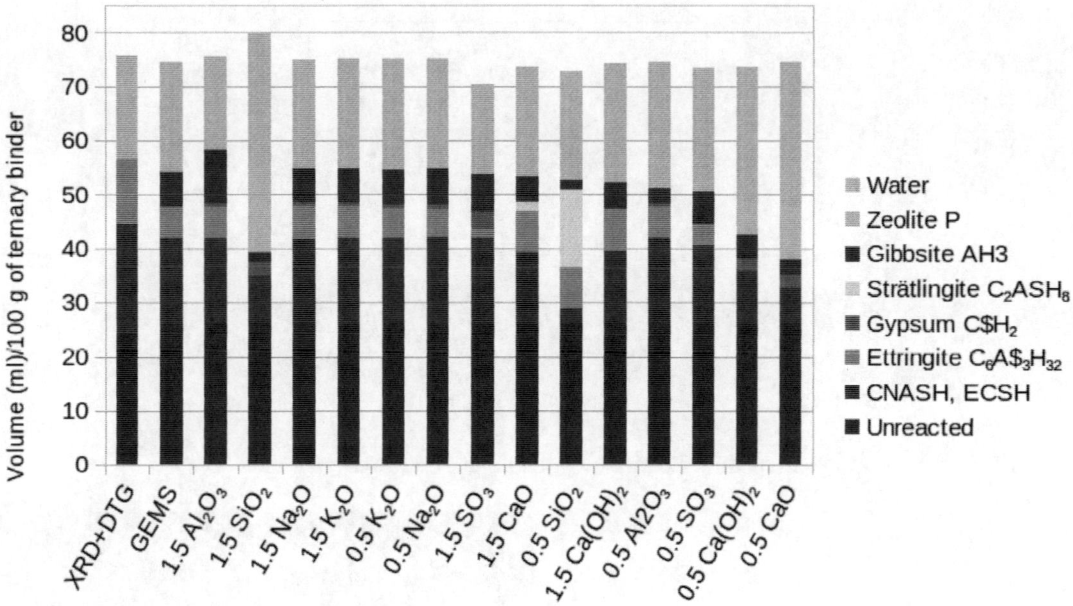

Figure 2. GEMS results and the sensitivity analysis of ternary binder.

Predictions of GEM-Selektor are summarized in Fig. 2. The column 'XRD+DTG' represents identified phases from Fig. 1 at the age of 60 days under sealed curing. The column 'GEMS' gives phase assemblage without further adjustments. In this particular case, 14% less volume in C-S-H + ettringite is predicted, however, a new gibbsite phase is revealed unnoticed previously. This might stem from gibbsite's amorphous nature when Na concentration is low.

Sorfix robustness is assessed by variations around default oxide composition by +50% and -50%. The ternary systems remain relatively stable in the ranges of oxide availability; C-S-H and ettringite still present the two main hydration products. Different hydrates contribute to space-filling properties, which guarantees certain engineering properties, for example elasticity or strength.

Hydration phases form at excess / deficiency of considered oxides. Zeolite P originates in CaO deficient systems, which is stimulated by low FBC or $Ca(OH)_2$. Deficiency in SiO_2 leads to strätlingite instead of C-S-H. Abundance of SiO_2 creates Zeolite P on behalf of aluminates.

3.3. Mock-up shotcrete

The mock-up shotcrete started from 2 m^3 wet-mix concrete, summarized in Table 3 [10]. A truss-tunnel frame imitated a tunnel profile, the height was approximately 5 m, see Fig. 4 left. Shotcrete exhibited fallout under 10%, which signalizes good adhesion and workability. Desired J2 curve was fulfilled, needle penetration test was used for early age assessment while drilled cores led to 28 day compressive strength, see Fig. 3. The mock-up test proved anticipated behavior in terms of workability, spraying and strength gain.

3.4. Tunnel lining in Prague metro D

Construction of the Metro D line in Prague began with a geological survey in June 2019. Shotcrete similar to the mock-up experiment has been used for primary lining and temporary structures, see Fig. 4 right. Up today, over 1000 tons of Sorfix replaced CEM I, saving over 700 tons of CO_2. This was calculated from substituted clinker CO_2 direct emissions [11] and

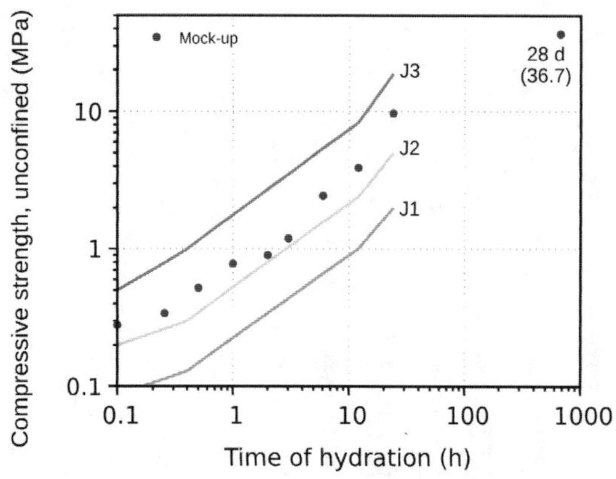

Figure 3. Strength evolution in the mock-up test.

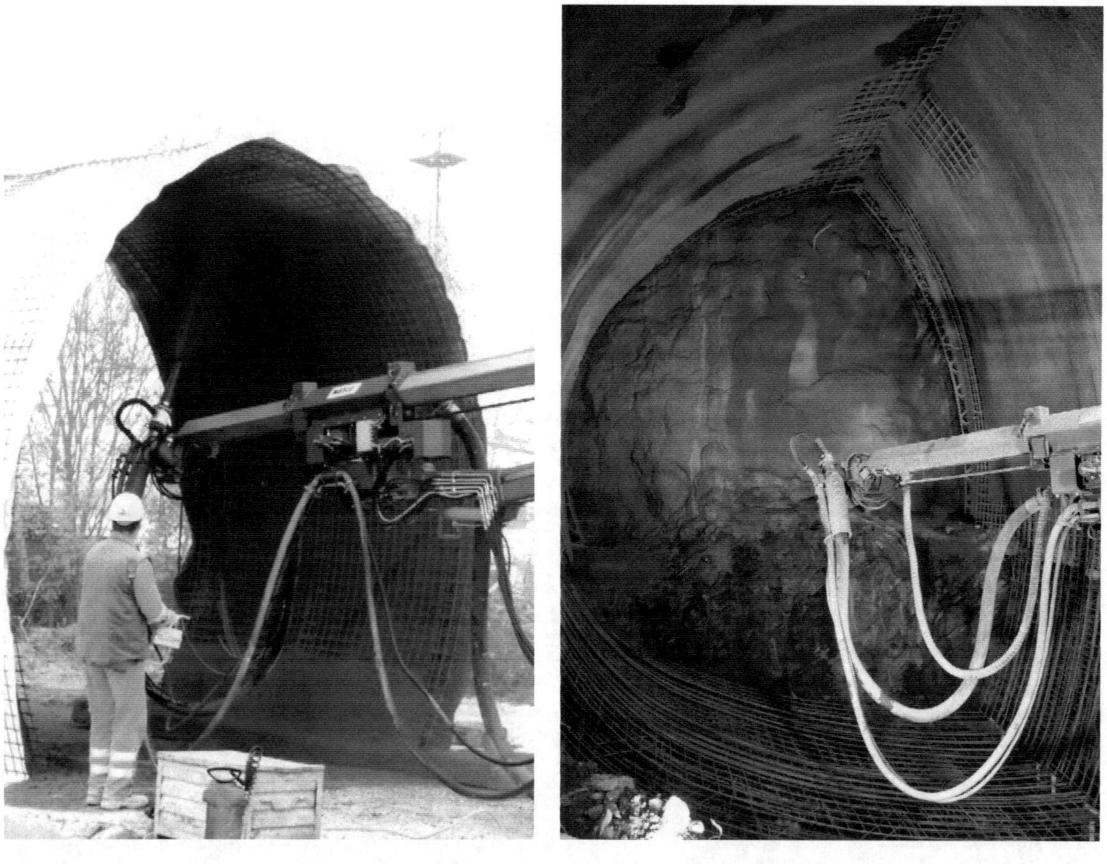

Figure 4. Mock-up experiment (left) and primary lining of metro D (right).

additional limestone decomposition as 1000*0.612/0.783 - 130*44/74 = 704 t, electricity for grinding is assumed to remain the same. Only minor technological adjustments were necessary, for example excessive transport time of fresh mix led to small slump loss which could be corrected by plasticizer or water addition.

4. Conclusions

The paper presented design of a ternary binder named Sorfix, yielding mainly ettringite and C-S-H due to hydration. Thermodynamical modeling proved robust capillary-filling capabilities even under large oxide variations. The advantage of early-age ettringite formation was utilized in a shotcrete, where Sorfix replaced 45% of Portland cement. The shotcrete successfully fulfilled technological and material requirements in a mock-up experiment and in Prague metro D's primary lining, saving over 700 tons of CO_2 and utilizing over 1000 tons of coal combustion products.

5. Acknowledgment

We gratefully acknowledge the financial support from The Technology Agency of the Czech Republic (TA CR) under the project Trend FW01010195, the Czech Science Foundation under the project 21-03118S and from the Grant Agency of the Czech Technical University in Prague No. SGS20/107/OHK1/2T/11.

References

[1] Robl T, Oberlink A and Jones M 2017 *Coal Combustion Products (CCP's): Characteristics, Utilization and Beneficiation. 1st ed.* (Woodhead Publishing Limited)
[2] Heidrich C, Feuerborn H J and Weir A 2013 *Proceedings of World of Coal Ash*
[3] Li X G, Chen Q B, Huang K Z, Ma B G and Wu B 2012 *Construction and Building Materials* **36** 182–187
[4] Lea F M 2019 *Lea's chemistry of cement and concrete* fifth edition. ed (Oxford, England: Butterworth-Heinemann) ISBN 0-08-100795-7
[5] Hlaváček P, Šulc R, Šmilauer V, Rößler C and Snop R 2018 *Cement and Concrete Composites* **90** 100–107 ISSN 0958-9465 URL `https://www.sciencedirect.com/science/article/pii/S0958946518302245`
[6] Škvára F, Šulc R, Snop R, Peterová A and Šídlová M 2018 *Cement and Concrete Composites* **93** 118–126 ISSN 0958-9465 URL `https://www.sciencedirect.com/science/article/pii/S0958946518300957`
[7] Jennings H M 2008 *Cem. Concr. Res.* **38** 275–289
[8] Hlobil M, Šmilauer V and Chanvillard G 2016 *Cement and Concrete Research* **83** 188 – 202
[9] Lothenbach B, Kulik D A, Matschei T, Balonis M, Baquerizo L, Dilnesa B, Miron G D and Myers R J 2019 *Cement and Concrete Research* **115** 472–506 ISSN 0008-8846 URL `https://www.sciencedirect.com/science/article/pii/S0008884617312073`
[10] Sovják R, Pešková S, Šmilauer V, Mára M, Růžička P, Černá Vydrová L and Konvalinka P 2019 *Case Studies in Construction Materials* **11** 1–12 ISSN 2214-5095 URL `https://www.sciencedirect.com/science/article/pii/S2214509518304200`
[11] Henrich C 2012 Sustainability - oddíl věnovaný České republice / slovensku 2012 Tech. rep. Buzzi Unicem

IOP Conf. Series: Materials Science and Engineering 1205 (2021) 012005 doi:10.1088/1757-899X/1205/1/012005

Use of clay from a local source for calcination and subsequent preparation of the mixed cement

B Smirova[1], **M Sedlacik**[1] **and R Novotny**[1]

[1] Brno University of Technology, Faculty of Chemistry, Materials Research Centre, Purkyňova 464/118, Brno Cz-612 00, Czech Republic

Email: Barbora.smirova@vut.cz

Abstract: This paper deals with the use of calcinated clay and micronized limestone as supplementary cementitious materials (SMCs) for preparation of blended Portland cement CEM II/B-M (Q-LL). Clay used in this study was calcinated at 700°C and pozzolanic activity after calcination was assessed using accelerated R3 pozzolanic test and modified Chapelle test. The influence of calcinated clay and limestone addition on mechanical properties and hydration process was investigated and an optimal ratio for 35% clinker replacement was found. Initial decrease of mechanical strength at early ages, caused by SCM addition, was almost compensated during maturation of the binder.

1. Introduction

Portland cement, with annual production of 4.1 billion tons worldwide, is responsible for 5–7% of total anthropogenic CO_2 emissions and 7% of total energy consumption. Increasing environmental concerns and focus on sustainable development have created considerable pressure on cement industry [1, 2]. One way of reducing the environmental impact is a partial replacement of clinker by supplementary cementitious materials (SMCs). Conventional SMCs include mainly pozzolanic materials. Most commonly used SMCs are granulated blast furnace slag and fly ash from coal power plants. However, due to high price and relatively low amount of blast furnace slag and decreasing supply of fly ash tied to decommissioning of coal power plants, alternative SMCs are required in order to satisfy the demand for these materials [3].

Clays are highly abundant and accessible worldwide. Calcination of clays between 550–900°C leads to dehydroxylation of clay minerals and the formation of partially disordered silica and alumina rich phases, making them a suitable candidate for use as SMC. Furthermore, compared to conventional SMCs, they can be supplied locally and independently on other industrial sectors. This eliminates the long-distance transport, commonly required with the use of i.e. blast furnace slag and provides long term stability of supply required for the sustainability of production [4]. A positive effect of calcinated clays on hydration process is caused mainly by filler effect in early stages and pozzolanic reaction later [5].

Addition of limestone powder to cement mixture influences hydration process via multitude of effects. Due to particles being smaller than cement particles, it has a filler effect causing a decrease of porosity and an increase in compressive strength [6]. Limestone particles also act as nucleation centres, accelerating hydration of C_3S and increasing the amount of hydration products at early ages. The reported chemical effect of limestone on hydration is limited in Portland cement due to the limited amount of aluminate phases in the clinker. However, it becomes more pronounced in combination with

SMCs containing active alumina and includes interaction with aluminate phases, stabilization of ettringite and formation of mono and hemicarbonate phases [7].

In this study, cement mixtures containing different calcinated clay to limestone ratios were prepared and characterized, with the aim of investigating the influence of calcinated clay and limestone addition on hydration process and strength development of cement mixtures.

2. Materials and methods

Portland cement (Českomoravský cement a.s., plant Mokrá, Czech Republic) with resulted specific surface area 391 m^2/kg (determined via Blaine method EN 196-6), gypsum (commercially purchased, pure; Penta, Czech Republic), used as setting regulator, clay calcinated at 700°C (Poštorná, Czech Republic), limestone (LB Cemix, s.r.o., plant Kotouč Štramberk, Czech Republic) sieved under 63 μm, coarse aggregate (Filtrační písky, spol. s r.o., Czech Republic), finely ground quartz sand (Sklopísek Střeleč a.s., Czech Republic) and deionised water were used for cement preparation. The cement composition is summarized in Table 1.

Table 1. Composition (g) of measured samples.

Sample	Cement	Calcined clay	Limestone	Finely ground sand	Gypsum
1	100.0	0.0	0.0	0.0	0.0
2	66.6	30.0	0.0	1.5	1.9
3	66.6	25.0	5.0	1.5	1.9
4	66.6	20.0	10.0	1.5	1.9
5	66.6	15.0	15.0	1.5	1.9

Samples intended for mechanical testing were prepared according to standard ČSN EN 196-1. The homogenized mixture was placed into moulds with the dimensions 4 × 4 × 16 cm. After 24 hours, the experimental specimens were demoulded and water curing conditions were applied under the controlled temperature of 20°C. Mechanical tests were performed after 2, 7 and 28 days.

The particle size of used raw materials was assessed by laser diffractometry (Helos KR, Sympatec) and is shown in Figure 1–Figure 2. Hydration process was analysed by isothermal calorimeter TAM air (TA Instruments) at isothermal temperature 25°C. Analysed samples (5g) were prepared according to standard ČSN EN 196-11. Measurements were performed with a reference sample. Chemical composition was evaluated using X-ray Fluorescence (XRF) analyser Vanta. The instrument analyses elements from magnesium to uranium. Elements lighter than magnesium in silicate materials are defined as a loss on ignition (LOI). X-ray Diffraction (XRD) was used to characterize formed crystalline phases. Phase composition analysis was carried out on Bruker D8 Advance diffractometer with Cu anode ($\lambda K\alpha$=1,54184 Å) and variable divergence apertures at Θ-Θ reflective Bragg-Brentano parafocusing geometry. Determination of amorphous phase content was carried out using 20 wt. % CaF_2 addition as a standard.

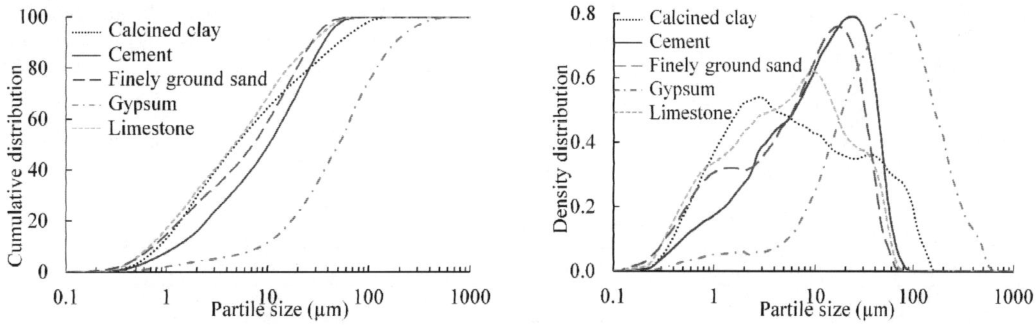

Figure 1. Particle size distribution of raw materials.

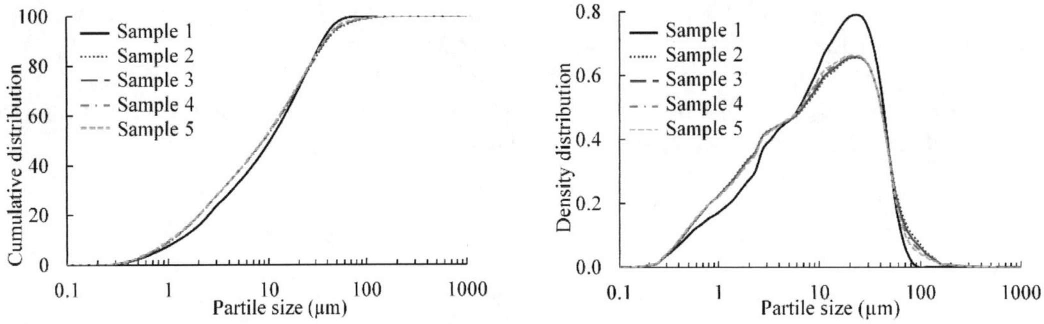

Figure 2. Particle size distribution of measured samples.

The pozzolanic activity of calcined clay was determined using two methods. The first method was the accelerated R^3 test [8]. A solid part of the used mixture contained 1 g calcined clay and 3 g $Ca(OH)_2$. Water to solid ratio of the used mixture was 1.2. Evolved heat from the sample was measured using isothermal calorimeter TAM air (TA Instruments) at isothermal temperature 25°C. Heat flow was measured for 168 hours. The first hour of the measurement was used for temperature calibration and CaO reaction, and the pozzolanic activity of the clay was determined from the heat evolved during 140 hours. The time of measurement was adjusted to account for the temperature difference, according to hydration heat of microsilica at 20°C and 25°C.

Pozzolanic activity of calcined clay was also assessed by modified Chapelle test according to NF P18-513 standard, Annex A. This test allows quantification of $Ca(OH)_2$ fixed by 0.4 g of calcined clay when mixed with 0.8 g of CaO and 100 ml of distilled CO_2 free water. The suspension was continuously stirred in an Erlenmeyer flask at 85±5°C for 16 hours. Erlenmeyer flask was then cooled to ambient temperature with tap water, 24 g of sucrose were added and the suspension was stirred with a magnetic bar for an additional 15 min and filtered. 10 ml of filtrate was titrated with HCl 0.15 M. Blank samples without calcined clay were prepared using the same method [9].

Formed hydration products were analysed using TG-DTA analysis (SDT650, TA Instruments). Samples, used in evolved heat measurement using an isothermal calorimeter, were crushed to 1–2 mm pieces. Hydration process was stopped by washing the samples 2× with isopropanol and 2× with diethylether and drying the samples at 45°C. Measurement was carried out up to 1000°C with heating rate 20°C/min under dried air atmosphere and sample size 40–50 mg.

3. Results and discussion

3.1. Characterization of clay particle

The chemical composition of calcinated clay can be seen in Table 2. Oxides contained in the clay in major quantities are silica and alumina. Other elements are only present in minor quantities. Silica to alumina ratio is of great importance in the use of calcinated clay in blended cements. The amount of silica influences the amount and type of formed hydration products. Alumina rich calcined clay increases the Al-uptake in C-S-H and the aluminate containing hydrates [10]. The loss on ignition value indicates that clay has lower carbonaceous matter and higher mineral matter contents. This confirms the chemical analysis of clay, where there is little MgO, CaO. The mineralogical composition of used clay can be seen in Table 3. The main identified minerals were quartz, illite, kaolinite and feldspars. The clay sample also contained over 30 % of an amorphous phase.

Table 2. Chemical analysis of clay (XRF).

Chemical composition	MgO	CaO	Al_2O_3	SiO_2	K_2O	TiO_2	Fe_2O_3	LOI (900 °C)
Weight (%)	0.9	0.5	19.9	61.8	2.0	1.1	4.2	9.5

Table 3. Chemical analysis of clay (XRD).

Mineral	Quartz	Chlorite	Biotite	Kaolinite	Illite	Albite	Orthoclase	Amorphous
Weight (%)	21.2	7.3	2.4	6.9	15.1	4.5	10.2	32.3

Pozzolanic activity values of clay calcinated at 700°C determined using accelerated R^3 test and modified Chapelle test according to standard NF P18-513 are shown in table 4, together with the pozzolanic activity of commonly used pozzolanic materials determined using the same methods. XRD analysis of clay showed the absence of limestone in clay. Therefore, correction of pozzolanic activity due to CaO formed during calcination was not needed.

Table 4. Pozzolanic activity of calcinated clay and commonly used pozzolans.

Raw materials	Quartz - blank	Metakaolin - Elkem	Silicafume - Imeta4	Slag - Štramberk	Calcined clay - Poštorná
R^3 (J/g)	17	222	193	109	202
Chapelle test ($mg_{Ca(OH)_2}/g_{pozzolanic}$)	127	1243	878	489	654

3.2. Isothermal calorimetry

The heat flow curve during hydration is shown in Figure 3, and the main kinetic events related to hydration reactions are highlighted. Portland cement's typical heat flow curve (sample 1) is compared with samples containing calcined clay and limestone in different ratios. The main hydration peak occurs after the end of the induction period. It is associated with the fast dissolution of alite and precipitation of C-S-H and C-(A)-S-H. Ettringite is also continuously precipitating during this time, and sulfate ions are adsorbed in C-(A)-S-H [11].

The presence of active alumina from calcined clay caused an increase in heat flow during the induction period. In samples 2–5, the maximum at the main hydration peak was reached faster. The presence of calcined clay and limestone accelerates the hydration of clinker phases due to filler effect [12]. This is caused by additional surface available for nucleation and growth of hydration products [13, 14]. According to Berodier and Scrivener, limestone also contributes to hydration of cementitious systems via the release of additional Ca^{2+} ions. Calcium ions accelerate the hydration of the clinker phases and shorten the induction time [15]. However, the height of the main hydration (first) peak in samples with calcined clay decreased by 20–25% compared to sample 1 serving as a reference sample.

A distinct aluminate hydration peak is observed for all samples. However, broadened and delayed aluminate hydration peak was observed in blend limestone and calcined clay. It has been reported that the amount of alumina available for the formation of the carboaluminate phase controls the reactivity of calcite [16, 17]. The studies suggest that other factors, in addition to the aluminate content, could affect the kinetics of calcite dissolution in low clinker factor cements [18]. In measured samples with calcined clay and limestone, the pozzolanic activity was affected, and the depletion of the sulphate content was slowed down [18]. The heat flow after 75 h in the cement with calcined clay was higher due to the ongoing pozzolanic reaction.

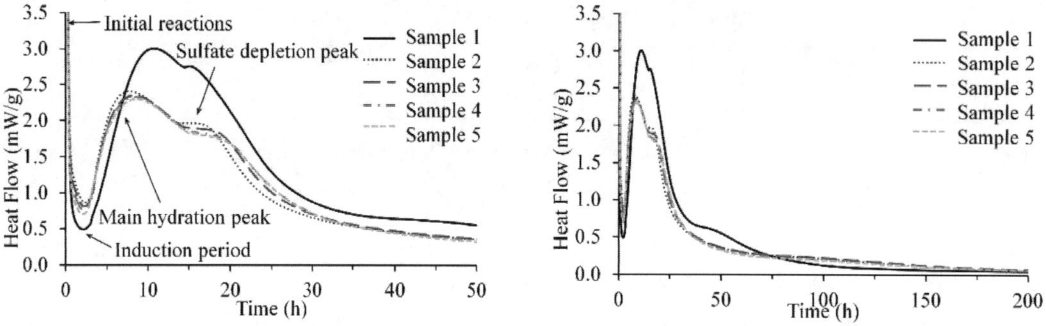

Figure 3. Heat release per gram of solid.

As can be seen from the total hydration heat of prepared samples (Figure 4), the substitution of clinker with calcined clay and limestone significantly influenced the hydration process. Cumulative hydration heat after 48 h, observed for the sample 1 (without calcined clay and limestone), was 259 J/g. The addition of calcined clay and limestone caused a decrease of evolved hydration heat by about 18 % (210–215 J/g). This difference became progressively smaller at later stages of hydration. Total hydration heat after 168 h in sample 1 was 331 J/g and only about 8–14% lower in other samples, which can be attributed to the pozzolanic reaction during later hydration stages. Out of clay containing samples, the highest heat was released by the sample with 5% limestone and 25% calcined clay. These observations are in agreement with other similar studies dealing with calcinated clay and limestone cement mixtures and can be explained by filler effect and pozzolanic reaction [19, 20].

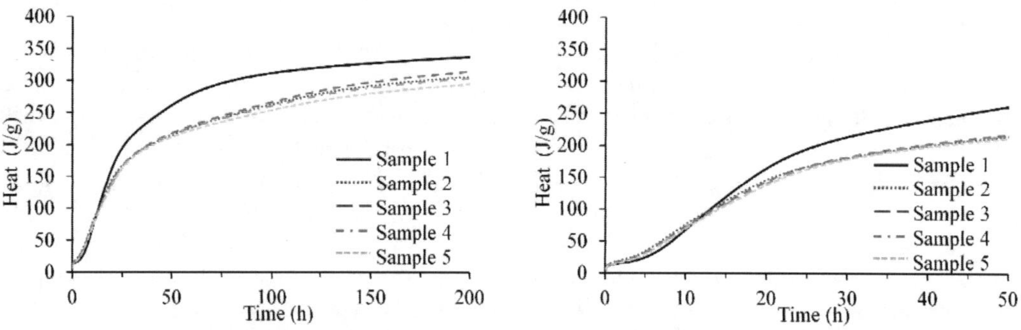

Figure 4. Cumulative heat released per gram of solid.

3.3. Testing of mechanical parameters

Typically, the degree of hydration of Portland cement is about 50% after one day of hydration. After 28 days, the degree of hydration typically increases to around 80%, and most of the unhydrated material is the slow-reacting belite phase and large cement grains [21]. Mechanical parameters of prepared samples with calcined clay and limestone were compared to Portland cement. The resulting compressive and flexural strengths at 1, 7 and 28 days are shown in Figure 5.

The addition of calcined clay caused significant decrease of compressive strength especially at early ages. Compressive strength values are relatively similar among samples 2–4. Ever bigger decrease was observable in sample 5. The difference in compressive strength compared to reference sample 1 after 2 days was 32–34% and 42 % for sample 5 containing 15% limestone and 15% calcined clay. After 7 days of the water curing conditions, the difference of compressive strength values decreased to only 26–29%, compared to the reference sample. One exception being sample 5 achieving only 60% of the compressive strength of the reference sample. After 28 days the compressive strength of clay containing samples were comparable to those of the sample 1. This trend is a clear illustration of the effect of ongoing pozzolanic reaction due to the calcined clay addition.

Studies have shown that the amorphous content in cement affects the mechanical properties. The limited amount of amorphous phase also reduces the reaction between calcined clay and limestone to form mono/hemicarboaluminate hydrates. However, the reduction of stable mono/hemicarboaluminate contributes to reduced compressive strength [22].

The lowest values of compressive strength after 28 days were achieved in sample 5, containing 15% calcined clay and the highest values in sample 3 with 25% calcined clay. The effect of pozzolanic activity can be seen here. At a later time was a grater increase in the compressive strength. The contribution of pozzolanic reaction to compressive strength becomes significant mainly at later stages of hydration process. [23].

A similar trend was also observed for the development of flexural strength. However, after 28 days flexural strength of the samples reached values similar to those of the reference sample. Furthermore, flexural strength loss caused by the substitution of cement was fully compensated after 28 days in sample 4, containing 20% of calcined clay and 10% of limestone. 2:1 ratio between calcinated clay and limestone was also found as optimal by other similar studies [22].

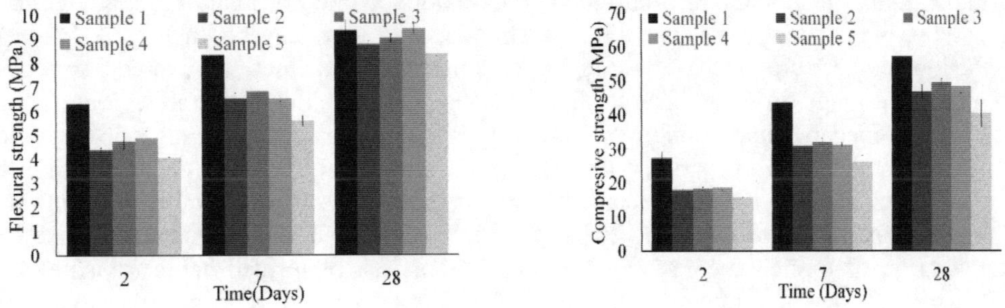

Figure 5. Mechanical parameters of measured samples.

3.4. *DTA analysis of hydration products*

Figure 6 shows results of TG/DTA analysis and DTG curves for measured samples after 8 days of hydration. TG/DTA curves show 3 main mass loss areas with corresponding endothermic peaks. Peaks in temperature range around 50–160°C are attributed to the decomposition of ettringite (Aft), C-S-H and C-A-S-H phases. Hemicarbonate aluminate hydrate (HC) decomposes at 140–200°C Characteristic peak of portlandite (CH) decomposition can be seen in the temperature range between 440–530°C. Decomposition of carbonates generally takes place between 625–875°C [24, 25].

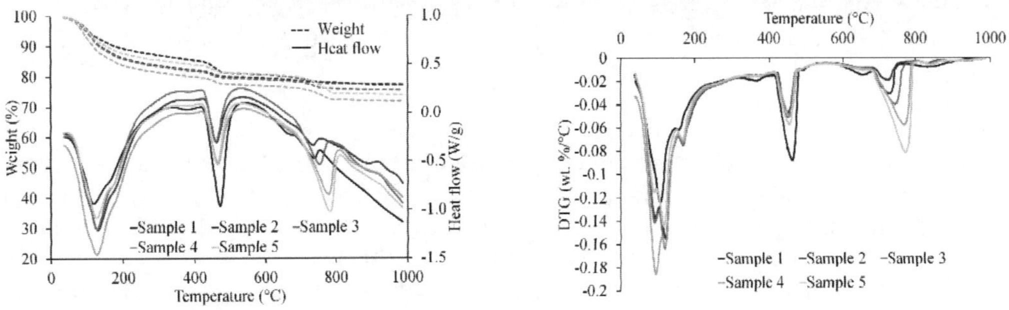

Figure 6. TG/DTA and DTG curves of hydrated cement samples.

Due to the complex nature of hydrated cement pastes and overlap of decomposition temperature ranges of different phases, quantification and identification of emerging phases using TG/DTA analysis can be difficult. In this study, the temperature range 50–400°C was assigned to mass loss caused by dehydration of ettringite, C-S-H, C-A-S-H and carboaluminate phases (AFm), 400–500°C to dehydroxylation of CH and 600–900°C to the decomposition of carbonates. Mass loss in these ranges is reported in Table 5.

Table 5. Mass loss of hydrated cement samples determined by thermal analysis.

Sample	Mass loss (wt. %)		
	Dehydration (C-S-H, C-A-S-H, Aft, Afm)	Dehydroxylation (CH)	Decarbonation (Carbonates)
1	14,1	4,2	3,4
2	17,4	2,2	2,0
3	17,7	2,4	3,4
4	19,2	2,4	5,0
5	15,7	2,7	6,6

With the increasing amount of calcined clay added to the mixture, the amount of CH in hydrated cement decreased. This trend can be partially attributed to the lower amount of cement in the paste and partially to the pozzolanic reaction of calcined clay, as can be seen with the increase of remaining portlandite, with the increasing amount of limestone in the mixture. While DTG curve of the reference PC sample only shows two distinctive peaks within 50–200°C temperature range, with a small shoulder observable on the low temperature range of the first peak, samples with calcined show three distinctive peaks in the temperature range 50–200°C. According to Boháč [25] the low temperature peak found around 90°C can be assigned to dehydration of amorphous C-S-H. This peak becomes most pronounced in sample 4 with calcined clay to limestone ratio 2:1. Temperature shift of C-S-H decomposition to lower temperature compared to Portland cement sample possibly suggests lower crystallinity of pozzolanic reaction products. Peak around 120°C can then be attributed to ettringite decomposition and further dehydration of formed C-S-H [26]. Decomposition peak of formed hemicarbonate was observed at 170°C [4, 27]. Higher amount of formed C-S-H and C-A-S-H provides further evidence of good pozzolanic activity of calcined clay and synergistic effect of calcined clay and limestone on cement hydration. These findings are in good agreement with available literature [4–6, 25].

The last significant mass loss was related to the decomposition of carbonate phases. Substitution of cement by SMC caused a significant drop in carbonate content in sample 2. Limestone addition manifested itself by an increase of carbonate content and also shifting the decomposition to higher temperatures in samples 3–5, suggesting different type, structure or crystallinity degree of added limestone. Carbonate peaks observable in samples 1–2 can be likely attributed to decomposition of the micro-filler present in commercially produced cement and carbonation product. The reference sample also showed a small additional peak at around 650°C, barely observable in other samples. This fact suggests the presence of poorly crystalline carbonation products, which generally decompose at this temperature and readily react with calcined clays, forming carboaluminate phases in samples 2–5 [26].

4. Conclusions

Blended Portland cements, to meeting the requirements set by standard EN-197-1 – CEM II/B-M, were successfully prepared by partial substitution of clinker by locally sourced calcined clay and limestone. The optimal calcined clay to limestone ratio, with 30 % clinker substitution, was found to be 2:1. The addition of calcined clay and limestone lead to increased heat flow during the induction period and earlier maximum on the main hydration peak due to active alumina content in the clay and filler effect. Heat flow after 90 h of hydration was higher than Portland cement due to the pozzolanic reaction of calcined clay. The addition of calcined clay and limestone lead to a lower amount of portlandite, increased amount of C-S-H and C-A-S-H and formation of carboaluminate hydrates. A significant increase in the amount of hydration products confirmed via thermal analysis further confirms the synergistic effect of calcined clay and limestone on Portland cement hydration.

Clay and limestone addition caused a decrease of initial compressive and flexural strength. However, after maturation, prepared blended cements reached long term strengths comparable to those of Portland cement.

Acknowledgements
This outcome has been achieved with the financial support by the project: TJ04000383 – "Use of Calcined Natural Clay Raw Materials in Portland Blended Cements" and using equipment that was funded by the project: GA19-16646S" The elimination of the negative impact of zinc in Portland cement by accelerating concrete admixtures", with financial support from the Czech science foundation.

References

[1] *Cement production in the United States and worldwide from 2010 to 2020* 2020 (*Online* https://www.statista.com/statistics/219343/cement-production-worldwide/)
[2] *Technology Roadmap - Low-Carbon Transition in the Cement Industry* 2018 (*Online* ww.iea.org/reports/technology-roadmap-low-carbon-transition-in-the-cement- industry)
[3] Jaskulski R, Jóźwiak-Niedźwiedzka D and Yakymechko Y 2020 Calcined Clay as Supplementary Cementitious Material *Materials* **13**
[4] Tironi A, Trezza A M, Scian N A, Irassar F E and Martirena F 2014 Thermal analysis to assess pozzolanic activity of calcined kaolinitic clays *J.Therm. Anal. Calorim* **117** pp 547–556
[5] Alujas A, Fernández R, Quintana R, Scrivener L K and Martirena F 2015 Pozzolanic reactivity of low grade kaolinitic clays *Appl. Clay Sci.* **108** pp 94–101
[6] Li Q, Trezza A M, Scian N A, Irassar F E and Martirena F 2014 The microstructure and mechanical properties of cementitious materials comprised of limestone 2014 *Ceramics - Silikaty* **117** pp 356–364
[7] Wang D, Shi C, Farzadnia N, Shi Z, Jia H and Ou Z 2018 A review on use of limestone powder in cement-based materials *Constr. Build. Mater.* **181** pp 659–672
[8] Avet F, Snellings R, Alujas Diaz A, Ben Haha M and Scrivener K 2016 Development of a new rapid, relevant and reliable (R3) test method to evaluate the pozzolanic reactivity of calcined kaolinitic clays *Cem. Concr. Res.* **85** pp 1–11
[9] Metakaolin - measuring the total quantity of fixed Calcium Hydroxide (Chapelle test modified) (*Online* http://www.metakaolin.info/quality-criteria/chapelle-test.html)
[10] Lothenbach B, Scrivener K and Hooton R D 2011 Supplementary cementitious materials *Cem. Concr. Res.* **41** pp 1244–1256
[11] Quennoz A and Scrivener K 2013 Interactions between alite and C3A-gypsum hydrations in model cements *Cem. Concr. Res.* **44** pp 46–54
[12] Lothenbach B, Scrivener K and Hooton R D 2011 Supplementary cementitious materials *Cem. Concr. Res.* **41** pp 1244–1256
[13] Stark J 2004 Optimierte Bindemittelsysteme für die Betonindustrie Beton **54** pp 486–490
[14] Péra J, Husson S and Guilhot B 1999 Influence of finely ground limestone on cement hydration *Cem.Concr. Compos.* **21** pp 99–105
[15] Berodier E, Scrivener K and Scherer G 2014 Understanding the Filler Effect on the Nucleation and Growth of C-S-H *J. Am. Ceram. Soc.* **97** pp 3764–3773
[16] Zajac M, Rossberg A, Le Saout G and Lothenbach B 2014 Influence of limestone and anhydrite on the hydration of Portland cements *Cem.Concr. Compos.* **46** pp 99–108
[17] Zajac M, Durdzinski P, Stabler C, Skocek J and Nied D 2018 Influence of calcium and magnesium carbonates on hydration kinetics, hydrate assemblage and microstructural development of metakaolin containing composite cements *Cem. Concr. Res.* **106** pp 91–102
[18] Krishnan S, Emmanuel C A and Bishnoi S 2019 Hydration and phase assemblage of ternary cements with calcined clay and limestone *Constr. Build. Mater.* **222** pp 64–72
[19] Huang W, Kazemi-Kamyab H, Sun W and Scrivener K 2017 Effect of cement substitution by limestone on the hydration and microstructural development of ultra-high performance concrete (UHPC) *Cem.Concr. Compos* **77** pp 86–101
[20] Du H and Pang S D 2020 High-performance concrete incorporating calcined kaolin clay and limestone as cement substitute *Constr. Build. Mater.* **264**
[21] Scrivener K and Nonat A 2011 Hydration of cementitious materials, present and future

IOP Conf. Series: Materials Science and Engineering 1205 (2021) 012005 doi:10.1088/1757-899X/1205/1/012005

Cem. Concr. Res. **41** pp 651–665

[22] Antoni M, Rossen J, Martirena F and Scrivener K 2012 Cement substitution by a combination of metakaolin and limestone *Cem. Concr. Res.* **42** pp 1579–1589

[23] Berodier E and Scrivener K 2015 Evolution of pore structure in blended systems *Cem. Concr. Res.* **73** pp 25–35,

[24] Kuliffayová M, Krajči L, Janotka I, Šmatko V, Jia H and Ou Z 2012 Thermal behaviour and characterization of cement composites with burnt kaolin sand *J. Therm. Anal. Calorim.* **108** pp 425–432

[25] Boháč M, Palou M, Novotný R, Másilko J, Šoukal F and Opravil T 2017 Influence of temperature on early hydration of Portland cement-metakaolin-slag system *J. Therm. Anal. Calorim.* **127** pp 309–318

[26] Gabrovšek R and Vuk T Kaučič 2006 Evaluation of the Hydration of Portland Cement Containing (Acta Chimica Slovenica)

[27] Alarcon-Ruiz L, Platret G, Massieu E and Ehrlacher 2005 A The use of thermal analysis in assessing the effect of temperature on a cement paste *Cem. Concr. Res.* **35** pp 609–613

IOP Conf. Series: Materials Science and Engineering 1205 (2021) 012006 doi:10.1088/1757-899X/1205/1/012006

Light autoclaved foam concrete with foam glass-based aggregate

P Bibora[1], V Prachar[1], O Svitak[1] and R. Zavrelova[1]

[1] Research Institute for Building Materials, Hnevkovskeho 30/65, 617 00 Brno, Czech Republic

Email: bibora@vustah.cz

Abstract. The presented research aims to develop new optimized composition of novel lightweight concrete with a very low bulk density. Low bulk density of developed lightweight concrete is achieved by using a combination of non-traditional lightweight artificial fillers, dispersed fibre reinforcement and pre-generated technical foam. The methodology of lightweight concrete test specimen production, based on technologies commonly applied for the production of lightweight concrete, foam concrete or autoclaved aerated concrete was also designed and verified. Not only the physical and mechanical parameters, but also the thermal insulation properties were verified on the produced test specimens of the developed lightweight concrete.

1. Introduction

Low bulk density is a characteristic feature of lightweight building materials. Lightweight building materials have good thermal insulation, ie. low thermal conductivity. On the other hand, these materials also have an unfavorably low heat accumulation. If we consider other conflicting properties such as strength and porosity, basis weight and soundproofing, sorption and sound absorption, we see that the problem of matching the low bulk density of these building materials with their other properties can be complicated [1].

Low bulk densities of lightweight concrete can be achieved in three ways:

a) Using aggregate voids
b) Using aggregate porosity
c) By creating pores in fine-grained cement, lime or lime-cement mortar

These methods can be combined with each other [1].

Lightweight concretes marked LC are lightened by cavities and pores in the concrete texture. They are used in structural-insulating, insulating structures and sometimes also for structural purposes, if the actual weight of the structure represents a decisive load. In other words, lightweight concrete is used for structural purposes only in the case of the minimum expected total load on the structure. Furthermore, lightweight concretes can be divided into dense lightweight concretes made of porous aggregate, interstitial concretes, aerated concretes and foam concretes. As light concretes, we commonly refer to a group of concretes with a bulk density of less than 2,000 kg/m^3, while it is currently possible to achieve values of significantly lower than 600 kg/m^3 by various production methods. In order to achieve such a low values of bulk densities of these lightweight concretes, it is necessary to replace all or a certain part of the natural aggregates with another input raw material. Artificial aggregates such as expanded perlite, polystyrene and various types of ceramic aggregates are currently used as the most common fillers for the production of lightweight concrete, due to the very limited possibilities of availability of natural lightweight aggregates. However, a lightweight concrete structure can also be achieved with the help of special foam additives, in this case we speak of so-called foam concrete, which is produced by implementation of a special foam or foam additive and related air pores into the cement matrix of concrete [2].

The term foam concrete is misleading in itself, as the majority of foam concrete does not contain any major aggregates, only fine sands in combination with cement, water and foam, so the final product should rather be referred to as foam mortar. In general, foam concrete is described as a material with an air content of more than 25%, which distinguishes it from other highly aerated materials. In its basic form, foam concrete is a mixture of sand, cement, water and a foaming agent or a pre-prepared synthetic or protein-based foam. The addition of pre-prepared foam to the base concrete mix reduces its bulk density, the more foam we add to the base mix, the lighter the final product will be. Depending on the application of foam concrete, part of the cement can be replaced or supplemented with other admixtures in the form of granulated blast furnace slag or fly ash. The filler in the form of sand can then be partially supplemented or replaced with ground limestone or crushed concrete. The foam that is added to the base mix is equally important and must be able to remain stable and not degrade during pumping, application and treatment of the concrete mix. This factor is especially important if the foam becomes the dominant component (more than 50% of the composition) of the foam concrete. This situation occurs especially with foam concrete with a bulk density of about 1,100 kg/m^3. Foam concretes with a lower bulk density must then be handled very carefully [3].

Lightweight concretes can be divided into structural or infill according to their strength characteristics. Marking of lightweight concretes is performed according to the ČSN EN 206-1 standard with an additional parameter D, which indicates their bulk density. The advantages of lightweight concrete are especially excellent thermal insulation capacity, low weight and frost resistance. The disadvantages may be lower resistance to mechanical stress and lower load-bearing capacity compared to conventional and structural concretes [2].

The aim of the presented research is the development and optimization of the composition of lightweight concrete using a combination of production technologies for lightweight concrete, foam concrete and autoclaved aerated concrete. Low bulk densities of developed concrete were achieved by a combination of lightweight aggregate based on foam glass and technical foam for the production of foam concrete. Sufficient mechanical parameters of the developed lightweight concrete were then supported by hydrothermal technology of curing in an autoclave and by the addition of dispersed fiber reinforcement. Fiber reinforcement is commonly used to provide enhanced toughness and ductility to brittle cementitious matrix [4]. Lightweight environmental concretes with physical-mechanical characteristics comparable to commonly introduced lightweight concretes were achieved. The intended use of this lightweight concrete is for infill non-load-bearing structures of buildings, floors and light monolithic partitions. The developed concrete mixture can also be used for the production of building elements in the form of blocks, insulating "sandwich" structures or composite building materials.

2. Experimental works

The proposed composition of the lightweight foam concrete was optimized to ensure the immediate stability of the fresh foam concrete mixture and to avoid a decrease in the mixture deposited in the molds due to the degradation of the pores in the foam concrete structure. For this reason, new input raw materials were successfully tested in the proposed recipes, especially light artificial fillers in the form of crushed waste foam glass of various fractions.

2.1. Mixture composition

The composition of the final mixture for the production of functional samples of lightweight concrete is shown in table 1. This is the mixture for which the best ratio of high strength and low bulk density was achieved.

Table 1. Composition of developed lightweight concrete mixture.

Composition	Quantity for 1 m^3
Cement CEM I 52.5 R	315 kg
Alumina Cement	56 kg
Mikrosilika	37 kg
Foam glass aggregate 8-16 mm	222 kg
Lithium Carbonate	0.4 kg
Tartaric Acid	0.4 kg
Water	222 kg
Technical Foam	370 l
Glass Fibres	0.8 kg

This mixture was used for research works aimed at verifying the laboratory and pilot production technology of developed lightweight concrete using hydrothermal curing of test specimens in an autoclave.

2.2. Production of test specimens, testing methods

Production of test specimens was carried out according to the requirements and statutes of the standard ČSN EN 12390-1 (731302) Testing of hardened concrete - Part 1: Shape, dimensions and other requirements for test specimens and molds. Lightweight concrete test specimens were produced - beams measuring 400 × 100 × 100 mm.

The weighed dry components - cement, microsilica, sand, light aggregate were placed in a homogenizer, mixed and homogenized for three minutes. The required amount of mixing water was weighed and the homogenized dry components were mixed thoroughly with water using a hand mixer for three to five minutes, depending on the appearance of the fresh mixture. The addition of fibres in a volume of 1% of the total dry mixture was added to the mixture at the end of the mixing, and stirring was continued for another two minutes to allow sufficient distribution of the fibres in the whole volume of the fresh mixture.

The laboratory foam generator was used for the production of technical foam. The laboratory foam generator is connected to a mobile air compressor, a liquid foaming additive is metered into the tank and then the required amount of technical foam is produced by means of a regulated air pressure. Prepared technical foam was further mixed with the fresh cement mixture at low speed, so that the foam structure and individual closed pores did not collapse. The volume of foam dosed into the fresh mixture was on average 1 l of foam per 5 kg of dry mixture.

The produced fresh mixture was placed in molds for the production of test specimens 400 × 100 × 100 mm. The test specimens were demolded and transferred to an autoclave.

The physical-mechanical parameters were tested and the influence of the autoclaving process on the resulting strength and bulk density of the test specimens was evaluated after 7 and 28 days of wet storage of the test specimens. Specimens were tested for physical-mechanical parameters according to the standards listed below.

- EN 12390-7. Testing hardened concrete - Part 7: Density of hardened concrete
- EN 12390-5. Testing hardened concrete - Part 5: Flexural strength of test specimens
- EN 12390-3. Testing hardened concrete - Part 3: Compressive strength of test specimens

The technology of lightweight concrete production was modified by a new technological step in the form of hydrothermal hardening of lightweight concrete test specimens in an autoclave and subsequently the influence of temperature and pressure on the resulting physical and mechanical properties of autoclaved foamed concrete was observed and then pilot plant production of lightweight foam concrete test beams was carried out.

2.3. Pilot production of the developed lightweight concrete

Verification of the possibility of pilot production of lightweight foam concrete with artificial aggregates based on foam glass was performed as part of research work. The aim was to verify the production of lightweight foam concrete for future products of larger formats. Test specimens of lightweight foam concrete measuring 400 × 100 × 100 mm were made for this purpose. Test specimens were cured by a hydrothermal process in a large autoclave during the pilot production process. Conditions for hydrothermal curing in the form of temperature, pressure and time of storage of manufactured specimens in the autoclave were as follows. The temperature of 190°C, the pressure of 1.3 MPa and the storage time of the test specimens in the autoclave was 24 hours. After hydrothermal curing in autoclave, the lightweight foam concrete test specimens were stored in a laboratory humid environment for another 26 days, after which time the beams were tested for physical and mechanical properties - flexural strength, compressive strength and bulk density.

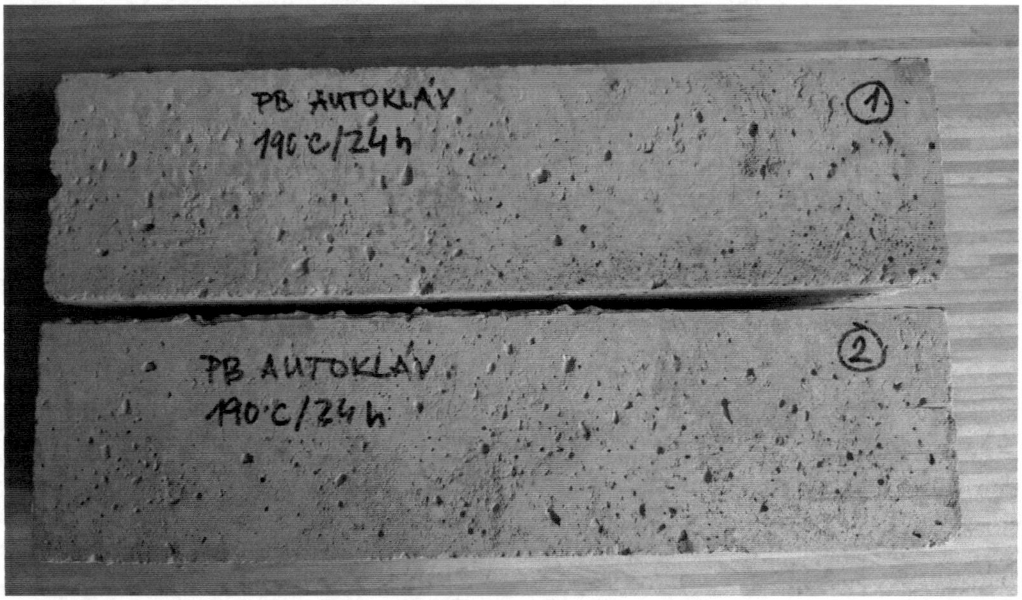

Figure 1. Test specimens of light foam concrete after completion of hydrothermal curing in an autoclave for 24 hours.

Figure 1 shows test specimens of lightweight foam concrete with artificial aggregate placed in a large autoclave before the start of the pilot phase of hydrothermal curing at a temperature of 190°C, a pressure of 1.3 MPa for 24 hours. Figure 1 shows the test specimens after the completion of the hydrothermal curing process. As can be seen in the figure, the integrity and appearance of the test specimens were not disturbed due to the effect of increased temperature and pressure.

3. Results and Discussion

3.1. Thermal insulation parameters of developed lightweight concrete

The produced test specimens of lightweight concrete with artificial aggregate showed relatively high porosity and low bulk densities of test specimens in the range of 600–800 kg / m^3. Therefore,

IOP Conf. Series: Materials Science and Engineering 1205 (2021) 012006 doi:10.1088/1757-899X/1205/1/012006

the measurement of thermal insulation properties was performed on the samples. The measurement was performed with an ISOMET instrument using an attached probe with a measurement range of 0.04–0.3 $W \cdot m^{-1} \cdot K^{-1}$. The probe was always placed in the centre of the sample. The measurement was performed on 3 samples of foam concrete, first in the direction of compaction and then in the perpendicular direction. The measurement with the ISOMET 2114 instrument is based on the analysis of the temperature response of the analyzed material to heat flow pulses. The heat flux is excited by the electric heating of a resistance heater inserted in the probe, which is in direct thermal contact with the test sample. The evaluation of thermal conductivity and volume heat capacity is based on the results of temperature records of regularly repeated measurements.

The following tables show the measured values of thermal insulation properties of developed lightweight concrete with artificial aggregate.

Table 2. Thermal insulation parameters of developed lightweight concrete.

Thermal insulation parameter	Value	Average values of all measurements
Thermal conductivity coefficient λ	$W \cdot m^{-1} \cdot K^{-1}$	0.1014
Volumetric heat capacity cρ	$10^{6} \cdot J \cdot m^{-3} \cdot K^{-1}$	0.4435
Thermal conductivity coefficient a	$10^{-6} \cdot m^{2} \cdot s^{-1}$	0.2335

As can be seen from table 2, the coefficient of thermal conductivity λ of the developed lightweight concrete is equal to the average value of 0.1014 $W \cdot m^{-1} \cdot K^{-1}$. If we compare this value with the coefficients of thermal conductivity λ of other light building materials [5], then we can state that the thermal insulation properties of developed lightweight concrete appear to be very satisfactory, with a better value of coefficient λ than other light building materials and concretes such as aerated concrete or lightweight concrete with expanded clay aggregate. A comparison of the values of the thermal conductivity coefficient λ for different materials, including the developed lightweight concrete, is shown in table 3.

Table 3. Thermal conductivity coefficient λ ($W \cdot m^{-1} \cdot K^{-1}$) for different materials.

Material	Thermal conductivity coefficient λ ($W \cdot m^{-1} \cdot K^{-1}$)
Developed lightweight concrete	**0.10–0.12**
Fly ash based aerated concrete	0.18–0.20*
Lightweight concrete with slag pumice	0.55–0.84*
Lightweight concrete with expanded clay	0.28–1.3*
Lightweight concrete with Perlite aggregate	0.10–0.16*
Lightweight concrete with slag	0.52–1.00*

*Values of thermal conductivity coefficient of the other lightweight construction materials. [5]

The coefficient of thermal conductivity λ of the developed lightweight foam concrete with artificial aggregate made of foam glass reaches better values than the values λ of the other lightweight concretes mentioned. [5] Only light concretes with light artificial aggregate Perlite reach comparable values.

3.2. Physical-mechanical parameters of developed lightweight concrete

The achieved values of autoclaved test specimens were compared with reference test specimens for which hydrothermal curing in an autoclave was not used and subsequently the influence of the autoclaving process on the resulting parameters was evaluated.

IOP Conf. Series: Materials Science and Engineering 1205 (2021) 012006 doi:10.1088/1757-899X/1205/1/012006

The resulting values of physical-mechanical parameters of lightweight concrete test specimens, hardened in an autoclave are shown in table 5, values of physical-mechanical parameters of non-autoclaved reference test specimens are shown in table 4.

Table 4. Values of physical and mechanical properties of reference test specimens.

Sample	Bulk density (kg/m^3)	Flexural strength (MPa)	Compressive strength (MPa)	
1	715	0.62	2.77	2.82
2	718	0.65	2.60	2.90
Ø	**717**	**0.63**	**2.77**	

Table 5. Values of physical and mechanical properties of autoclaved test specimens.

Sample	Bulk density (kg/m^3)	Flexural strength (MPa)	Compressive strength (MPa)	
1	660	0.67	2.66	2.86
2	687	0.80	3.27	2.99
Ø	**647**	**0.74**	**2.95**	

Table 5 shows the results of autoclaved test specimens produced by a hydrothermal curing process during the verification pilot plant production phase. The test specimens were cured in an autoclave at a temperature of 190°C, a pressure of 1.3 MPa and for 24 hours. If we compare the individual average values of physical and mechanical parameters of autoclaved and non-autoclaved test specimens, given in table 4 and 5, then it can be stated, that due to hydrothermal hardening of test specimens in the autoclave there was a slight decrease in the values of bulk densities of autoclaved test specimens with a simultaneous slight increase in strength characteristics. The positive influence of the hydrothermal process in the autoclave on the resulting physically mechanical and probably also thermal insulation parameters of the test specimens was verified.

Table 6 shows a comparison of the achieved parameters of the developed concretes with the typical properties of common foam concretes depending on the required bulk density [3].

Table 6. Comparison of the achieved parameters of the developed light concrete with the typical properties of common foam concretes.

Material	Bulk density (kg/m^3)	Compressive strength (MPa)	Thermal conductivity ($W \cdot m^{-1} \cdot K^{-1}$)
Developed concrete	**647**	**2.95**	**0.1-0.12**
Foamed concrete 400	400*	0.5–1.0*	0.1*
Foamed concrete 600	600*	1.0–1.5*	0.11*
Foamed concrete 800	800*	1.5–2.0*	0.17–0.23*
Foamed concrete 1,000	1,000*	2.5–3.0*	0.23–0.30*

*Typical properties of foamed concrete [3].

Based on the findings above, it can be stated, that the developed lightweight concrete with the bulk density value of about 650 kg/m^3, reached values of compressive strength comparable to compressive strength values typical for foamed concretes with a bulk density of about 1,000 kg/m^3. The lower bulk density of the developed lightweight concrete had a positive effect on the values of thermal conductivity. As can be seen from table 6, conventional foam concretes with comparable values of compressive strengths reach bulk densities of about 1,000 kg / m^3. At this bulk density, the typical value of foam concrete thermal conductivity coefficient λ is equal to the range 0.23–0.30 $W \cdot m^{-1} \cdot K^{-1}$. The value of the thermal conductivity coefficient λ of developed lightweight concrete was equal to the range 0.1–0.12 $W \cdot m^{-1} \cdot K^{-1}$ and the value of bulk density was around 650 kg/m^3. The achieved results proved that the test beams of developed lightweight concrete had significantly lower bulk density, higher strength and better thermal conductivity, compared to conventional foam concrete.

4. Conclusion

Implemented research work has shown that it is possible to realize the production of lightweight foam concrete with artificial waste aggregates based on foam glass and fibre reinforcement, even in the case of pilot production of composite elements of larger dimensions. It has been shown that the use of the hydrothermal curing process of the designed lightweight foam concrete in an autoclave leads to increased values of strength characteristics of specimens. However, the use of hydrothermal hardening of the developed foam concrete in practice will depend mainly on the economic aspect in the form of energy prices necessary for the implementation of this proposed production process.

The thermal insulation properties of the developed foam concretes, with very satisfactory results, were achieved. In particular, the achieved values of the coefficient of thermal conductivity λ have shown that the developed lightweight foam concrete achieves better thermal insulation parameters than in the case of compared lightweight concretes with artificial aggregates or typical foam concretes. It is thus possible to assume the successful application of lightweight foam concrete, e.g. in the form of lightweight insulating fillings for ceilings, thermal insulation partitions and panels, or lightweight insulating composite prefabricates.

During the described research, a proven technology for the production of lightweight concrete with artificial aggregates and fibre reinforcement was developed. This result was fulfilled in the form of a Technological Regulation describing a proven technology for the production of lightweight concrete with artificial aggregates and fibre reinforcement.

Acknowledgments

This outcome has been achieved with the institutional support for the long-term development of research organizations by the Ministry of Industry and Trade of the Czech Republic.

References

[1] Drochytka R and Matulova P 2006 *Lightweight building materials, Module M01* (Technical University Brno)
[2] Pytlik P 2000 *Concrete technology* (Technical University Brno)
[3] Dhir R K, Newlands M D and McCarthy A 2005 Use od Foamed Concrete in Construction *Thomas Telford Publishing* (London)
[4] Drdlova M, Popovic M and Koutny O 2018 Blast resistance of hybrid fibre reinforced concrete containing polyvinyl alcohol, polypropylene and steel fibres with various shape parameters, et al *European Physical Journal-Special Topics* Vol. 227 Issue 1–2 pp 111–126
[5] https://stavba.tzb-info.cz/

Impact of the curing conditions and carbon dioxide ingress on heavyweight concrete

J Cepcianska[1], J Dragomirova[1], E Kuzielova[1, 2], M Zemlicka[1] and M T Palou[1, 2]

[1] Institute of Construction and Architecture, Slovak Academy of Sciences, Dúbravská cesta 9, SK-845 03 Bratislava, Slovak Republic
[2] Faculty of Chemical and Food Technology, Slovak University of Technology, Radlinského 9, SK-812 37 Bratislava, Slovak Republic

Email: jana.cepcianska@savba.sk

Abstract. The present work investigates the influence of curing conditions on the mechanical and physical properties of heavyweight concrete. The prismatic bars of 40 mm × 40 mm × 160 mm dimension were cured in a climatic chamber (relative humidity 30%, average temperature 26°C), wet (100% of humidity, average room temperature 26°C) and CO_2 chamber-wet (relative humidity 90%, average temperature 50°C and average CO_2 concentration 20 %) conditions for 2, 7, 28 and 90 days. Density, compressive strength, dynamic modulus of elasticity, and longitudinal shrinkage were determined at different ages of curing. Mercury Intrusion Porosimetry was used to analyze and determine the influence of carbonation on pore structure evolution. Samples cured under CO_2- wet conditions showed a higher compressive strength (54.05, 66.83, 84.98, 96.35 MPa) compared to that of the samples exposed to wet (45.49, 65.87, 78.91, 93.80 MPa) and dry (39.62, 46.52, 48.45, 45.28 MPa) conditions at all ages. The dynamic modulus of elasticity of CO_2-wet cured samples (53.02, 51.48, 59.24, 67.60 GPa) was lower than that of samples cured in wet conditions (59.82, 66.76, 78.84, 80.27, GPa), but higher than that of dry-cured samples (45.74, 45.73, 43.91, 44.62 GPa). The density of the samples exposed to all curing conditions was higher than 3800 kg/m^3. Carbonation led to a decrease in total porosity (from 10% to 20%) and an increase in density (from 320 to 390 kg/m^3). Also, the time and curing conditions have strongly influenced the pore structure. The precipitation of calcium carbonate in the matrix of concrete and the acceleration of hydration reaction under wet conditions has led to a decrease in porosity.

1. Introduction

Although ordinary concrete is widely used as the main building material, heavyweight concrete plays a specific role of structural and functional material in the construction of nuclear power plants (NPPs), medical units, particle accelerator facilities or storage bunkers of nuclear wastes where construction is exposed to strong neutron or gamma-ray radiation [1–3]. Therefore, the development of heavyweight concrete, which can meet the functional and structure (mechanic) demands of a radiation shielding material, with proven low activation and high mechanical properties is highly required [3]. The manufacture of heavyweight concrete uses heavy natural aggregates such as barites or magnetite [4] or manufactured aggregates such as iron or lead shot. The heavy aggregates containing many heavy elements are crucial in improving concrete shielding properties and hence contribute greatly to the attenuation of photons and neutrons [5]. Generally, concrete with specific gravities higher that 2600 kg m^{-3} is called heavyweight concrete according to EN [6] and aggregates with specific gravity higher than 3000 kg.m^{-3} are considered as heavyweight aggregates. Indeed, heavyweight concrete can undergo the same curing effects of temperatures and carbonation as ordinary concrete. It is manufactured using ordinary Portland cement or its blends as binder [1, 3–4]. Therefore, the hydration of cement phases in heavyweight concrete generates by the same manner different hydrous water-insoluble products with the capacity to blend the fillers, strengthen the structure and perform the durability of concrete over the time. The main products of cement hydration are calcium hydroxide $Ca(OH)_2$, calcium silicate hydrates $C_xS_yH_z$ (with different ratios), ettringite monosulfate or other calcium aluminate hydrates and in some cases calcium aluminosilicate hydrate. Calcium hydroxide $Ca(OH)_2$, calcium silicate hydrates $C_xS_yH_z$ are sensible to the action of carbon dioxide and produce calcium carbonate –phenomenon known as carbonation or carbon sequestration [7, 8].

In addition, unhydrated tricalcium silicate and dicalcium silicate (C_3S and C_2S) can also react with CO_2 in the presence of humidity to form calcium carbonate. The main carbonation reactions can be summarized as followed:

$$CO_2 + Ca(OH)_2 = CaCO_3 + H_2O \quad (1)$$

$$CSH + CO_2 = 2\,CaCO_3 + SiO_2aq + H_2O \quad (2)$$

where equation (2) describes total carbonation of C-S-H.

The carbonation mechanism, as reported by [8] can be divide into 8 stages: (1) diffusion of the gaseous $CO_2(g)$ trough external layer surrounding the hardened concrete; (2) diffusion of $CO_2(g)$ into the concrete pore solution; (3) solvation of $CO_2(g)$ to $CO_2(aq)$ in the pore water;(4) hydration of $CO_2(aq)$ to $H_2CO_3(aq)$; (5) ionization of H_2CO_3 to H^+, HCO_3^- and CO_3^{2-} (6) dissolution of calcium containing phases to Ca^{2+}; (7) nucleation of $CaCO_3(s)$ and (8) precipitation of solid phases.

Unfortunately, CO_2 lowers the pH of the alkaline environment surrounding steel reinforcements in concrete, this process reduces the level of a protective passive layer on the surface of steel reinforcement which makes it vulnerable for corrosion [9–11]. Moreover, in the presence of water excess, carbonation can produce carbonic acid and consequently a soluble $Ca(HCO_3)_2$ is formed.

$$CO_2 + H_2O = H_2CO_3 \quad (3)$$

$$H_2CO_3 + Ca(OH)_2 = CaCO_3 + 2H_2O \quad (4)$$

$$CaCO_3 + H_2CO_3 = Ca(HCO_3)_2 \quad (5)$$

Ions of HCO_3^- can then diffuse inward the sample and react there with the OH^- ions to form the $CaCO_3$. Dissolution of $CaCO_3$ to form Ca^{2+} and HCO_3^- can appear only when pH value sufficiently drops. This process is supposed to take place only in the carbonation zone in the moist samples.

The effect of carbonation on the compressive strength of concrete was deeply studied by authors [10, 11]. Carbonation of fresh concrete can increase its compressive strength with time. Indeed, the ingress of CO_2 trough fresh concrete is more accentuated compared to hardened concrete due to the slow diffusion rate trough the dense microstructure.

1.1. Carbonation and shrinkage

Shrinkage can be classified into 4 groups depending on curing conditions [12, 13]: (1) plastic shrinkage, (2) autogeneous shrinkage/ self-shrinkage due to self-dessification, (3) drying shrinkage, and (4) carbonation shrinkage. Autogenous shrinkage is defined as a concrete volume contraction caused by the reduction of internal moisture due to consumption of water by internal chemical and structural reactions of the concrete components. It occurs without moisture transfer to the environment and contributes significantly to concrete cracking when the water–cement (*w/c*) ratio is less than 0.4. Self-shrinkage is caused by the loss of water when the water is consumed by cement hydration. Drying shrinkage occurs in environment with a lower relative humidity [12, 13]. As for shrinkage carbonation, it is above all the process resulting from chemical reactions and the precipitation of fine grains of calcium carbonate in a porous structure inside the concrete matrix and which induces the change in volume.

The main hypotheses proposed to explain the mechanism of CO_2-induced shrinkage were reported by [14, 15] who stated that, under partially saturated conditions, shrinkage is due to the pressure exerted by the gas on the dissolved molecules of $Ca(OH)_2$ in highly "compressible fresh cement paste". Although this idea is refuted by other authors who claimed that this hypothesis is only valid for calcium hydroxide, it can be thought that this phenomenon may occur at an earlier hydration period in fresh concrete when partial pressure of CO_2 can act on the component and cause the volume change.

The carbonation of C-S-H phases is hypothetically occurring via a topochemical mechanism in which no shrinkage was observed. Calcium carbonate is formed mainly in outer product regions as microcrystalline vaterite or calcite. Reardon et al. [14] are of the similar idea postulating that the carbonation of C-S-H occurs through decomposition by a pseudomorphic reaction and hence, does not affect the overall shrinkage. Contrary to this idea, author [15] has considered shrinkage of C-S-H through decalcification or loss of calcium by reaction. Indeed, carbonation remove calcium ions from C-S-H and changes the C/S ratio. When the ratio C/S is reduced below 1.2, the pastes shrank significantly and irreversibly. Authors of these findings called this phenomenon "decalcification shrinkage", that is due to the structure changes in C–S–H at Ca/Si ~ 1.2 and eventually to the decomposition of C–S–H into silica gel. That why blended cement paste exhibited greater decalcification shrinkage than the pure cement pastes due to its lower initial Ca/Si ratio for C–S–H gel [14].

1.2. Effect of carbonation on the pore structure of the hardening concrete

Another area of interesting research on concrete carbonation is the change in pore structure. The works of Claisse and Liu [16, 17] have studied the effect of carbonation on permeability and pore volume of carbonated concrete cured under wet and dry conditions and exposed in a carbonation chamber for up to 140 days. They pointed out that carbonation reduces the permeability and total porosity. It is evident that the precipitation of fine crystals of vaterite or aragonite in pore contributes to the reduction of total porosity and also the diffusion rate of gas, water and heat.

Although the carbonation reaction reduces the total porosity of OPC-based concretes, its significant impact on increase of the proportion of capillary pores (pore size greater than 30 nm) can be noticed on the pore size distribution. The increase in capillary porosity promotes the diffusivity of ions through the cement matrix. It has been reported that the specific surfaces of OPC paste samples are likely to decrease with an increase in the degree of carbonation [18].

It has been reported a strong correlation between the surface area measured by nitrogen BET and the degree of carbonation as determined by thermogravimetric analysis [18, 19].

The information on the carbonation shrinkage of heavyweight concrete is limited. Therefore, the present work was undertaken to understand the effect of three different curing regimes on physical and mechanical characteristics of heavyweight concrete. The material composition corresponds to concrete, but the measurements were done using standard for mortars. The experimental investigation includes long-term shrinkage, bulk density, compressive strength, dynamic modulus of elasticity and pore structure using mercury intrusion porosimetry (MIP).

2. Experimental

2.1. Used materials:

- Cement CEM I Extra 42.5 R; producer: Cement plant CRH a.s., Rohožník, Slovak Republic. Density: 3,143 kg/m^3; specific surface: 4,341 cm^2/g
- Barite aggregate; producer: SABAR s.r.o., Markušovce, Slovak Republic. Density: 3 845 kg/m^3
- Cast iron aggregate; producer: IKB Slévárna Písečná s.r.o., Czech Republic. Grain size 22 and 32 mm. Density: 7 600 kg/m^3

2.2. Concrete setting retarder based on Lignosulfonates; producer:

Stachema Bratislava, a.s., Slovak Republic.

2.3. Concrete composition

Composition of prepared concrete samples is depicted in table 1, in which the designations for cement, superplasticizer and mixing water have been used.

Table 1. Composition of prepared samples (in wt.%).

Material	Content
Barite	55.06
Cast Iron	31.82
Cement, CEM I	9.06
Superplasticizer, SP	0.12
Water, W	3.94
Total	100

2.4. Preparation of aggregate

Barite aggregate was crushed and sieved into different size fractions and remixed to obtain aggregate mixtures with defined particle size distribution falling into region of granulometric curves. Grain size composition was set to obtain grading curves of aggregates with largest grain of 16 mm.

Table 2. Grading of aggregate with the largest grain size of 16 mm.

Fraction size [mm]	Aggregate grain-size composition [%]	Content in concrete [%]
0/0.063	0.25	0.137
0.063/0.125	0.125	0.068
0.125/0.25	11.025	6.070
0.25/0.5	4.56	2.510
0.5/1	4.56	2.510
1.0/2.0	15.2	8.369
2.0/4.0	9.77	5.379
4.0/8.0	21.93	12.074
8.0/16.0	32.57	17.93
Total	99.99	55.054

2.5. Samples preparation and curing conditions

The manufacture of heavyweight concrete samples (trivial named EDU) was done in 3 phases: (1) the aggregates were mixed with a small quantity of water to fill the open pores before cement was added and the mixture was mixed in a semi-dry state during 5 min. (2) Then, half of the water and superplasticizer blend was added into the mixture which was then mixed for 8 min. (3) In the last stage, the remaining water and superplasticizer blend was added and mixed for 5 min. After mixing the concrete mixture was placed into the prism molds (40 × 40 × 160 mm). The molds then were covered with a plastic foil in order to avoid evaporation of surface water and were left to harden at a room temperature. After 24 h, the concretes specimens were demolded and subjected to three different curing regimes: climatic chamber (30 % of relative humidity, average temperature 26°C), humid curing (100 % r. h., average room temperature 26°C) and CO_2-chamber-wet (90 % r. h. at average temperature 50°C, average CO_2 concentration 20%) for 7, 28 and 90 days.

2.6. Test methods

- The compressive strength of specimens cured under different environments was measured according with the standard, EN 196-1 (2000) at the ages of 7, 28 and 90 days using a 1000 kN capacity mortar compression machine. The average compressive strength test results are calculated from six individual measurements.
- The dynamic modulus of elasticity was determined using Ultrasonic Meter Matest device. The measurement was performed on 40 × 40 × 160 mm prism specimens, which were afterwards used for the destructive compressive strength determination. The specimen was clamped at the middle section. The electromagnetic exciter (emitter) probe and the receiver probe were placed against each other on the two end faces of the sample at the mid-section. The exciter induced longitudinal pulses of sound into the concrete and the time the sound takes to travel from the transmitter probe to the receiver probe through the material was measured. Knowing the resonance frequency, the dynamic modulus was calculated using the following equation:

$$E_d = 4.n^2.L^2.\rho.10^{-15} \quad (6)$$

  where E_d is the dynamic modulus of elasticity (GPa), n is resonant frequency (Hz), L is length of specimen (mm); ρ is the density of specimen (kg/m^3).
- Mercury intrusion porosimeter Quantachrome Poremaster 60GT (Quantachrome UK Limited) was used for the estimation of the curing environment on the pore structure parameters of the heavyweight concretes at 2, 7, 28 and 90 days. About 2 grams of samples without coarse aggregates used for the tests, were soaked in acetone and diethyl ether. The samples were dried in a vacuum atmosphere for 2 h before testing. The maximum applied pressure of mercury was 414 MPa, equivalent to a Washburn pore radius of 1.8 nm according to equation (7) [20]:

$$r = -\frac{2\gamma cos\theta}{P} \quad (7)$$

  where r is the pore radius, γ is the surface tension of mercury (N/m), θ is the contact angle between mercury and the solid materials and P is the applied pressure (Washburn, 1921, Carlos, 1998).

3. Results and Discussion

3.1. Density

The density of heavyweight concrete after 2, 7, 28 and 90 days are shown in figure 1. The bulk density of EDU heavyweight concrete is higher than 3900 kg.m^{-3}. The influence of curing environment has a negligible influence on bulk density because of high density aggregates. The negligible variation of bulk density with time and curing conditions resulted from measurement errors as it can be observed with the intensity of standard deviations.

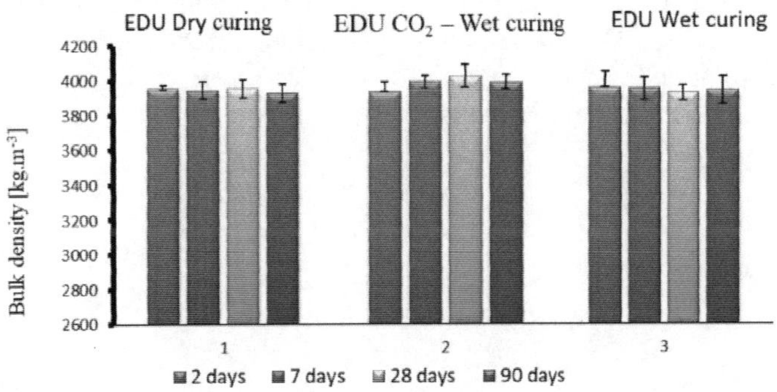

Figure 1. Bulk density of EDU heavyweight concrete (dried samples).

3.2. Compressive strength

The compressive strength of hardened EDU heavyweight concrete cured under the three different environments is shown in figure 2. At all time of curing, compressive strength is found dependent on environment. Under dry curing conditions, the development of compression strength is negligible due to loss of water by evaporation. It is a general rule that the compressive strength strongly depends on the degree of hydration which runs only in the presence of water. 90-day compressive strength exceeding 45 MPa increased only by 21% compared to the 2-day one.

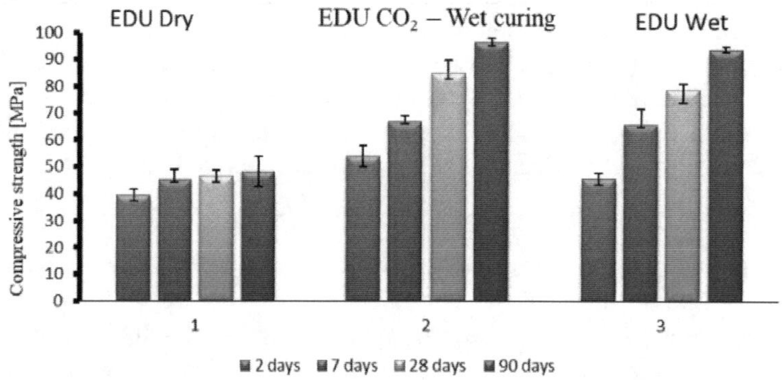

Figure 2. Compressive strength of heavyweight concrete EDU (dried samples).

The compressive strength evolved positively over time marked by a significant difference values measured at 2, 7, 28 and 90 days under wet and CO_2-wet curing environments (figure 2). It is evident that under these conditions the hydration of the cement proceeded and hence increased the compressive strength. 90-day compressive strength increased by about 106% and 78% for samples cured under wet and CO_2-wet curings respectively in comparisson with the samples cured for 2 days. Nevertheless, it should be noted that the values of compressive strength of samples cured under CO_2-wet conditions are higher than that of samples exposed to wet environment. This difference is higher at the beginning than at the later time. However, it should be noted that this beneficial effect of CO_2 on the mechanical properties of OPC based systems is observed when the fresh concrete mix is subjected to carbonation. During this phase, CO_2 under pressure diffuses easily into the concrete bulk and supports the homogeneous formation of calcium carbonate in concrete matrix. The precipitation of calcium carbonate in micro and macropore contributes to the reduction of total porosity and hence, to the increase of compressive strength [13, 15, 21]. When the hardened (or matured) concrete samples are subjected to carbonation, the CO_2 penetration depth is mostly limited to the outer layer of the samples because of the slow diffusion rate through the dense microstructure. In such condition, the increase in strength of OPC-based systems due to the carbonation is not significant. These findings were reported by also [7]. The enhanced compressive strength can be explained by the degree of hydration. Indeed, it was reported [22–23] that carbonation accelerates the hydration of tricalcium silicate. Hydration is a complex physical and chemical process. The consumption of calcium hydroxide by carbonation can induce acceleration of C_3S dissolution to keep the alkalinity of liquid phase. In the meantime, carbonation can decalcify CSH producing poorly calcium silicate gel with a lower content of calcium.

3.3. *Dynamic modulus of elasticity (DME)*

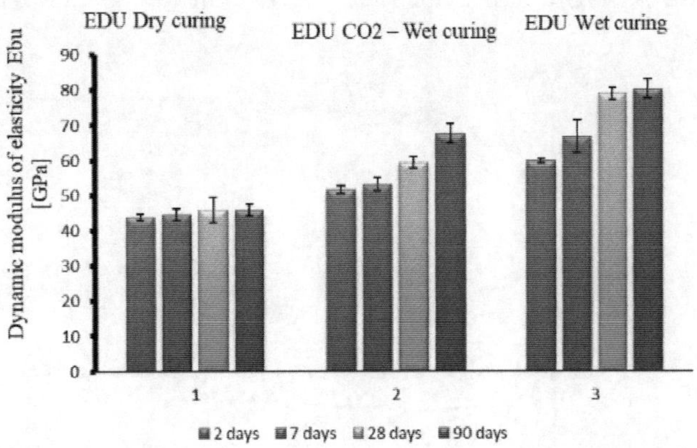

Figure 3. Dynamic modulus of elasticity (DME).

The DME of prepared samples cured in dry environment increased very slightly from the beginning to 90 days. DME is an important characteristics of concrete and can be used to compute the dynamic Young's Modulus (E) [4]. The higher the value of DME, the stiffer concrete is. Also, it characterizes the heterogeneity of structure such as aggregate, water content, air voids, transition zone, and porosity. It depends on concrete age, curing regime and material composition. As it can be observed in figure 3, DMEincreased significantly in wet and CO_2-wet curing regimes. Unfortunately, the values of DME of A EDU heavyweight concrete cured in CO_2-wet environment, are slightly lower than that of EDU cured in wet environment. It can be expected that the precipitation of huge amount of calcium carbonate crystals may reduce the stiffness of concrete. An increase of DME depends the concrete maturity which is directly dependent of degree of hydration. It can be related to the shrinkage also.

3.4. *Shrinkage characteristics*

The typical shrinkage strain behavior of the three EDU heavyweight concretes is plotted in figure 4. Samples underwent volume changes because of physical and chemical interaction with the internal and external environment. It is worth noting that the curing conditions significantly affected the magnitude of concrete shrinkage because the flux of moisture out of the pore structure of concrete is mainly related to the restraining effect of concrete maturity. As dry curing ran at r. h. below 30% humidity, it is evident that drying shrinkage occurred. The mechanism of dry shrinkage has been reported as the effect of capillary stress when water evaporates from the capillary pores. Then, the tension is transferred to the walls of pores causing thus drying shrinkage [12]. Wet cured and dry cured heavyweight concrete samples have opposite behavior. While in dry conditions, samples showed important shrinkage due to evaporation of water, in wet conditions samples expanded slightly. In these conditions, the formation of new hydration products was supported by 100% humidity which keep almost constant quantity of water in the concrete matrix to ensure further hydration. A low content of water in the concrete matrix leads to an excessively dry condition for the ionization of CO_2 and prevents the dissolution of calcium hydroxide, while a high humidity inhibits the diffusion of dissolved CO_2 through open surface pores. This is why the carbonation tests on mortars or concrete are carried out in conditions of humidity not exceeding 70%. The course of CO_2-wet cured samples is interesting. The carbonation shrinkage is subject to different interpretations. During the first hours of curing, $Ca(OH)_2$ submitted to the pressure of CO_2 diffusion can be compressed leading to intensive shrinkage. The shrinkage is observed during the first 11 days. In the meantime, CSH was exposed to the action of CO_2 and its decalcification leads to an excessively dry condition. The decalcification of CSH structure causes the volume change known as "decalcification shrinkage", predominantly at the earlier period of hydration. Acceleration of C_3S hydration and decalcification of CSH are controversely and this phenomenon can explain lower shrinkage of CO_2-wet cured heavyweight concrete compared to wet cured one. On the other hand, crystallization of calcium carbonate can compensate the effect of carbonation shrinkage.

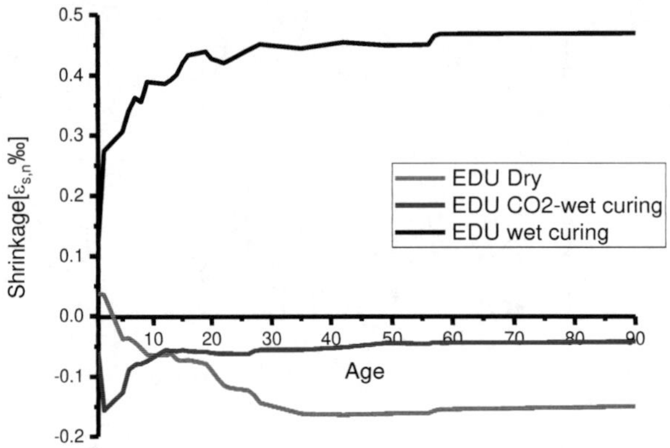

Figure 4. Continual shrinkage of EDU heavyweight concrete.

3.5. *Pore structure analysis*

The visual evolution of concrete carbonation observed by application of phenolphthalein solution on samples of age 90-days is depicted in figure 5. Practically, total volume carbonation as consequence of CO_2 diffusion into the sample was observed on the sample stored in CO_2 chamber. Samples stored in a wet environment and in climatic chamber showed no signs of carbonation.

Figure 5. Carbonation test. From the left concrete sample cured in climatic chamber, wet and CO_2 chamber.

The present work prefers to interpret the influence of carbonation on concrete through in-depth analysis of MIP results. Characteristics, such as total porosity with interparticle and intraparticle porosity components, pore size distribution, bulk density, total surface area were determined. These parameters have provided therefore the relevant information regarding the influence of curing environments on the pore structure of heavyweight concrete.

Total surface area is the sum of surface area of all pores and voids filled up to pressure P. Total porosity is the sum of interparticle porosity (spaces between particles) and intraparticle porosity (within particles).

These pore parameters result from the following equations (Carlos A 1998 26) [25]:

$$\text{Interparticle porosity (\%)} = 100\frac{V_v}{V_b} \tag{8}$$

$$\text{Intraparticle porosity (\%)} = 100\frac{V_t-V_v}{V_b} \tag{9}$$

$$\text{Mercury intrusion (total) porosity (\%)} = 100\frac{V_t}{V_b} \tag{10}$$

where V_b is the bulk volume of the sample, V_v is the volume of mercury intruded up to the interparticle filling limit and V_t is the total volume of mercury intruded up to the maximum pressure.

The classification of pores proposed by Metha and Monteirio [25] was used to divide the pores into particular categories according to their diameter as follows: gel pores (< 4.5 nm), mesopores (4.5–50 nm), middle capillary pores (50–10^2 nm), large capillary pores (10^2–10^4 nm), macropores (> 10^4 nm). The characteristics of pore structure are reported in table 3. It is evident that the total porosity depended on curing conditions and on the curing time. Dry curing caused a substantial increase in total porosity when compared with wet and CO_2-wet curing due to the lack of water for advanced hydration.

Table 3. Characteristics of pore structure of EDU heavyweight concrete.

Curing condition	Dry curing				CO_2-Wet curing				Wet curing			
Curing period[day]	2	7	28	90	2	7	28	90	2	7	28	90
Total porosity[%]	10.766	10.02	9.147	9.062	7.028	7.694	6.144	4.102	6.366	5.37	6.1933	4.057
Interparticle porosity[%]	0.96	0.633	0.683	0.58	0.378	1.098	0.915	0.851	0.815	0.72	0.8305	0.857
Intraparticle porosity[%]	9.806	9.387	8.514	8.482	6.65	6.596	5.229	3.251	5.551	4.65	5.3628	3.2
Total surface Area [m^2/g]	1.488	1.283	1.299	1.303	3.107	3.254	3.483	3.856	0.951	0.928	2.560	1.287
Bulk density[g/cm^3]	3.909	3.981	3.895	3.785	3.955	4.299	4.430	4.345	4.004	4.064	4.227	4.337

In such conditions, water evaporated and let empty space or voids. Since hydration cannot continue, the initial large pores are not filled with hydrated products. The higher total porosity explains the lower value of compressive strength and DME of EDU heavyweight concretes cured under dry conditions. Surprisingly, the total porosity of samples cured in CO_2-wet condition is slightly higher than that of samples exposed to wet conditions. The main influence of carbonation can be observed with the values of total surface area. The explanation of this finding escape still to our understanding, but it could be related to the presence of gel pores as reported by pore size distribution (PSD) at figure 6. However, it can be seen that the intraparticle porosity which is the main component of the total porosity is higher in samples CO_2-wet cured than that cured in wet conditions.

3.6. *Influence of curing in dry curing regime*

Other relevant demonstration of the influence of curing conditions and curing time on pore structure is illustrated by pore size distribution and cumulative pore volume shown in figures 5–10. The cumulative pore volume expresses the summation of mercury volume intruded into the pores and interparticle voids versus the applied pressure. These two characteristics are uniform with samples cured in dry conditions (figure 5).

The total porosity is constituted of middle capillary pores (50–10^2 nm) and large capillary pores (10^2–10^4 nm). After 7 days, one noted a negligible change in pore structure. This stagnation is due to the lack of water responsible for hydration of cement phases to produce nano size CSH with possibility to fill large pores. If he concentration of CO_2 in air is negligible, therefore its effect on carbonation of sample is also negligible.

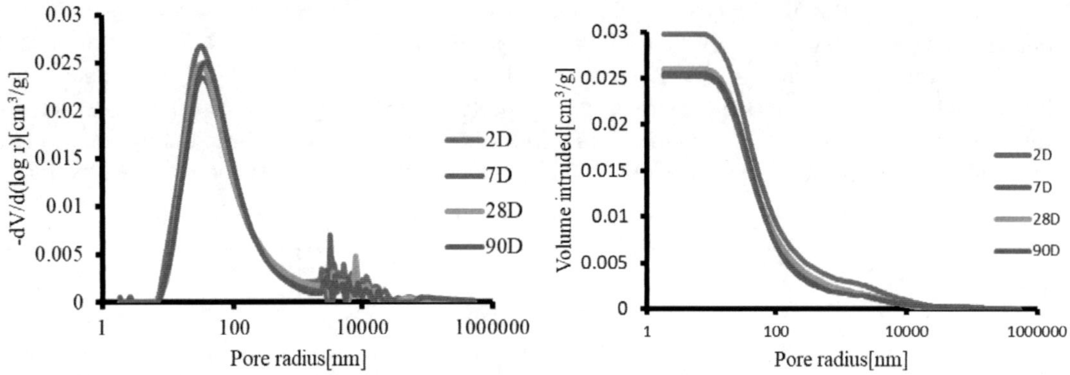

Figure 6. Characteristics of pore structure of dry cured EDU heavyweight concrete.

3.7. *Influence of curing in CO_2-wet curing regime*

When cured in wet and CO_2-wet conditions, the curve of PSD (figure 6) showed the presence of gel pores (< 4.5 nm) and mesopores (4.5–50 nm) represented by curves having the form of a zigzag. The presence of gel pores resulted from the formation of CSH. The pore size distribution in ranged between gel pore, middle capillary pores (50–10^2 nm) with a negligible portion of large capillary pores (10^2–10^4 nm).

The presence of macro pores is not observed. The volume of intruded mercury is reduced and we observe an increase of middle capillary pores (50–10^2 nm) after 2, 7, 28 days of curing.

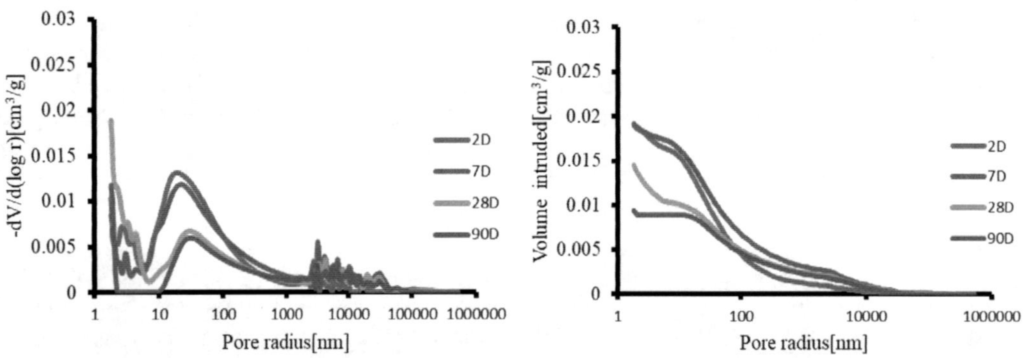

Figure 7. Characteristics of pore structure of CO_2-wet cured EDU heavyweight concrete.

3.8. *Influence of curing in wet curing regime*

The wet curing conditions are known as standard ones for evaluation of the mechanical and physical properties of mortars and concrete.

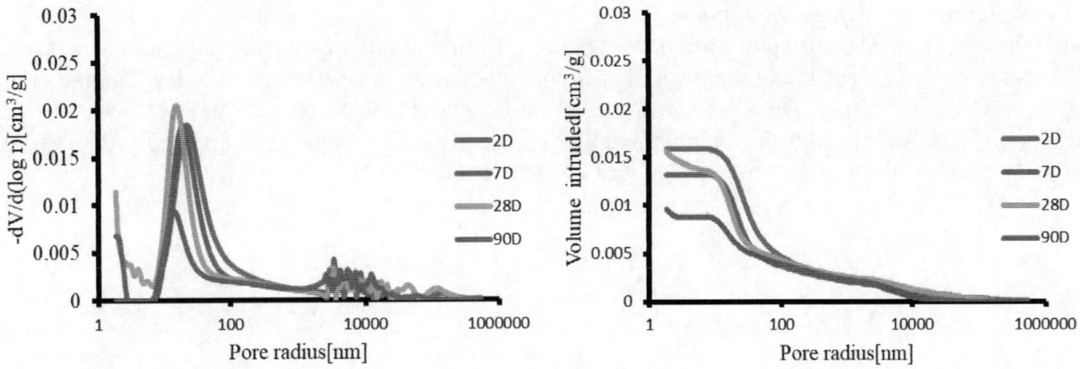

Figure 8. Characteristics of pore structure of wet cured EDU heavyweight concrete.

These conditions prevent the samples from drying out and ensure the continuity of the hydration which is responsible for the development of mechanical and physical properties including pore structure. The important presence of middle capillary pores (50–10^2 nm) illustrates the degree of hydration under these conditions (figure 8). One can observe the presence of gel pores (< 4.5 nm) and mesopores (4.5–50 nm) from the age of 28 days, while these pores are found in samples cured under CO_2-wet from the beginning.

3.9. *Influence of curing environment on the pore characteristics*

Figures 9–12 illustrate the effect of different curing environment on the pore structure of EDU heavyweight concrete at 2, 7, 28 and 90days. These figures clearly show the effect of conditions on pore size distribution and volume intruded. Dry conditions are distinguished by the high values of volume intruded and the positioning of PSD in the region of micropores with important portion of capillary pores. As previously reported, the stagnation of the pore structure development is caused by the lack of water.

3.10. *Influence of 2-day curing regimes*

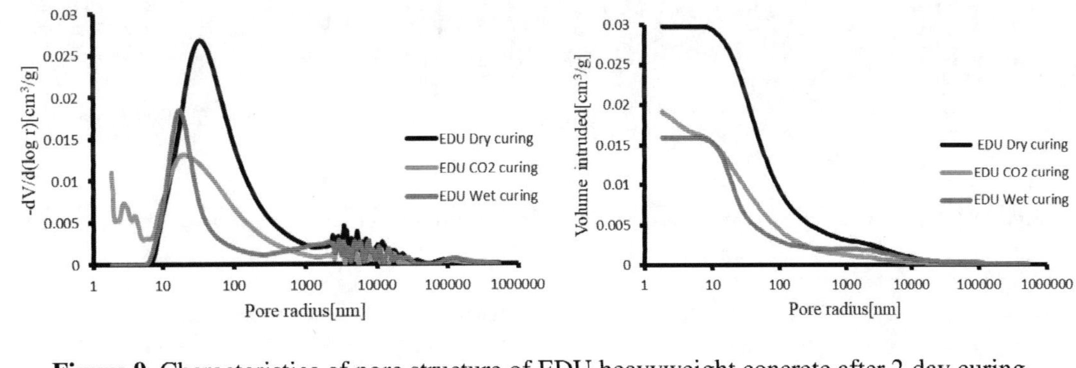

Figure 9. Characteristics of pore structure of EDU heavyweight concrete after 2-day curing.

After 2 day-curing, one notes the displacement of PSD towards finer pore structure with apparition of gel pore in sample CO_2-cured (figure 9). The assumption of CO_2 penetration into the depth of the material is usually limited to the carbonation front. CO_2 resp. CO_3^{2-} reaction with calcium hydroxide in the alkaline environment of the cement matrix is relatively fast and shifted to the side of the reaction products. The formation of these products can lead to reduction of pore sizes [27].

3.11. Influence of 7-day curing regimes

The 7-day curing under different environments did not bring about any substantial change compared with 2-day curing except that the portion of gel pores increased in CO_2-cured samples. Sample cured under wet conditions had the lowest volume intruded, a slight displacement of PSD towards finer structure including mostly the middle capillary pores (50–10^2 nm) and large capillary pores (10^2–10^4 nm), but the presence of gel pore was not detected (figure 10).

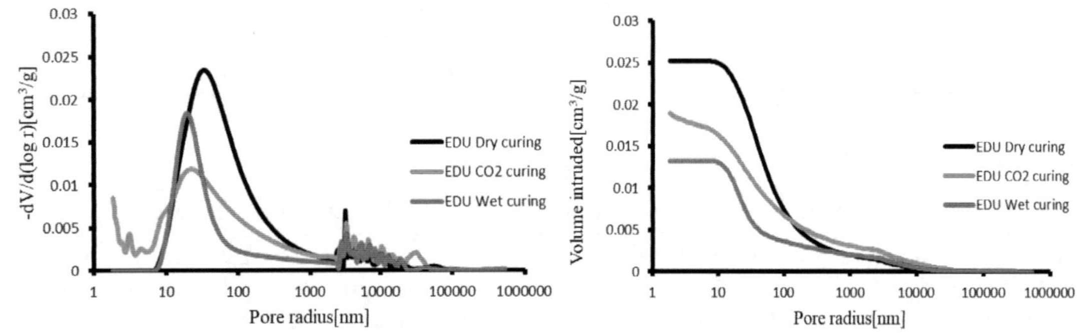

Figure 10. Characteristics of pore structure of EDU heavyweight concrete after 7-day curing.

3.12. Influence of 28-day curing regimes

The main difference with curing conditions was observed in figure 10 with the apparition and extension of gel pore portion. The hydration and carbonation effect was evidenced by pore refinement of samples cured under wet and CO2-wet conditions. At this stage, volume intruded of both samples was reduced compared to dry cured sample. The refinement was possible thanks to the progress in hydration and carbonation. The compressive strength of the samples in this time also increased (figures 1, 2).

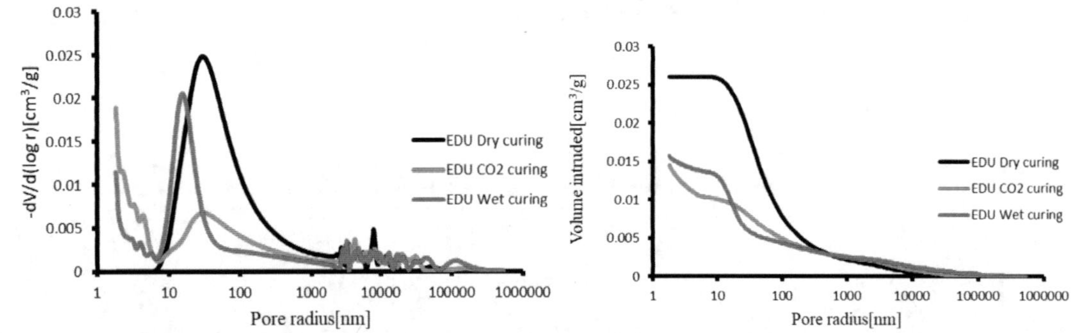

Figure 11. Characteristics of pore structure of EDU heavyweight concrete after 28-day curing.

3.13. Influence of 90-day curing regimes

90-day curing brought a significant change in pore structure. The porosity of wet and CO_2-wet cured EDU heavyweight concretes decreased. The values of total volume intruded for both samples are similar. Also, a drastic reduction of the intensity of PSD peak of samples cured under wet and CO_2-wet conditions was observed. In the meantime, the portion of gel pore increased. The 90-day curing increased the portion of gel both in wet and CO_2-cured samples. Hydration continued causing the formation of a higher amount of CSH (figure 12).

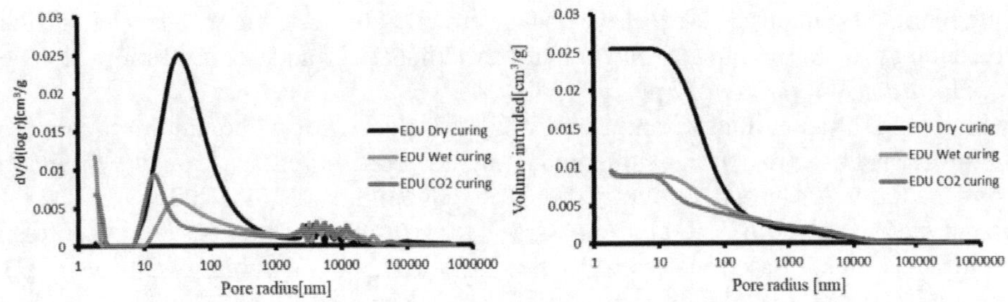

Figure 12. Characteristics of pore structure of EDU heavyweight concrete after 90-day curing.

4. Conclusion

Heavyweight concrete was manufactured and the prepared samples were subjected to curing in three different curing environments: dry curing, CO_2-wet curing and wet curing. The effect of carbonation on mechanical and physical properties of samples was studied. Curing in dry condition did not ensure the increase of the compressive strength and dynamic modulus of elasticity at later time. CO_2-wet curing together with wet curing has a positive impact on compressive strength increase. Ultimately, wet treatment results in the highest sample strengths. Also the dynamic modulus of samples cured in CO_2-wet and wet environment reached higher values in comparison with samples cured in climatic chamber. The pore structure has not been refined. Curing of samples in the CO_2-wet environment has led to the increase in compressive strength, even compared to the standard curing conditions. It affected the shrinkage, specific surface and the portion of gel pore from the beginning. The wet condition (standard) has refined the pores structure with the apparition of gel pore after 28 days.

Acknowledgement

This work was supported by courtesy of APVV-19-0490, APVV-15-0631, Slovak Grant Agency VEGA No. 2/0032/21 and 2/0017/21.

References

[1] Ouda A 2015 Development of high performance heavy density concrete using different aggregates for Gamma-Ray shielding *Prog.Nucl. Energy* **79** pp 48–55

[2] Kyoungsoo P, Hyung_tae K, Tae-Hyun K and Eunso Ch 2016 Effect of neutron irradiation on response of reinforced concrete members for nuclear power plants. *Nuclear Engineering and Design* **310** pp 15–26

[3] Dragomirová J and Palou MT 2019 Development of High-Compressive Heavyweight Concrete Based on Portland Cement and Supplementary Cementitious *Materials Materials Science Forum* **955** pp 44–49

[4] Dragomirová J, Palou M T, Kuzielová E, Žemlička M, Novotný R and Gméling K 2020 Optimization of cementitious composite for heavyweight concrete preparation using conduction calorimetry *Jour. Therm. Anal. Calorim* **142** pp 255–266

[5] Dragomirová J, Palou M T, Gméling K, Szilágyi V, Harsányi I and Szentmiklósi L 2019 Design of heavyweight concrete used in radiation protection based on complete NAA, PGAA and XRF results; activation and physical properties In Brittle Matrix Composites 12 *Proc. 12th Int. Symp. Brittle Matrix Composites* (Warsaw : Institute of Fundamental Technological Research) pp 195–210

[6] STN EN 206+A1: 2017 Concrete_ Part 1: Concrete. Specification, performance, production and conformity STN (Bratislava)

[7] Groves G W, Rodway D I and Richardson I G 1990 The carbonation of hardened cement pastes *Adv. Cem. Res.* **3** pp 117–125

[8] Jang J, Kim G M, Kim H J and Lee H K 2016 Review on recent advances in CO_2 utilization and sequestration technologies in cement-based materials *Constr. Build. Mater.* **127** pp 762–773

[9] Fern´andez B M, Simons S J R, Hills C D and Carey P J 2004 A review of accelerated carbonation technology in the treatment of cement-based materials and sequestration of CO_2 *Journal of Hazardous Materials* **112** pp 193–205

[10] Liu R, Jiang L, Xu J, Xiong C and Song Z 2014 Influence of carbonation on chloride induced reinforcement corrosion in simulated concrete pore solutions *Constr. Build. Mater.* **56** pp 16–20 (*Online* http://dx.doi.org/10.1016/ j.conbuildmat.2014.01.030)

[11] Marangu J M, Thiong'o J K and Wachira J M 2019 Review of Carbonation Resistance in Hydrated Cement Based Materials, Review Article *Hindawi Journal of Chemistry* **ID 8489671** pp 1–6 (*Online* https://doi.org/10.1155/2019/8489671)

[12] Mushtaq S M, Siddique R, Goyal S and Kaur K 2021 Experimental studies and drying shrinkage prediction model for concrete containing waste foundry sand *Cleaner Engineering and Technology* **2** pp 100071

[13] Cheung A K F and Leung C K Y 2011 Shrinkage reduction of high strength fiber reinforced cementitious composites (HSFRCC) with various water-to-binder ratios *Cement & Concrete Composites* **33** pp 661–667

[14] Reardon J, James E and Abouchar B R 1989 High pressure carbonation of cementitious grout *Cem. Concr. Res.* **19** pp 385–399

[15] Chen J J, Thomas J J and Jennings H M 2006 Decalcification shrinkage of cement paste *Cem. Concr. Res.* **36** pp 801–809 (*Online* http://dx.doi.org/10.1016/j. cemconres.2005.11.003)

[16] Claisse P A, Elsayad H I and Shaaban I G 1999 Permeability and pore volume of carbonated concrete *ACI Mater. J.* **96** pp 378–381

[17] Liu L, Jiang J, Xu C and Xiong Z S 2014 Influence of carbonation on chloride induced reinforcement corrosion in simulated concrete pore solutions *Constr. Build. Mater.* **56** pp 16–20 (*Online* http://dx.doi.org/10.1016/ j.conbuildmat.2014.01.030)

[18] Ngala V T and Page C L 1997 Effects of carbonation on pore structure and diffusional properties of hydrated cement pastes *Cem. Concr. Res.* **27** pp 995–1007

[19] Pihlajavaara S E 1968 Some results of the effect of carbonation on the porosity and pore size distribution of cement paste *Mater. Constr.* pp 521–526

[20] Washburn E W 1921 Note on a Method of Determining the Distribution of Pore Sizes in a Porous Material. Proc. Nation. Acad. Sc. USA(PNAS) **7** **(4)** pp 115–116 (*Online* https://doi.org/ 10.1073/pnas.7.4.115)

[21] Zhenghu L, Yinchuan G, Zhihui C, Aiqin S, Xiao Q, Jingyu Y and Ming Z 2019 Research on shrinkage development and fracture properties of internal curing pavement concrete based on humidity compensation *Constr. Build. Mater* **203** pp 417–43

[22] Berger R L, Young J F and Leung K 1972 Acceleration of hydration of calcium silicates by carbon dioxide treatment *Nat. Phys. Sci.* **240** pp 16–18

[23] Young J F, Berger R L and Breese J 1974 Accelerated curing of compacted calcium silicate mortars on exposure to CO_2 *J. Am. Ceram. Soc.* **57** pp 394–397 (*Online* http://dx.doi.org/ 10.1111/j.1151-2916.1974.tb11420.x)

[24] Sevelsted T F and Skibsted J 2015 Carbonation of C–S–H and C–A–S–H samples studied by 13C, 27Al and 29Si MAS NMR spectroscopy *Cem. Concr. Res.* **71** pp 56–65 (*Online* http://dx.doi.org/10.1016/j.cemconres.2015.01.019)

[25] Carlos A. León y León 1998 New perspectives in mercury porosimetry *Advances in Colloid and Interface Science* 76 –77 pp 341–372 (*Online* https ://www.scien cedir ect.com/journ al/advan ces-in-collo id-and-inter face-science/vol/76/suppl /C)

[26] Mehta P K and Monteiro P J 2017 *Concrete Microstructure, Properties and Materials* (McGraw-Hill Education: New York, NY, USA) (*Online* accessengineeringlibrary.com)

[27] In-Seok Yoon and Chun-Ho Chang 2020 Time Evolution of CO_2 Diffusivity of Carbonated Concrete *Appl. Sci.* 10 pp 8910

Impact performance of rubber reinforced concrete with waste steel fibres

M Drdlova[1], P Bibora[1] and V Prachař[1]

[1] Research Institute for Building Materials, Hněvkovského 65, Brno 61700, Czech Republic

Email: drdlova@vustah.cz

Abstract. This study introduces cementitious composite with rubber granulate and waste steel fibres as a new material for construction industry with an enhanced energy absorption capability and impact toughness. Detailed research on physico-mechanical properties of high-performance concrete with waste steel fibres and partial replacement of the aggregates by rubber granulate was performed, with emphasis on impact energy absorption potential. Different aggregate replacement ratios (0–30% wt.) and fibre amount (0–3% wt.) were investigated. The influence of rubber sizes, rubber content and steel fibre content on the mechanical parameters of the rubberized concrete at both quasistatic and dynamic loads was evaluated and discussed. With increasing amount of rubber granulate, the concrete suffered from reduction of its mechanical parameters – compressive and flexural strength, however the energy dissipation capability showed rising trend. This study demonstrated the potential of rubberized concrete with waste steel fibres for use in structures with higher impact resistance requirements.

1. Introduction

In 2016, around 3.5 million tonnes of waste tires were produced in EU [1]. High percentage was recycled and used for the production of playground and sport pitches tiles, in hot mix asphalt as a highway construction material, subgrade insulation, sound absorbing barriers, lightweight fill material, drainage materials, flowable fills, and road embankments or energy recovery, but still about 160,000 tonnes of tires (about 4.6%) were landfilled. Taken into account the numbers of landfilled tires, it is clear, that there is a wide space for new applications with waste tires content. A tire consists of rubber elastomer, carbon black, metal, textile fabric and additives. The amount of each constituent varies depending on tire type and producer, but approximately the tire contains 47% rubber, 22% carbon black, 17% steel cords, 5% fabrics, and the remaining percentage consists of some other minor additives, as reported by Evans [2]. The possibility of both rubber-based material and steel wire are currently widely tested to partially replace the raw materials for production of several cement-based composites. Concrete is the most spread material in the construction industry. Optimizing its composition towards a greener and more cost-effective variants while maintaining its physico-mechanical parameters is a global challenge, significant to society. Due to the high strength, uniform properties, elasticity, and inert behaviour towards the cement matrix, rubber from tire waste appears to be a potentially suitable component for substituting part of coarse and fine aggregates. Up to now, published studies have mainly focused on evaluating the mechanical parameters of rubberized concrete at quasistatic load. The substitution of aggregate by rubber recyclate usually results in a reduction of the mechanical parameters of concrete, especially its compressive and flexural strength. However, to some extent, it may shift the properties of concrete towards a higher ability to dissipate impact energy [1]. Siddika [1] and Gereges [3] reported in their studies a significant reduction in the compressive strength of concrete when aggregate was replaced by rubber granulate. (Up to 63% reduction in compressive strength when 20% aggregate was substituted has been observed). According to the study of Thomas and Gupta [4], the optimal rate of replacement of fine aggregate with rubber granulate is 12.5%, in terms of maintaining the strength of concrete, resistance to water absorption and carbonation. Dynamic properties of standard concrete and fibre reinforced concrete have been widely studied in the last decades, e.g. [5-8], but not much data has been published yet regarding the dynamic properties of concrete with rubber particles. It is generally known that materials subjected to impact loading must exhibit high impact energy absorption capacity and toughness. The research work carried out so far also shows that the addition of crushed rubber to concrete increases toughness and impact resistance [9-12]. For example, the performance of the concrete with waste tire particles content has been studied by Atahan [12], who carried out crash tests on barriers made of rubber concrete. His results indicate that the presence of the rubber particles improves absorption of the vehicle's kinetic energy during the crash to the barrier and reported as the most effective the replacement ratio of the aggregates by the rubber particles between 20-40% [12]. The behaviour of rubberized concrete at high strain rates using Split Hopkinson pressure bar was investigated by Liu et al. [9]. The results published in his study show that the energy absorption capacity increases with the increasing rate of rubber granules in the concrete, but only up to 10% of the rubber content. Above this amount, the absorption capacity decreases again. A decrease of impact resistance when a higher percentage of aggregates is replaced by rubber granulate was also reported by Atahan [13], who observed that 100% replacement of aggregate with rubber recyclate resulted in a significant reduction of impact load resistance (72%). On the other hand, Corinaldesi and Donnini [14] and Al-Tayeb [15] observed improved impact energy absorption when 50% sand was replaced by rubber granulate. The majority of studies mentioned above have confirmed that the presence of rubber particles can increase the energy dissipation capability of concrete and its impact resistance, which can be crucial in some applications such as road construction (restraint systems, guardrails). Rubberized concrete, therefore, appears to be a suitable candidate for impact load applications. Still, more data is needed regarding the energy dissipation capability and impact toughness of rubberized concrete with the addition of commercial and waste steel fibres; studies available in this field are scarce and results are not entirely consistent. The main goal of the presented research is thus to enhance the knowledge in this field. The obtained results will be used in the ongoing development of elements for safety enhancement, e.g., in the field of transport or critical infrastructures protection.

2. Experimental program

2.1. Materials and specimen preparation

Cement CEM 42.5 R (700 kg/m^3), silicious sand (624 kg/m^3; Střelec quarry) with grain size of 0–2 mm, coarse aggregates with grain size of 2–4 and 4–8 mm (298 and 446 kg/m^3, granodiorite, Olbramovice quarry), rubber granulate with grain size of 0–1, 1–3 and 3–6 mm (processed and supplied by RPG Recycling, see figure 1, left) and water (189 kg/m^3) were the main components of the reference concrete mixture R0. Superplasticizer Glenium 422 produced by BASF, was added. The proportion of superplasticizer was varied in range of 15–20 kg/m^3, as to keep similar and sufficient workability for all mixes. Silica fume Elkem 940U (150 kg/m^3), produced by BASF, was used to add fine fractions to achieve the maximum compaction of the mixture and to exploit its pozzolanic properties. The waste steel fibres from tires, without further classification, were added as the reinforcement, see figure 1 right (processed and supplied by RPG Recycling). The fibres shape parameters vary, the distribution of their diameter and length depicts figure 2. For comparison, steel fibres KrampeHarex 30/0.6DE (length 30 mm, diameter 0.6 mm), without coating, has been used. The mixes are summarized in table 1. In total, 10 mixes were prepared and investigated. Concrete mixes were prepared with water/cement ratio of 0.27; first the components were dry-mixed for 2–3 minutes in concrete mixer, then the water with superplasticizer was added and mixed for another 2 minutes. As the last step, the fibres were added to the mixture and homogenized for 60 seconds. Prepared mixtures were placed into steel moulds treated with releasing agent and vibrated on a table vibrator. After demoulding, the specimens were stored in water for 28 days. Beams with dimensions of 100 × 100 × 400 mm were prepared for the quasi-static tests. For dynamic tests, slabs of 400 × 400 × 40 mm, cubes of 150 mm edge and prismatic specimens with dimensions of 300 × 100 × 100 mm were prepared.

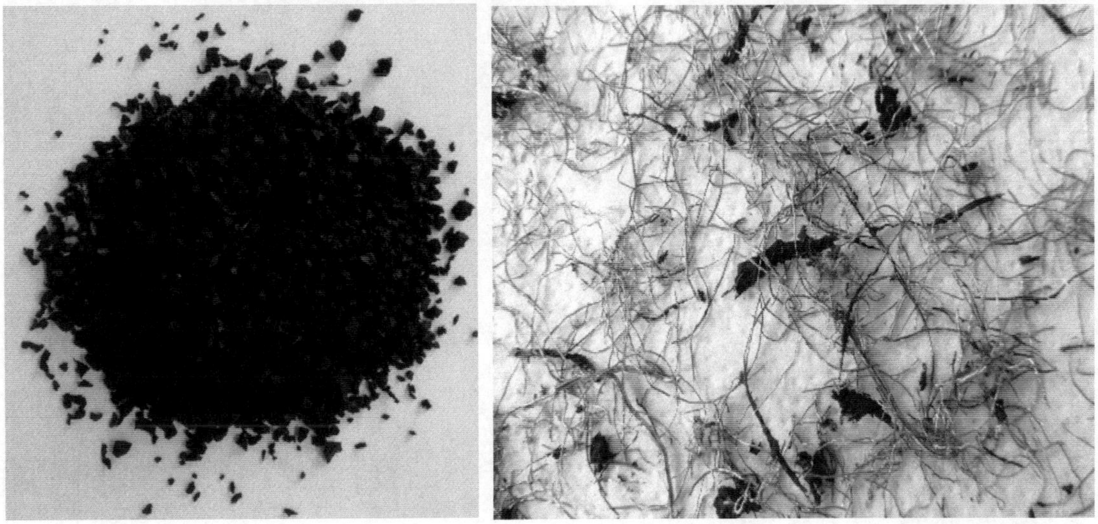

Figure 1. Coarse rubber particles used in the mixtures (left), unclassified waste steel fibres (right).

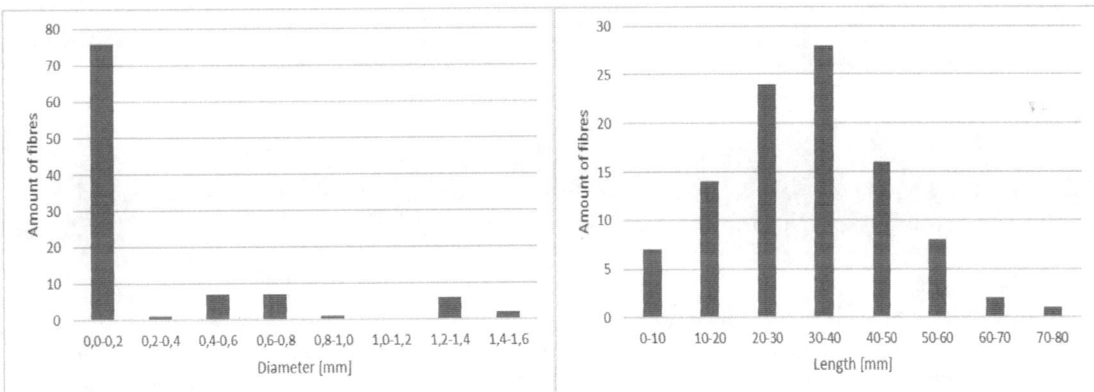

Figure 2. The distribution of fibre diameter and length.

2.2. Physico-mechanical parameters

Mechanical parameters were tested on a TIRAtest 2710, R58/02 press machine at a 5 mm/min loading rate. First, a flexural tensile strength test (three-point) was carried out, followed by a compressive strength test performed on the fractions of the test specimens. Both strength tests and bulk density determination were performed according to the standard ČSN EN 12390-3. Additionally, the water absorption was determined according to the ČSN 73 1325 standard.

Table 1. Description of the mixes.

Designation	Fibre content (vol. %)	Total vol. % of aggregate replacement	Replacement percentage (vol. %)		
			Rubber 0–1 mm	Rubber 1–3 mm	Rubber 3–6 mm
R0	0	0	0	0	0
R10/10-10-10	0	10	10	10	10
R20/20-20-20	0	20	20	20	20
R30/30-30-30	0	30	30	30	30
R20/44-0-0	0	20	44	0	0
R20/0-0-62	0	20	0	0	62
FR20/1	1.0	20	20	20	20
FR20/2	2.0	20	20	20	20
FR20/3	3.0	20	20	20	20
FR20/3KH	3.0	20	20	20	20

2.3. Impact resistance drop test

The impact resistance drop test device, including the test methodology, was developed in Research Institute for Building materials. It consists of a solid steel stand with dimensions of 400 × 400 mm on which the test specimen is fixed. The stand is connected with the support structure with two brackets through which a vertical steel guiding rod with a spherical ending is passed (see figure. 3). The steel rod is equipped with a 4 kg weight that can move freely vertically along the rod. The weight is repeatedly dropped from a height of 700 mm to the centre of the test specimen, and the number of blows required to produce the first visible crack (so called initial crack) and the ultimate failure is recorded. The total absorbed energy for both stages is then calculated using the following formula:

$$E = m.a_g.h.x \tag{1}$$

Where:

E...Total absorbed energy [J]; m...Weight of the impactor [kg]; a_g...Gravity acceleration [$m.s^{-2}$]; h...Height of drop [m]; x...Number of the drops [-].

2.4. Rebound test

Rebound test has been carried out on cubes with the edge length of 150 mm. A steel ball weighing 0.5 kg was used for the measurements. During the test, the ball was released from a certain height (1000 mm) towards the centre of the test specimen. After rebounding from the specimen, the reached height was recorded using the Nikon J5 high-speed camera. The absorbed energy was calculated as the difference of the potential energies of the steel ball at the release and after the rebound, see equation 2:

$$E = E_{p,in} - E_{p,\,r}, \tag{2}$$

Where:

$$E_{p,\,r}, = m.\, a_g.h_r \tag{3}$$

$$E_{p,in} = m.\, a_g.h_{in} \tag{4}$$

and m...Weight of the impactor [g]; a_g...Gravity acceleration [$m.s^{-2}$]; h_{in}...Height of drop - initial [m]; h_r...Height of drop - rebound [m]

2.5. Energy dissipation under dynamic compression tests

Dynamic compression tests were performed on the specimens with 0, 10, 20 and 30% replacement of the natural aggregates with rubber particles (specimens R0, R10/10-10-10, R20/20-20-20, R30/30-30-30) with the aim to determine and compare the energy dissipated in viscoelastic regime. The maximum applied load was 30% of the compressive elastic limit of the concrete in order to comply with the testing condition in the viscoelastic range. To determine the dissipated energy, following calculations according to the methodology described in [16] were used:

$$W=\int 0\pi 2\omega\sigma d\varepsilon dtdt \quad (5)$$

where ω is the angular frequency (rad/s) of the applied load, σ is compressive stress, ε is unitary deformation. Substituting $\sigma=\sigma_0\sin(\omega t)$, $\varepsilon=\varepsilon_0\cos(\omega t-\delta)$, the following can be obtained:

$$W=\sigma_0\varepsilon_0\cos\delta 2+\pi\sin\delta 4 \quad (6)$$

Where σ_0 is the amplitude of the compressive stress, ε_0 is the maximum amplitude of the unitary deformation and δ is the phase shift coefficient that expresses the delay between strain and stress.

From the point of view of the whole cycle, the retained elastic energy is zero. But designating as ΔW the elastic energy dissipated in a cycle and as Wa the maximum density of elastic stored energy during the load stage, the equation (7) can be obtained [16]:

$$\Delta WWa=2\pi\tan\delta. \quad (7)$$

It can also be expressed in the next form:

$$\Delta W=\pi\sigma_0\varepsilon_0\sin(\delta)=\pi\varepsilon_0^2E'' \quad (8)$$

The relation $\Delta W/W$a is denominated as specific loss [16]. It shows the percentage of the dissipated energy with regard to maximum elastic energy stored during the load. Two frequencies, 10 and 20 Hz, were used for tests.

3. Result and discussion

3.1. Physico-mechanical properties

The results of determined physico-mechanical properties are summarized in table 2. The compressive strength was negatively affected by replacement of the natural aggregates with rubber granulate, even though the replacement levels were not very high. The values of compressive strength dropped from 120.1 to 53.8 MPa (30% replacement), which is almost 55% reduction. Similar reduction of compressive strength was observed also by Alsaif et al [17], Gonen [18] and Gupta et al [19]. This is probably due to the different parameters of the cement matrix and the rubber particles, with the rubber being significantly more elastic. When loaded, cracks then form at the interface between the rubber and the matrix, which subsequently reduce the load-bearing capacity of the concrete. Additionally, the replacement of high-strength aggregate by rubber with low load bearing capacity must reduce the total load-carrying potential of the whole composite. The other reason of the compressive strength reduction is the weaker chemical interaction and adhesion between the rubber particles and cement matrix and generally more porous structure of rubberized concrete. This can partly be mitigated by surface treatment of rubber granulate, e.g. by NaOH, H_2SO_4, $Ca(OH)_2$, methanol, ethanol, silane coupling agent, acetone or CH_3COOH [20] and Mohammadi [21]. Further addition of steel fibres did not improve the compressive strength, see table 2.

Table 2. Properties of the designed steel fibre reinforced rubberized concrete.

Designation	Bulk density (kg.m^{-3})	Compressive strength (MPa)	Flexural strength (MPa)	Water absorption (%)
R0	2,422	120.1	11.9	2.5
R10/10-10-10	2,270	102.0	11.3	3.7
R20/20-20-20	2,130	75.1	9.8	4.2
R30/30-30-30	2,091	53.8	6.0	5.1
R20/44-0-0	2,165	68.4	9.3	4.5
R20/0-0-62	2,149	86.4	8.6	4.6
FR20/1	2,193	75.2	16.3	4.3
FR20/2	2,260	72.5	17.3	4.5
FR20/3	2,321	71.0	18.9	4.9
FR20/3KH	2,315	72.1	18.1	4.9

The results show that also the size of rubber particles affects the compressive strength of rubberized concrete. Higher compressive strength was achieved in the case of concretes with coarser fraction of rubber granulates, which is in coherence with findings of Liu [9], but in contradiction with results of study published by Gonen [18]. This divergence may be attributed to different overall composition of the tested concretes, as well as different aggregate gradation. Flexural strength was also reduced when the natural aggregates was replaced by rubber, but the level of reduction was generally lower, in particular in the case of 10% replacement. Other authors reported similar trends for comparable replacement ratios [9]. The opposite trend was reported by Jokar [22] for content of rubber aggregate limited to 5%.

The addition of steel fibres both commercial and waste improved the flexural strength, to similar extent. Rubberized concrete presents higher water absorption value; the water absorption value increases with rising rubber particle replacement ratio and is not affected by particle size.

3.2. *Impact resistance drop test*

The number of impacts required to produce the first crack and complete failure of the slab specimen was evaluated as part of the impact energy absorption rate assessment. The corresponding energy was then calculated according to equation 1. The impact drop test device and the specimen after the test are depicted in figure 3. Test results (mean value of 3 slabs) for all specimens are summarized in table 3.

Table 3. Impact drop test results.

Designation	Number of blows at first crack (-)	Number of blows for ultimate failure (-)	Impact energy at first crack (J)	Impact energy at ultimate failure (J)
R0	8	11	220	302
R10/10-10-10	13	16	357	439
R20/20-20-20	14	18	384	494
R30/30-30-30	16	20	440	549
R20/44-0-0	13	19	357	522
R20/0-0-62	12	17	384	467
FR20/1	91	120	2,500	3,297
FR20/2	95	244	2,610	6,704
FR20/3	104	299	2,858	8,215
FR20/3KH	120	330	3,297	9,067

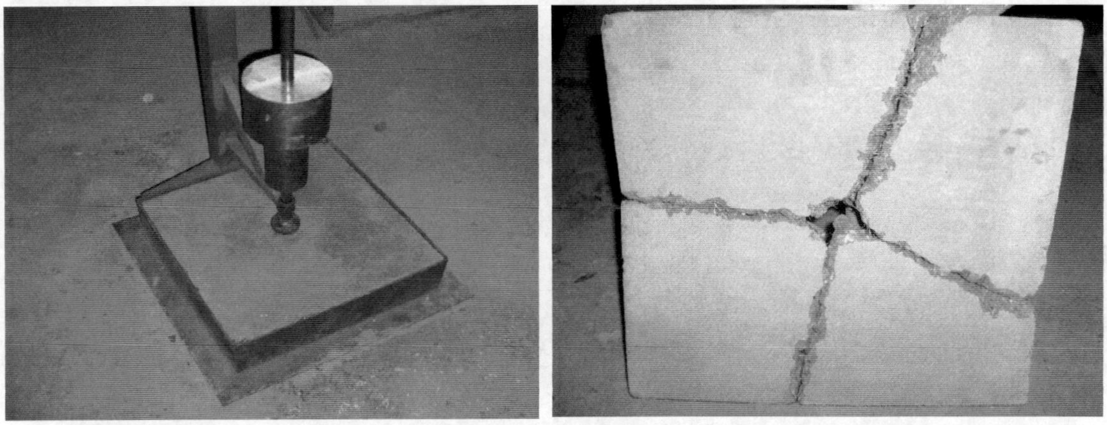

Figure 3. Impact drop test device (left) and slab specimen FR20/3KH after the test (right).

The impact resistance of rubberized concrete, for both first crack and ultimate failure, is higher compared to the specimen without rubber particles. According to Munoz-Sanchez [23], the higher toughness of rubberized concrete is produced from rubber's ability to absorb high tensile loads. Rubberized concrete showed better resistance to crack control, because it has overall better ductility compared to concrete with only natural aggregates. Rubber particles are able to absorb sudden shock because of its nature, this cannot be achieved by natural aggregates because of their brittle nature [1]. With 30% replacement of natural aggregate, the impact energy at ultimate failure was 81% higher compared to the reference specimen. The effect of rubber particle size cannot be properly assessed, as the results of impact energy at first crack and ultimate failure are not coherent – further research in this field is required. Even though the addition of rubber particles enhances the impact resistance, the most significant improvement of impact resistance was achieved with addition of fibres. The highest values of impact energy were achieved for specimen FR20/3KH with 3 vol.% of commercial fibres (3,297 J at first crack and 9,067 J at ultimate failure), but competitive result was achieved also in the case of specimen FR20/3 with 3 vol.% of waste steel fibres (2,858 J and 8,215 J, respectively). From the perspective of drop impact, the presence of fibres, not rubber particles, has determinative influence on the final impact resistance of the concrete, which is in accordance with the findings of Medina [24].

3.3. Rebound test and energy absorption

Test results from rebound test are summarized in Figure 4. It was observed, that increasing amount of rubber particles enhances the amount of absorbed energy – with 0, 10, 20 and 30% replacement, 6.3%, 7.3% and 11.2% increase was detected. During the impact load, some of the energy which the rubberized concrete is exposed to, can be absorbed by the rubber particles; this brings higher impact resistance in comparison with the standard concrete. Similar method to evaluate the impact energy absorption (rebound test) was also performed by Gupta [19] and Medina [24] with consistent results of rising absorption capacity with increasing replacement level up to 25% [19] and 30% [24].

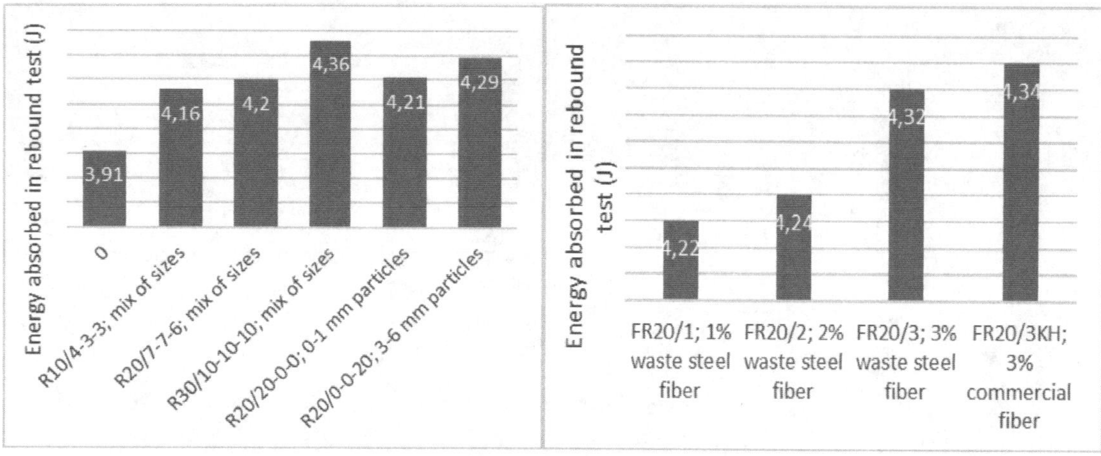

Figure 4. Results of rebound test of concrete with different amount and size of rubber particles (left) and with addition of steel fibres (right).

3.4. Energy dissipation under dynamic compression tests

The ΔW/Wa values calculated (according to the methodology described in 2.5) for tested specimens are given in figure 5. The graph summarizes the percentage of dissipated energy (at frequencies of 10 and 20 Hz) as a function of replacement ratio of rubber particles. The dependence on rubber particle amount and frequency can be stated. At frequency of 20 Hz, larger dissipated energy was observed compared to 10 Hz. With increasing content of rubber particles, the dissipated energy was rising. The amount of dissipated energy was calculated 26–29% (20 Hz) and 22–26% (10 Hz), which makes rubberized concrete promising material for absorbing energy under dynamic actions. 5, 10 and 20 Hz frequency measurements on concrete with 3.5 and 5% of rubber fibres were performed by Hernandez-Olivares [16], who also reported better values of dissipated energy for specimens with higher rubber content and dependency of the dissipated energy on the frequency.

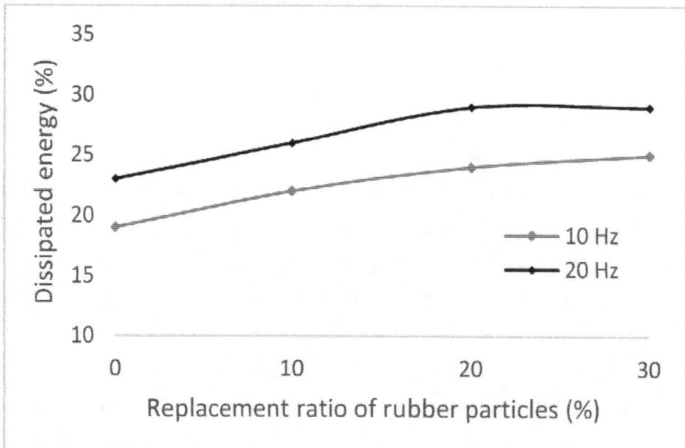

Figure 5. Energy dissipated within dynamic compressive strength test.

4. Conclusions

Mechanical parameters of rubberized concrete with waste steel fibres, with emphasis on impact energy absorption potential, were evaluated, using the drop test, rebound test and energy dissipation under dynamic compression test. The influence of addition of steel fibres and amount of the rubber particles on the mechanical properties of the material was also investigated.

The following conclusions can be drawn:

- Replacing part of the aggregate with rubber granulate reduces the compressive strength of the concrete. The rubberized concrete with 20 and 30% replacement ratio dropped about 37 and 55%, respectively, in the compressive strength. This mainly can be attributed to decreased amount of high-strength natural aggregates and the weaker bonding between the rubber and the cement.

- The addition of fibers, both waste, and reference, does not affect the compressive strength of the concrete with rubber particles. However, they have a significant improving effect on the flexural tensile strength and overall impact resistance.
- Rubberized concrete presents higher water absorption value; the water absorption value increases with rising rubber particle replacement ratio and is not affected by particle size.
- Evaluating the overall impact resistance assessed by impact drop test, the impact resistance was significantly higher influenced by the addition of steel fibres then by addition of rubber particles.
- The amount of energy dissipated by rubber concrete during the dynamic compressive test was calculated 26–29% (for 20 Hz frequency) and 22–26% (for 10 Hz frequency). Within the investigated range of rubber particles amount, the rising trend of dissipated energy was observed with increasing amount of rubber particles in the concrete mixture.
- The effect of rubber particle size on energy absorption was also investigated, but with inconsistent results. Further research will be performed in this field.
- According to the aforementioned findings, rubberized concrete with addition of waste steel fibres shows better energy absorbing potential compared to similar concrete without rubber and fibre presence; thus, has potential to be used in applications where energy absorption capability, deformability and toughness are of importance. Other advantage is lower price of the composite and environmental benefit of waste utilization and raw material saving.

Acknowledgments
The presented research is part of the development process of elements for safety enhancement in the civilian sector. The authors would like to thank to the Ministry of the Interior of Czech Republic, project No. VI20192022129, for financial support.

References

[1] Siddika et. al 2019 Properties and utilizations of waste tire rubber in concrete: A review *Construction and Building Materials* (224) pp 711–731
[2] Evans A and Evans E 2006 Composition of a tyre: Typical components, *The waste and resources action Programme report.*
[3] Gerges N N, Issa C A and Fawaz S A 2018 Rubber concrete: Mechanical and dynamical properties *Case Studies in Construction Materials* (9) e00184
[4] Thomas B S and Gupta R C 2015 Long term behaviour of cement concrete containing discarded tire rubber *Journal of Clean Production* **102** pp 78–87
[5] Drdlova M, Cechmanek R and Ridky R 2015 Blast impact behaviour of concrete with different fibre reinforcement Conference: *International Conference DYMAT*, Web of Conferences Volume 94)
[6] Bibora P, Drdlova M, Prachar V et al. 2017 UHPC for Blast and Ballistic Protection, Explosion Testing and Composition Optimization Conference, IOP Conference Series – Materials Science and Engineering Volume 251) DOI: 10.1088/175-899X/251/1/012004
[7] Frank M, Drdlova M, Prachar V et al 2016 Effect of Filler on the Mechanical and Dynamic Properties of Impact Energy-absobing Materials *Polymers and Polymer composites* **24** (1) pp 1–6
[8] Drdlova M, Buchar J, Ridky R et al 2015 Blast resistance characteristics of concrete with different types of fibre reinforcement *Structural Concrete* **16** (4) pp 508–517
[9] Liu F, Chen G, Li L and Guo Y 2012 Study of impact performance of rubber reinforced concrete *Construction and Building Materials* **36** pp 604–616
[10] Eldin N N and Senouci A B 1993 *ASCE J. Materials in Civil Engineering* **5** pp 478-487

[11] Topcu I B 1995 The properties of rubberized concretes *Cement and Concrete Research* **25** (2) pp 304–310
[12] Atahan A and Sevim U K 2008 Testing and comparison of concrete barriers containing shredded waste tire chips *Materials Letters* **62** pp 3754–3757
[13] Atahan A O and Yucel A O 2012 Crumb rubber in concrete: Static and dynamic evaluation *Construction and Building Materials* **36** pp 617–622
[14] Corinaldesi V and Donnini J 2018 New Trends Eco-Efficient Recycled Concrete, *book chapter 4 Waste rubber aggregates*
[15] Al-Tayeb M M, Abu Bakar B H, Akil H M and Ismail H 2013 *Experimental Mechanics* **53** pp 377
[16] Hernández-Olivares F, Barluenga G, Bollati M and Witoszek B 2002 Static and dynamic behaviour of recycled tyre rubber-filled concrete *Cement and Concrete Research* **32** pp 1587–1596
[17] Alsaif A, Koutas L, Bernal S A, Guadagnini M and Pilakoutas K 2018 Mechanical performance of steel fibre reinforced rubberised concrete for flexible concrete pavements *Construction and Building Materials* (172) pp 533–543
[18] Gonen T 2018 Freezing-thawing and impact resistance of concretes containing waste crumb rubbers *Construction and Building Materials* **177** pp 436–442
[19] Gupta T, Siddique S, Sharma R K and Chaudhary S 2019 Behaviour of waste rubber powder and hybrid rubber concrete in aggressive environment *Construction and Building Materials* **217** pp 283–291
[20] Li Y, Zhang S, Wang R and Dang F 2019 Potential use of waste tire rubber as aggregate in cement concrete – a comprehensive review *Construction and Building Materials* **225** pp 1183–1201
[21] Mohammadi I, Khabbaz H and Vessalas K 2016 Enhancing mechanical performance of rubberised concrete pavements with sodium hydroxide treatment *Materials and Structures* **49** pp 813
[22] Jokar F, Khorram M, Karimi G and Hataf N 2019 Experimental investigation of mechanical properties of crumbed rubber concrete containing natural zeolite Construction and Building materials (208) pp 651–658
[23] Munoz-Sanchez B, Arevalos-Caballero M J and Pacheco-Menor M C 2017 Influence of acetic acid and calcium hydroxide treatments of rubber waste on the properties of rubberized mortars *Materials and Structures* **50** pp 75
[24] Medina N F, Medina D F, Hernandes-Olivares F and Navacerrada M A 2017 Mechanical and thermal properties of concrete incorporating rubber and fibres from tyre recycling *Construction and Building Materials* **144** pp 563–573

Lime-based mortars with various binder composition: characterization and freeze-thaw resistance assessment

D Janotova[1] **and Z Slizkova**[1]

[1] Institute of Theoretical and Applied Mechanics of the Czech Academy of Sciences, Prosecká 809/76, 190 00 Prague, Czech Republic

Email: janotova@itam.cas.cz

Abstract. The study focused on lime mortars with different lime binder types regarding the frost attack effects on their microstructure and mechanical characteristics. The performances of studied mortars in hardened state was significantly influenced by the amount of mixing water and by curing conditions. Inhomogeneous microstructure was observed inside the 360 days old specimens with dimensions 40 × 40 × 160 mm in the case of all lime mortars types. The different state of the binder hardening with respect to various distances from the specimen surface and the different behaviour of matured outer part and immature inner part of lime mortar specimens influenced performed tests and reflected in all results. After 10 and 20 freeze cycles, respectively, the compressive strength of all lime mortar specimens with the hydraulic binder component increased, indicating a beneficial effect of the water on the hydration previously unreacted hydraulic binders. On the other hand, the flexural strength of the frost-aged specimens decreased significantly, indicating the drastic procedure of the test performed concerning lime mortars characteristics, especially when the pure air lime binder was used. Improvement of the testing procedure especially for lime mortars, which are characterized by slow hardening, was recommended.

1. Introduction

Recently, lime has become one of the main materials used in the conservation and restoration of historic buildings. Lime plasters, renders and masonry mortars are commonly utilized in the repair of cultural heritage buildings, because of their compatibility with traditional masonry fabrics [1]. The quality of a lime binder fundamentally affects the performance and durability of the lime mortar manufactured from it, particularly in cases when the building construction containing lime mortar is exposed to frequent water penetration and freezing-thawing events. Frost damage is a crucial physical cause of the decay of building materials. This phenomenon is of great importance in countries where near-zero temperatures conditions are frequent. The intensity and duration of freezing, the cyclic action, as well as interstitial moisture determine the severity of the effect [2].

Generally, non-hydraulic and hydraulic lime binders – depending on the chemical composition of the limestone from which the lime was burned – have been used in lime mortars produced throughout history and also today. The lime binder in mortars is often modified by admixtures or additives for various reasons. Concerning moisture and frost exposure of building constructions, pozzolanic materials were often added to the air lime binder to improve the performance of the prepared mortar. Later, especially in the last century, Portland cement was often used as the hydraulic admixture added to air lime binders in repair mortars meant to be applied on structures affected by high moisture and cold climatic conditions.

The experiment presented in this article masonries focuses on the resistance of lime mortars prepared using various lime based binders to freezing-thawing cycles. The binder types for the experiment were chosen considering the raw materials available for use in a repair mortar to be designed today. The study is intended to provide knowledge needed for a well-advised design of repair mortars for traditional masonries in various exposure conditions and related frost resistance requirements.

2. Materials and methods

Six different mortar mixtures were prepared. A hydrated air lime powder CL 90 (Čerťák®, Vápenka Čertovy schody, a.s.), a metakaolin admixture (Mefisto L05, České lupkové závody, a.s.), two types of natural hydraulic limes, NHL3.5 (Calcidur®, Zement- und Kalkwerke Otterbein) and NHL5 (Hydradur®, Zement- und Kalkwerke Otterbein) and a white Portland cement (HET, CEM I 52.5 R, Aalborg Portland A/S, Denmark) were used as the binders. A pure silica sand with controlled particle size distribution 0–4 mm supplied by Provodín Sands a.s. was used as the aggregate. Table 1 shows the weight ratios of binder to aggregate and water to dry-solids used in the preparation of the mortar specimens.

Table 1. Mortars composition and the water to dry solid ratio.

Mortar	**L**	**LM**	**NHL3.5**	**NHL5**	**CL1**	**CL2**
Composition weight parts	CL90:ag 1:3	CL90:M:ag (0.75:0.25):3	NHL3.5:ag 1:3	NHL5:ag 1:3	CL90:wPC:ag (0.9:1):15	CL90:wPC:ag (0.5:1):10
w/ds	0.26	0.24	0.17	0.17	0.14	0.13

Notes: CL90 = hydrated lime powder; M = metakaolin; NHL = natural hydraulic lime; wPC = white Portland cement, ag = aggregate; w/ds = water to dry solids weight ratio.

Mortar mixtures were prepared using the desirable amount of kneading water to obtain good workability and comparable consistency in all the mortars, 170±5 mm [3], measured using the flow table test [4]. The binder and dry aggregate were mixed for three minutes at low speed using a laboratory mixer MATEST-E093. Water was then added and the mixtures were blended for another 1.5 min. The fresh mortars were mechanically compacted into prismatic casts of dimensions 40 × 40 × 160 mm. The specimens were left inside the moulds for one day and were then stored for a further six days at 90±5% relative humidity at room temperature 20±5°C. The mortar prisms were then stored for 360 days under controlled conditions at a temperature of 20±5°C at 60±10% relative humidity, and placed on grid-lined shelves to provide air flow. The 60% RH value was chosen with respect to the average relative air humidity in Prague during the construction season [5].

2.1. Methods of Testing Hardened Mortar Samples

The open porosity of the mortar specimens was determined by means of hydrostatic weighing under low pressure after total immersion in water for 24h according to the procedure described in [6].

Pore size distribution was analysed using mercury intrusion porosimetry (Poremaster PM 60-13, Quantachrome). The pressure applied in the measurement (0.0055 to 200 MPa) corresponds to pore sizes with a diameter of 258 μm to 6 nm. Two specimens of each mortar type were analysed.

The morphologies of the mortar specimens after the freezing test were observed by optical microscopy (Olympus BX53M). One thin section from each hardened mortar category (40 × 40 mm) was prepared.

Flexural strength was determined based on the three-point flexural test, and the compressive test was done with half of the remaining samples obtained from the flexural test, according to [7].

Dynamic modulus of elasticity was obtained based on determination of longitudinal resonance frequency using the ultrasonic device USG 40 (Krompholz Geotron Elektronik, FRG, 250 kHz, USG-T transmitter and USE-T receiver).

Water absorption coefficient (kg.m^2.hod$^{-1/2}$) was measured in a free-water intake experiment [8]. The three halves of the 4 × 4 × 16 cm samples that remained after the mechanical tests were immersed in 1 mm of water (using glass rods) inside a covered box to maintain constant hygrothermal conditions and to limit water evaporation from the samples.

2.2. Frost aging test

The frost resistance test was performed according to the Czech standard procedure [9]. Three specimens of standard dimensions 40 × 40 × 160 mm and an age of 360 days were initially dried to a constant mass at 60°C. The specimens were then immersed in water at an ambient temperature of 20±5°C for 24 h and then exposed to freezing at -20±5°C in a freezing box for a period of four hours (figure 1). The specimens were then thawed in water at an ambient temperature of 20±5°C for at least two hours before performing another cycle. This group of the frost exposed specimens was labelled "F".

During the freeze-thaw procedure the water absorption was monitored by weighing the thawed specimens on hydrostatic scales to follow changes in water uptake (in w.%). The total loss of mass of the material remaining in the vessels used for thawing the samples was also recorded. Morphological changes by photographic recording, weight variations of the specimens and ultrasonic wave velocity propagation were measured during the ageing tests.

The frost resistance test was terminated when the samples (F) showed moderate to severe degradation patterns. The samples were then dried at 60°C to constant weight and subjected to the dynamic modulus of elasticity, the flexural and the compressive strength characterizations. The halves of the specimens remaining after the mechanical tests were used to determine changes in the water transport properties of mortars that could indicate their porosity modification due to frost loading: capillary absorption coefficient [10], and water uptake at saturation by immersion under reduced pressure [6].

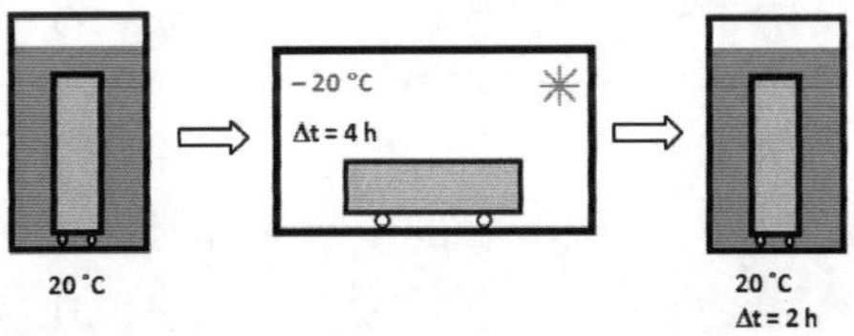

Figure 1. Scheme represents the steps for performing one cycle of the frost ageing test.

3. Results and discussion

3.1. Fresh mortar properties

The mixing water amounts needed to prepare fresh mortar of consistency 170 mm (flow table) are summarised in table 1 for all types of tested mortar. The largest amount was needed to the lime mortar L due to the large specific surface area of the hydrated lime powder particles. The rapid drying of the mixing water caused the formation of shrinking cracks in the hardened L mortar which displayed in the pore size distribution within the pore size range from 7 to 217 μm (figure 2). The substitution of 25-w% of lime hydrate by metakaolin decreased the water consumption in the lime-metakaolin mortar LM by less than 8% compared to the L mortar. Similar water consumption in the mortar with the same type of metakaolin was also reported in [11]. The reason for the lower water consumption is probably the fact that the specific surface area of the metakaolin (S_{BET} Mefisto L05 = 15.41 m^2.g^{-1}) is slightly smaller compared to the specific surface area of lime hydrate (S_{BET} CL90 = 16.84 m^2.g^{-1}). In the case of the NHL3.5 and NHL5 mortars, the kneading water consumption was 36% resp. 33% lower in comparison with air lime mortar L. This can also be explained by the lower specific surface area of the hydraulic lime particles [12]. The low shrinkage of that mortar can be attributed to the consumption of some water during hydration instead of evaporation. The low w/ds ratios of the CL1 a CL2 lime-cement mortars are assigned to the amount of cement used [13]. Since the aerial lime proportion is larger in the CL1 mortar, also the largest water demand was observed.

3.2. Hardened mortar properties

Table 2 summarizes the mortars properties in 360 days of age. The water accessible porosity was 32% for the lime mortar L and its pore size distribution curve showed the presence of two peaks (figure 2). The first peak in the range 0.1–1 µm was assigned to the binder porosity while the second in 10–100 µm range to shrinkage cracks. Slightly higher open porosity, of 34%, was for the lime-pozzolan LM mortar (table 2), which can be explained by the porous character of the pozzolanic reaction products, mainly the C-S-H gels [14], [15]. The gel′ pores, smaller than 0.1 um, are shown in the distribution curve of this mortar (figure 2). LM mortar did not develop shrinkage cracks like L mortar which is probably related to the improvement of the mechanical properties granted by the rapid pozzolanic reaction [17]. Hydraulic limes mortars had porosity approximately 28% and the pore size distribution of these mortars was unimodal. The wide peak of pores in the range 0.5–1 µm is attributable to portlandite reaction while the 0.01–0.2 µm porosity fraction to hydraulic phases [18]. The addition of the white portland cement to the lime in CL1 and CL2 mortars visibly shifts the pores sizes into a lower range and macropores occurring in the pure air lime mortar L have completely disappeared. This tendency is more noticeable in CL2 with higher Portland cement amount.

Table 2. Properties of the hardened mortar with 360 days of age. (*These values represent the average of at least three measurements.)*

Mortar	Open porosity [%-v]	Water absorption coefficient [$kg \cdot m^{-2} \cdot h^{-1/2}$]	Flexural strength [MPa]	Compressive strength [MPa]
L	31.4 (±0.3)	32.2 (±1.4)	0.6 (±0.1)	1.8 (±0.2)
LM	34.1 (±0.2)	8.6 (±1.4)	1.9 (±0.4)	5.9 (±0.4)
NHL3.5	25.9 (±0.4)	16.4 (±0.8)	1.0 (±0.17)	6.8 (±0.3)
NHL5	26.8 (±0.2)	18.9 (±0.3)	0.8 (±0.3)	6.1 (±0.7)
CL1	23.3 (±0.5)	10.9 (±1.1)	2.7 (±0.3)	7.9 (±0.7)
CL2	20.5 (±0.3)	4.6 (±0.4)	4.8 (±0.4)	19.6 (±1.9)

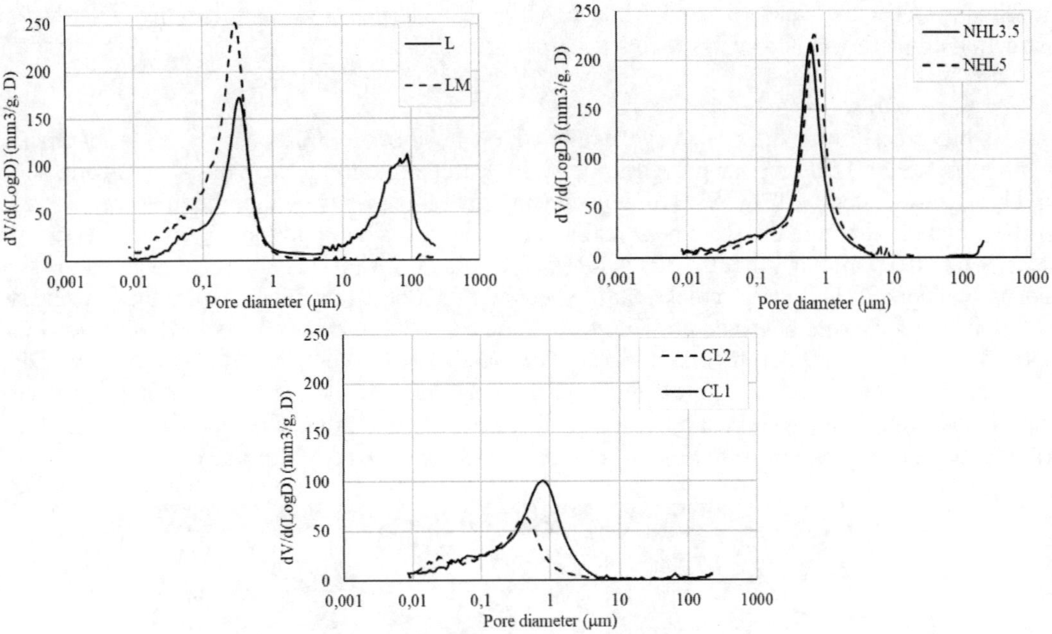

Figure 2. Mercury intrusion curves show the relative pore size distribution of the mortar after 360 days of curing.

The capillary water uptake behaviour of the studied mortars strongly relates to the pore size values determined by mercury intrusion porosimetry (figure 2). The highest water absorption coefficient, of 32 $kg.m^{-2}.h^{-1/2}$, was measured for the air lime mortar L (table 2). This value correlates well with the presence of pores between 10 and 100 μm. Conversely, the LM mortar had the capillary absorption coefficient nearly 4 times lower in comparison with L. This is in accordance with the observed pores size distribution curve with the main peak shifted towards smaller pore diameters which are not capillary active. Similar capillary absorption rate was also recorded on similar composition lime-metakaolin mortar [14]. The main reason for the slow liquid water transport in lime-metakaolin mortar is probably the effect of CSH gels that can bond by van der Waals forces more water molecules in the pore walls than calcium carbonate [14]. [11] reported reduction of the capillary absorption coefficient of lime-pozzolana mortars with the increasing amount of metakaolin. The mortars with hydraulic lime NHL3.5 and NHL5 also showed low capillary absorption coefficient (16.4 and 18.9 $kg.m^{-2}.h^{-1/2}$) corresponding to the presence of hydration products in the microstructure [16]. The slowest capillary transport was determined for lime-cement mortars CL1 and CL2 (10.9 and 4.6 $kg.m^{-2}.h^{-1/2}$). Hydrated phases affect the capillary pores with a decrease in the total porosity and a shift to finer pores [20].

Table 2 shows the flexural and compressive strength of the samples with 360 days of age. The lowest values of flexural and compressive strength obtained for the lime mortar L can be attributed to the shrinkage cracks. Relatively low measured strengths of the natural hydraulic lime mortars NHL3.5 and NHL5 was probably caused by a combination of two factors which negatively influenced the strengthening: 1) insufficient amount of kneading water used for the fresh mortar preparation and 2) not enough wet curing conditions. [21] state that the optimum value of the mortar flow rate of lime NHL3.5 and NHL5 should approach 185 mm. [22] admits that the strength of hydraulic mortars may not always correspond to the classification according to EN 459-1. The values of both compressive and bending strength of the lime-pozzolana plaster LM containing metakaolin can be clearly attributed to the formation of C-S-H and hydrated calcium aluminate structures that have higher strength than calcium carbonate [14]. The highest values of strengths showed the lime-cement mortar CL2 (with a higher amount of white Portland cement), followed by the lime-cement mortar CL1. However, it must be considered that the strength of the repair mortar containing Portland cement may be too high compared to mechanical characteristics of used historic building materials and compatibility with historic construction properties should be assessed.

3.3. Resistance to frost aging

Some visible aspects of mortars damage exhibited in the form of surface spalling or microcrack formation (figure 3). Lime mortar proved to be highly susceptible to the freezing and thawing action after it broke down after the 4th cycle. Hydraulic lime mortars showed several longitudinal cracks already after the 5th cycle although the specimens maintained their cohesion up to the 10th cycle. Degradation was faster for NHL5 mortar. After the 3rd cycle, LM/F mortar also showed significant degradation signs as a dense network of fine surface cracks on the specimen's surface. Immediately after the freezing step very fine ice crystals were observed growing from these cracks. The cracks developed only along with the 40 × 160 mm side faces of the mortar prisms, i.e. parallel to the compaction plane, the top surface showing no signs of visually detectable crack development. After completion of the test and final drying, the specimens emitted a hollow sound when tapped. Cement-lime mortars CL1 and CL2 exhibited fine hairline fissures in several areas after the 8th and 15th cycle, respectively.

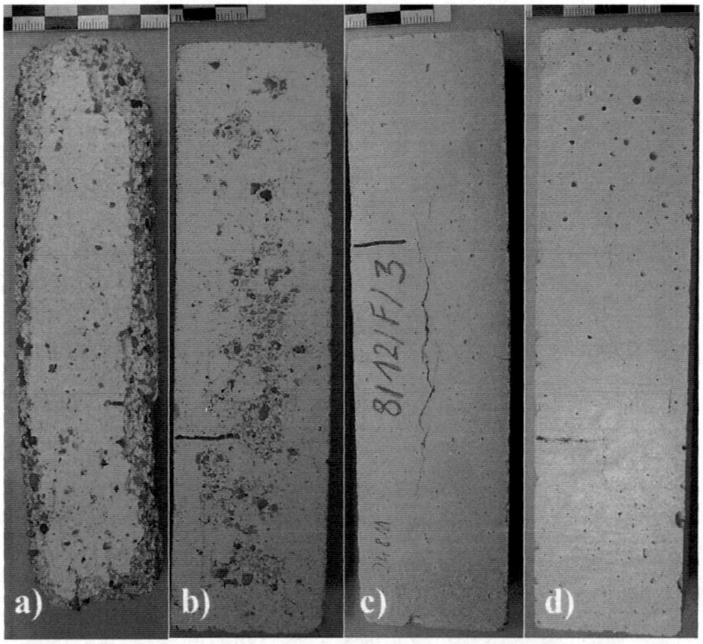

Figure 3. Degradation pattern of specimens after frost aging: a) progressive granular disintegration of L/F specimens after 3rd cycle; b) surface fissuration of LM/F after 10 cycles; c) deep fissures of NHL5 after 10 cycles; d) hairline cracks of CL2 after 20 cycles. Scale next to the specimens shows 1 mm intervals.

Table 3 shows changes of mortars properties after 10 resp. 20 cycles of freeze-thaw aging, except data for L mortar that was destroyed after four cycles. The compressive strength of all frozen specimens with the hydraulic binder increased, indicating a beneficial effect of the water on the hydration previously unreacted hydraulic binders. On the other hand, the flexural strength of the Frost-aged specimens decreased significantly after 10 and 20 freeze cycles, respectively, indicating the drastic character of the test performed. (E.g. The flexural strength of the 360-day-old LM mortar decreased by 74% compared to the strengths of the same old non-freezing set. The flexural strength of frozen mortar specimens NHL3.5, NHL5 and CL1 decreased almost by 40, 63 and 56% after 10 cy freezing, while the flexural strength of CL2 decreased by 69% after 20 freeze cycles. Based on a comparison of the resulting strength values, the most frost resistant/durable mortar appear to be NHL3.5, NHL5, CL1 and CL2.

Table 3. Change of mortars properties after freeze-thaw aging.
(*The down / up arrow symbol clearly indicates the trend of decrease / increase of the characteristic after the end of the freezing test.*)

Mortar (Nr. of cycles)	Capillary absorption coefficient [$kg \cdot m^{-2} \cdot h^{-1/2}$]	Open porosity [%]	Flexural strength [MPa]	Compressive strength [MPa]	Dynamic modulus of elasticity E_{dyn} [MPa]
L (4 cy)	-	-	-	-	-
LM (10 cy)	3.5% ↓	1.5% ↑	74% ↓	3.4% ↓	55.9% ↓
NHL3.5 (10 cy)	2.5% ↓	0.4% ↑	40% ↓	35.3% ↑	48.6% ↓
NHL5 (10 cy)	11% ↓	3.7% ↑	62.5% ↓	1.6% ↑	21.9% ↓
CL1 (10 cy)	16.5% ↓	0.4% ↑	55.6% ↓	20.3% ↑	55.6% ↓
CL2 (20 cy)	46% ↓	4% ↓	68.7% ↓	25.5% ↑	31.9% ↓

The lowest dynamic modulus loss was determined for LM and CL1 samples after ten cycles (reduction of 45%), followed by NHL3.5 (loss of 51%) and NHL5 (loss of 78%). CL2 mortar exhibited the dynamic modulus loss of 68% after 20 cycles. The lower reduction of dynamic modulus of elasticity obtained for LM and CL1 indicates a better mechanical durability of these mixtures against the freeze-thaw cycles. Cracks formed in frozen specimens work similarly to capillary active pores and are able to absorb additional amount of water [23]. However, in our case there was only a slight increase in open porosity (maximum NHL5 3.7%). In this case, the effect of storing samples in water prevailed, which had a favorable effect on the additional hydration of previously unreacted hydraulic binders as a result of which the capillary absorption coefficient significantly decreased. The additional hydration of the binder created smaller pores in the mortars which transport water more slowly. The largest capillary absorption coefficient increase is evident in CL1 and CL2 mortars with a Portland cement.

The phenolphthalein staining test (1% -vol. Phenolphthalein solution) revealed a surprising conclusion. Deep purple color indicates that the central portion of the Frost-aged specimens was not completely cured even after 360 days. Photographs of cross-sections of the Frost-aged specimens test in figure 4 indicate that 10–30% of the volume of the test specimens contained unreacted binder.

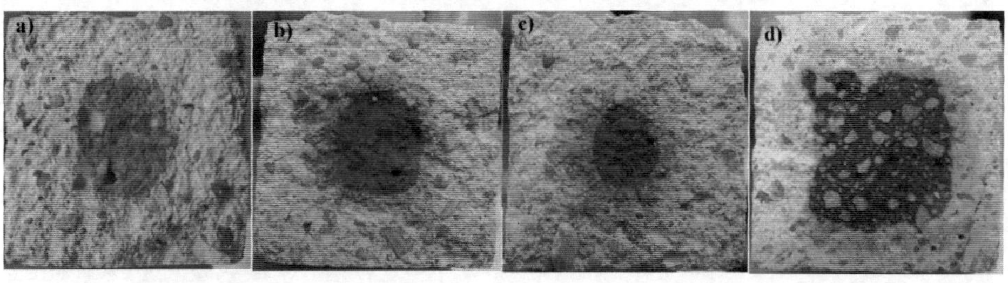

Figure 4. Cross-sections of the Frost-aged specimens. The purple color marks the part of the specimen area (4 × 4 cm) that contains the unreacted binder: a) LM, 10 cy; b) NHL3.5, 10 cy; c) NHL5, 10 cy; d) CL2, 20 cy.

IOP Conf. Series: Materials Science and Engineering 1205 (2021) 012009 doi:10.1088/1757-899X/1205/1/012009

4. Conclusion

Most cultural heritage structures are subjected to a range of environmental risks including frost damage. Repair mortars with appropriate microstructure and durability in relation to expected freeze-thaw loading should be designed and applied within the conservation interventions. The frost attack effects on microstructure and mechanical characteristics of lime mortars was studied for this purpose. In the same time, the used testing procedure based on saturation of the test specimens with water was assessed with respect to the observed lime mortars characteristics and behaviour.

Although samples of lime mortars (in the form of standard test samples 4 × 4 × 16 cm) were stored in regulated laboratory conditions recommended by the standard and subsequently in laboratory regulated conditions corresponding to real external conditions, the binder was not homogeneously matured even after one year (360 days), as is evident from the phenolphthalein test. The least cured is air lime mortar, however, lime-cement mortars also have immature centers of beams. It follows that lime mortars are tested in a different hardened state, which affects (distorts) the test result. The frost resistance test should be adapted for lime mortars, e.g. using smaller specimens.

The mortar with pure air lime binder showed significant shrinkage apparently due to the large amount of mixing water which resulted in a low mechanical strength (mainly bending strength) and low durability when the mortar was subjected to freeze-thaw cycles. It has been confirmed that the preparation of mortar with an air lime binder requires great care. The appropriate consistency should be achieved by kneading fresh mortar with a minimum amount of water. Subsequently, the lime mortar must be protected against rapid drying and the hardening mortar must be repeatedly moistened by gentle spraying with water. In order to create a microstructure of the mortar without cracks, and thus more resistant to frost.

Lime-pozzolanic mortar based on metakaolin admixture to pure air lime improved mechanical strength and did not produce shrinkage cracks. Due to the pozzolanic reaction of the binder and the formation of C-S-H gels, the pores in the porous structure predominate in the range of 0.01–1 μm, which limit capillary suction rate (the water absorption coefficient dropped to a quarter of that observed for pure air lime mortar). The compressive strength of the mortar increased, but the flexural strength decreased significantly after 10 freezing cycles. After the 3rd cycle, mortar showed significant degradation signs as a dense network of fine surface cracks on the specimen's surface. Even in this case, we can assume that the amount of mixing water required with respect to the required consistency of 170 mm was too high and that the reduction of mixing water would contribute to better frost resistance.

Hydraulic limes mortars possessed also lower porosity and lower water absorption coefficient compared to the air lime mortar. Contrary to expectations, low flexural strengths have been established for these mortars which has had a negative impact on frost resistance. Hydration of hydraulic binders has probably been negatively influenced either by the low relative humidity setting for a curing (90±5% for 1 week then 60±5% RH) or too much mixing water played a role. Hydraulic lime mortars showed several longitudinal cracks already after the 5th cycle although the specimens maintained their cohesion up to the 10th cycle.

Lime-cement mortars CL1 and CL2 achieved the highest compressive and flexural strengths. However, even after 360 days, the standard beams of lime-cement mortars were not cured until the middle of the beam. The porosity accessible to water is the lowest of the mortars compared. The porous structure is dominated by pores smaller than 0.1 μm, which do not transport liquid water, but can retain it on the surface of the pores through hydrogen bonds. There is also a significant slowing down of water transport through capillary pores in comparison with other studied mortars.

The compressive strength of all Frost-aged specimens with the hydraulic binder component increased, indicating a beneficial effect of the water on the hydration previously unreacted hydraulic binders. On the other hand, the flexural strength of the Frost-aged specimens decreased significantly after 10 and 20 freeze cycles, respectively, indicating the drastic nature of the test performed.

Acknowledgement
This research was supported by the CONSECH20 project ("CONSErvation of 20th century concrete Cultural Heritage in urban changing environments") supported by the JPI CH Heritage in Changing Environments.

References

[1] Pavía S, Fitzgerald B and Treacy E 2005 An assessment of lime mortars for masonry repair *Concrete Research in Ireland Colloquium, Ciaran McNally Ed. University College Dublin* pp 101–108
[2] Grossi C, Brimblecombe P and Ian H 2007 Predicting long term freeze-thaw risks on Europe built heritage and archaeological sites in a changing climate *The Science of the total environment* 377. 273-81. 10.1016/j.scitotenv.2007.02.014.
[3] EN 1015-2: 1999 Methods of test for mortar for masonry – Part 2: Bulk sampling of mortars and preparation of test mortars
[4] EN 1015-3: 1999 Methods of test for mortar for masonry – Part 3: Determination of consistence of fresh mortar (by flow table)
[5] Czech Hydrometeorologic Institute Average relative humidity data in Prague during the period from April to September 2011 reported on the website of the Czech Hydrometeorologic Institute (retrieved on June 2021) http://portal.chmi.cz/historicka-data/pocasi/mesicni-data
[6] EN 1936: 2007 Natural stone test methods. Determination of real density and apparent density and of total and open porosity
[7] EN 1015-11: 1999 Methods of test for mortar for masonry: Determination of flexural and compressive strength of hardened mortar
[8] EN 1925: 1999 Natural stone test methods. Determination of water absorption coefficient by capillarity
[9] ČSN 72 2452: 1968 Testing of frost resistance of mortar
[10] EN 1015-18: 2002 Methods of test for mortar for masonry. Determination of water absorption coefficient due to capillary action of hardened mortar
[11] Vejmelková E, Keppert M, Rovnaníková P, Kešner Z and Černý R 2012 Properties of lime composites containing a new type of pozzolana for the improvement of strength and durability *Composites Part B: Engineering* vol. **43** pp 3534–3540
[12] Silva B A, Ferreira Pinto A P and Gomes A 2014 Influence of natural hydraulic lime content on the properties of aerial lime-based mortars *Construction and Building Materials* vol. **15** pp 208–218
[13] Zhang D, Zhao J, Wang D, Xu Ch, Zhai M and Ma X 2018 Comparative study on the properties of three hydraulic lime mortar systems: Natural hydraulic lime mortar, cement-aerial lime-based mortar and slag-aerial lime-based mortar *Construction and Building Materials* vol. **186** 20. P. 42-52. https://doi.org/10.1016/j.conbuildmat.2018.07.053
[14] Černý R, Kunca A, Tydlitat V, Drchalova J and Rovnanikova P 2006 Effect of pozzolanic admixtures on mechanical, thermal and hygric properties of lime plasters *Constr Build Mater* vol. **20** pp 49–57
[15] Vejmelková E, Keppert M, Keršner Z, Rovnaníková P and Černý R 2012 Mechanical, fracture-mechanical, hydric, thermal, and durability properties of lime–metakaolin plaster for renovation of historic buildings *Construction and Building Materials* vol. **31** pp 22–28
[16] Arandigoyen M and Alvarez J I 2007 Pore structure and mechanical properties of cement-lime mortars *Cement and Concrete Research* vol. **37** pp 767–775 DOI:10.1016/ j.cemconres.2007.02.023
[17] Gameiro A, Santos Silva A, Veiga R and Velosa A 2012 Hydration products of lime–metakaolin pastes at ambient temperature with ageing *Thermochimica Acta* 535 pp 36–41
[18] Arizzi A, Viles H and Cultrone G 2012 Experimental testing of the durability of lime-based mortars used for rendering historic buildings *Construction and Building Materials* vol. **28**

pp 807–818

[19] Vejmelková E, Keppert M, Rovnaníkova P, Keršner Z and Černý R 2011 Properties of lime composites containing a new type of pozzolana for the improvement of strength and durability *Composites: Part B: Engineering* vol. **43** pp 3534–3540

[20] Cizer O 2009 Competition between carbonation and hydration on the hardening of calcium hydroxide and calcium silicate binders (PhD thesis, Katholieke Universiteit Leuven) ISBN 978-94-6018-055-2

[21] Hanley R and Pavia S 2008 A study of the workability of natural hydraulic lime mortars and its influence on strength *Materials and Structures* vol. **41** pp 373–381

[22] Figueiredo C, Lawrence M and Ball R 2016 Chemical and physical characterisation of three NHL 2 binders and the relationship with the mortar properties in L Villegas, I Lombillo, H Blanco & Y Boffill (eds), *REHABEND 2016 Euro-American Congress: Construction Pathology, Rehabilitation Technology And Heritage Management* (6th REHABEND Congress) C2.2.17, Santander, Spain.

[23] Klemm A J and Klemm P 1997 The effects of the alternate freezing and thawing cycles on the pore structure of cementitious composites modified by MHEC and PVA *Building and Environment* vol. **32** pp 509–512

IOP Conf. Series: Materials Science and Engineering 1205 (2021) 012010 doi:10.1088/1757-899X/1205/1/012010

Development of rapid set mortar for additive manufacturing

B Nespor[1], M Bohac[1] and M Nejedlik[1]

[1] Research Institute for Building Materials, Hnevkovskeho 65, Brno 61700, Czech Republic

Email: nespor@vustah.cz

Abstract. The study presents the development results of blended cement mortars for additive manufacturing. The goal was to achieve a balanced rheological behaviour of the fresh mixture through optimal granulometry, suitable grain shape and the choice of micro-admixtures. The article closely describes the granulometric and rheological characterization of the proposed ternary and quaternary blended mortar mixtures with various SCM's. The print quality and shape stability of the promising mortars and the researched 3DP mixes were evaluated during incremental layering for comparison. The performed experiments confirmed that the mixtures containing blast furnace slag, quartz powder and metakaolin are more plastic and allow faster layering in height. The slaked lime-enriched mixture has excellent plasticity and is able to resist deformation during overlaying, the printed layers are without surface defects. The small addition of polypropylene fibre (0.08 wt.%) improves the plasticity and shape retention ability.

1. Introduction

Concrete is the single most widely used material in the world, after water, with a usage of about 13 billion tons per year. A primary reason for its popularity as a building material stems from the fact that it goes naturally from a fluid to a solid state – being able to flow and fill a mould, and upon hardening sustains a load. This gives great flexibility in terms of material handling and placement, something that has yet to be fully exploited in the world of digital fabrication [1].

Digitization of production is one of the pillars of the industry 4.0 initiative, which is considered the fourth industrial revolution. This includes 3D printing (3DP), a subset of digital fabrication, and a technology that is still at the nascent stage of realization in architecture and construction, notwithstanding the growing attention in recent years [2–6]. The future seems to be in the connection of 3D printing and so-called 4D materials, i.e. materials with properties that change depending on the change of boundary conditions (e.g. time, temperature, stress, dynamic effects, etc.), or digital materials to enable user programming of functional properties [7–8].

Cement mortars for 3DP technology have the potential to help the necessary digitization of construction industry and thus enable the unlimited implementation of computer-optimized designs of structure elements and buildings. In the future, it is possible to anticipate the gradual replacement of a significant part of current construction technologies by additive manufacturing. The 3DP technology or additive manufacturing or also incremental production technology is a relatively new, progressive, rapidly developing area of development in the construction industry in recent years. Additive manufacturing allows the creation of architectural models, building elements and buildings based on a digital design using a machine 3D printer. A suitable material is gradually applied by the printhead layer by layer. It is a computer-automated manufacturing process based on a 3D model.

IOP Conf. Series: Materials Science and Engineering 1205 (2021) 012010 doi:10.1088/1757-899X/1205/1/012010

This technology significantly speeds up construction, reduces labour costs and enables creation of a wide range of modern architectural works of any shape without restriction.

By using progressive cement-based binders with SCM's (supplementary cementitious materials), CO_2 emissions and energy consumption in the production of Portland clinker can be reduced. A suitable ratio of ternary or quaternary binders can improve the granulometry of the densely filled polydisperse system, the rheology of the 3DP mortar and the printability interval. 3DP mortars should achieve relatively high flow limits and low plastic viscosities to achieve print quality and dimensional stability.

The use of recycled materials is also very promising, as they can make a significant contribution to saving raw materials and energy in conjunction with additive manufacturing technology. Utilization of the fine fraction of recycled material is often problematic and the use of recycled materials for additive manufacturing is a new solution for the recovery of this waste.

2. Experimental

2.1. Materials and methods

Limiting the use of chemical additives was a leitmotif in designing 3DP mixtures. The aim was to achieve a balanced rheological behaviour of the mortar mixture through optimal granulometry, suitable grain shape and the choice of micro-admixtures. The design of the composition of the quick-drying mortar mixture for 3D printing was based on a previous study; see Boháč et al [9], in which the role of various aggregates on the rheology of fresh 3D printing concrete was described. One of the key admixtures was the crushed recycled concrete from the recycling centre, which, with a suitable dose of fines, does not affect the plasticity of the mortar mixture, and at the same time, increases the required yield stress and adhesion of the fresh material.

The ternary blended binder of the reference mixture (TBM) is composed of Portland cement (C) CEM I 52.5 R (Cement Hranice), fly ash (FA) (Veolia Energie ČR, plant Přerov) and silica fume (SF) Elkem 971 (Elkem) in proportions 40:40:20 wt.%. Special quartz sand B30 (Tvarbet Moravia, sandpit Bzenec) of fraction below 1 mm with the fine fraction of recycled concrete aggregate (RCA) (Ditton) were used as aggregate. The unpolluted fine fraction of crushed RCA below 0.355 mm was used as a partial replacement (4 wt. %) for sand. The ratio of binder to aggregates was 1:1.7. Polycarboxylate superplasticizer (SP) Glenium ACE 446 (Basf) was used for all mixtures. The used water-binder ratio was 0.3 to 0.43 due to the use of fibre and concern for the various water demands of SCM's.

3DP mortars from literature review – SM1–5 [10–14] (see table 1), differed in the ratio of ternary binders, mainly by the increasing amount of Portland clinker, but also by the proportion of binder to aggregate, the dose of mixing water and plasticizer, or the use of reinforcement. These compositions were slightly modified from the original. A fine fraction of RCA was used as a partial replacement (4 wt. %) for sand. To eliminate cracks caused by plastic shrinkage and plastic settling, the reference mixture was enriched with polypropylene fibres with a length of 3, 6 or 8 mm (PFRM 1–5).

Table 1. Mix design of printing mortars.

Prescription	C	SCM		Aggregates		Reinforcement			SP	w/b
		FA	SF	Sand	RCA	3	6	8 mm		
3DP mixtures from literature review										
TBM [Bohac et al 2018]	40	40	20	163.2	6.8				1.0 (b)	0.3–0.4
SM 1 [Tay et al 2016]	50	38	12	153.6	6.4				0.3 (b)	0.4
SM 2 [Nerella et al 2019]	55	30	15	172.8	7.2				2.0 (b)	0.3
SM 3 [Nerella et al 2016]	55	22	23	152.6	6.4				1.3 (b)	0.3
SM 4 [Le et al 2012]	70	20	10	144.0	6.0			0.15 (b)	2.0 (b)	0.28
SM 5 [Anell 2015]	80	10	10	132.5	5.5			0.14 (b)	1.4 (b)	0.27
Fibre-reinforced mortars										
PFRM 1	40	40	20	163.2	6.8	0.08 (b)			1 (b)	0.37
PFRM 2	40	40	20	163.2	6.8		0.08 (b)		1 (b)	0.40
PFRM 3	40	40	20	163.2	6.8			0.08 (b)	1 (b)	0.39
PFRM 4	40	40	20	163.2	6.8			0.15 (b)	1 (b)	0.43
PFRM 5	50	12	38	170.9	7.1	0.14 (b)			0.5 (b)	0.40

The blended binder of the reference mortar was further modified with various admixtures. The main goal was to optimize print quality and shape stability of 3DP mixture and mainly to reduce the cost of the binder. Microsilica is replaced by ground granulated blast furnace slag SMŠ 400 (BFS) (Cemix, plant Kotouč Štramberk), finely ground limestone (L) Vitošov 7000 (Vápenka Vitošov), metakaolin (MK) Mefisto K05 (České lupkové závody, plant Nové Strašecí), quartz powder (MD) Mikro-Dorsilit 120 (Gebrüder Dorfner). Different doses of slaked lime (SL) High purity lime (Research institute for building materials) were tested to improve plasticity. The mix designs QBM 1–7 are given in table 2.

Table 2. Mix design of quaternary blended cement mortars for 3D printing.

Prescription	C	SCM							Aggregates		SP	w/b
		FA	SF	L	BFS	MK	MD	SL	Sand	RCA		
QBM 1	40	40	10	10					163.2	6.8	1 (b)	0.4
QBM 2	40	40	10		10				163.2	6.8	1 (b)	0.30.4
QBM 3	40	40	10			10			163.2	6.8	1 (b)	0.4
QBM 4	40	40	10				10		163.2	6.8	1 (b)	0.3–0.4
QBM 5	31	31	16					22	124.8	5.2	0.8 (b)	0,64*
QBM 6	35	35	18					12	144.0	6.0	0.9 (b)	0.59*
QBM 7	37	37	19					6	153.6	6.4	0.9 (b)	0.55*

* The water to binder ratio includes the water required for slaking of lime

2.2. Granulometric characterization

Coarse RCA was crushed and sieved to 0.355 mm (crushed RCA), the oversize fraction was grounded and sieved to 0.355 mm (ground RCA). The effect of preparation of RCA on the shape factor was observed with optical microscope Nikon Eclipse LV100ND in linearly polarized light at 500× magnification.

Particle size distribution (PSD) of cement, various SCM's, sand, crushed and ground RCA was determined by a sieve analysis (above 0.4 mm) and CILAS 920 laser particle-size analyser with a range of 0.3–400 μm. Before the measurement, particles were dispersed in isopropyl alcohol using sonication for 60 seconds.

The optimal dose of selected microsilica substitutes was designed using Elkem Materials Mixture Analyzer (EMMA). Input data from PSD were uploaded into EMMA particle packing. The cumulative

PSD curves of the quaternary blended cement mortars were compared with the ideal grain size curve according to the modified Andreassen model for ideal packing density with the distribution coefficient 0.38 as in equation (1).

$$CPFT = \frac{D_P^q - D_S^q}{D_L^q - D_S^q} 100 \qquad (1)$$

where, CPFT = the cumulative (volume) percent finer than, DP = the particle size, DS = smallest particle size, DL = largest particle size, q = the distribution coefficient or exponent.

The distribution coefficient (fine to coarse particle ratio) was proposed with concern for extrudability and shape-retention-ability of fresh mortars. Densely packed polydisperse systems have q = 0.2–04. If the exponent increases, it means an increase of the coarse materials, and if it decreases, the amount of the fine materials is increased [15]. Due to the slope of the cumulative curves of the studied mixtures, the distribution coefficient of 0.38 was chosen.

2.3. Rheological characterization

Each blended binder was thoroughly homogenized by a laboratory homogenizer for 30 minutes. Dry components (binder and aggregates) were dry-mixed for 2 minutes in the mixer, then water with superplasticizer was added and mixed for another 3 minutes.

Rheological characterization of mortars was carried out using a TA Instruments Discovery Hybrid Rheometer DHR-1 under standard laboratory conditions at 25°C. Flow characteristics were monitored in a range of 1–150 s^{-1} using the geometry of concentric cylinders (DIN). Yield stress and plastic viscosity were calculated from flow curves by Bingham model equation (2).

$$\tau = \tau_0 + \eta\dot{\gamma} \qquad (2)$$

where τ = shear stress (Pa), τ_0 = yield stress (Pa), η = viscosity (Pa.s), $\dot{\gamma}$ = shear rate (s^{-1}).

Thixotropy or rheopexy (time-dependent viscosity) were evaluated from hysteresis loops. All mixtures were pre-sheared at a constant shear rate of 10 s^{-1} for 10 s.

The shape stability of freshly printed mortars was simulated using a rheometer with parallel-plate geometry in the axial mode. The increase in axial force was monitored as the upper geometry approached. During one cycle, the geometry approaches 20 μm at a speed of 2 μm/s. The geometry approach was set in minute intervals. The number of cycles indicates stability when layering the material during 3D printing. More cycles mean less stability of the fresh mortar mixture.

Buildability (resistance to deformation) was assessed using a rheometer in oscillatory mode by amplitude sweep test in a log strain ramp from 0.001% to 1%, which was within the linear viscoelastic region (LVER). The LVER indicates the area in which the sample resists the applied deformation without destroying the structure. The LVER of storage modulus (G') and loss modulus (G”) was determined. The visco-elastic behaviour was characterized by the complex modulus (G*), which is expressed by the vector sum of individual moduli (3).

$$|G^*| = \sqrt{(G')^2 + (G'')^2} \qquad (3)$$

where: G* = complex modulus, G’ = storage modulus, G” = loss modulus.

The physical stability of the samples was monitored using critical strain (γ_c), which determines the upper limit of the LVER. The critical strain can be read from the stress-strain curve as the onset of the decrease in complex modulus.

2.4. Technological tests

Technological tests were performed on the optimized mortars. Determination of normal consistency (paste without liquefaction agent) and determination of initial and final setting time of mortars (paste with 1% liquefaction agent) was performed using a penetration test (Vicat) according to EN 196-3.

The consistency of fresh mortar was determined by Haegerman flow table, using a prescribed metal cone after the prescribed number of vertical shocks, by measuring the diameter of the spilt test specimen according to ČSN 72 2441.

Determination of compressive strength, flexural strength, tensile splitting strength was performed according to EN 196-1 (at the age of 1, 2, 7, 28 and 90 days). Moisture content y and water absorptivity were determined according to ČSN 73 1316.

Durability was determined by a standard freeze-thaw resistance test according to ČSN 73 1322 and an accelerated durability test. Weather changes simulation involves the cyclical alternation of summer and winter dominant factors. The weather model is based on the long-term trend of climate development in the Czech Republic (1961–2021). One model year of external environment simulation consists of cyclic water sprinkling and drying by radiant heat drying by radiant heat (18 cycles) a freezing and thawing (1 cycle). The test specimens (20 × 20 × 100 mm) were subjected to water sprinkling for 170 min. There was a 10-minute delay after each sprinkling, followed by radiant heat stress using 2400 W.m^{-2} IR lamps. The radiant heat was supplemented by xenon lamps, which supply a component of UV radiation with an output of about 200 W.m^{-2}. The surface temperature of the test specimens was about 70°C. The effect time of radiant heat and UV radiation was 170 minutes. There was a 10-minute break before the next water sprinkling cycle. After 18 cycles of the thunder-shower simulation, test specimens were saturated by water at a temperature of 20°C for 2 h. Subsequently, the samples were surface dried and frozen to -20°C for 2 h and then thawed in water (20°C) for another 2 h. After completion
of weathering cycles, changes in mechanical properties of reference and cyclically stressed samples were evaluated.

Determination of volume changes was performed using the beam method according to ČSN 73 1320. The test specimens (40 × 40 × 160 mm) were fitted with metal targets. The longitudinal strain was determined according to the formula (4).

$$\varepsilon_{s,n} = \frac{\Delta Z_n}{Z} \cdot 1000 \tag{4}$$

where: $\varepsilon_{s,n}$ = longitudinal strain (mm/m). Δz_n = change in length over original length (mm), z = original length (mm). Changes in length were measured after demoulding (1, 4, 14, 36, 39 days) and after determination of freeze-thaw resistance (50, 75, 100 freezing cycles) using a calliper.

2.5. Pilot plant experiments
The pilot plant verification of the designed mortars was carried out by Construction 3D printer "AMT" S-6045. Numerous printing experiments were performed using a model design of a circle with a diameter of 500 mm. The material was printed at a speed of 100 mm/s through the print head with a diameter of 40 mm. The layer height was 10 mm. The dimensional conformity of the printed layer with the model 3D design (print quality) and the ability to resist deformations during overlaying (shape stability) were evaluated by optical observation. Surface defects and cracking were also monitored.

3. Result and discussion

3.1. Granulometric characterization
The effect of preparation of unpolluted coarse fraction of RCA on the particle size and shape factor is summarized in figure 1. RCA was ground by crushing and grinding to a very similar fineness. The RCA particles are often non-spherical or irregularly shaped and have various shapes, including a rectangle, ellipse, or random shapes with sharp edges. The granulometry of the two RCA is very similar and the particle size does not have to be differentiated when assessing the rheological properties.

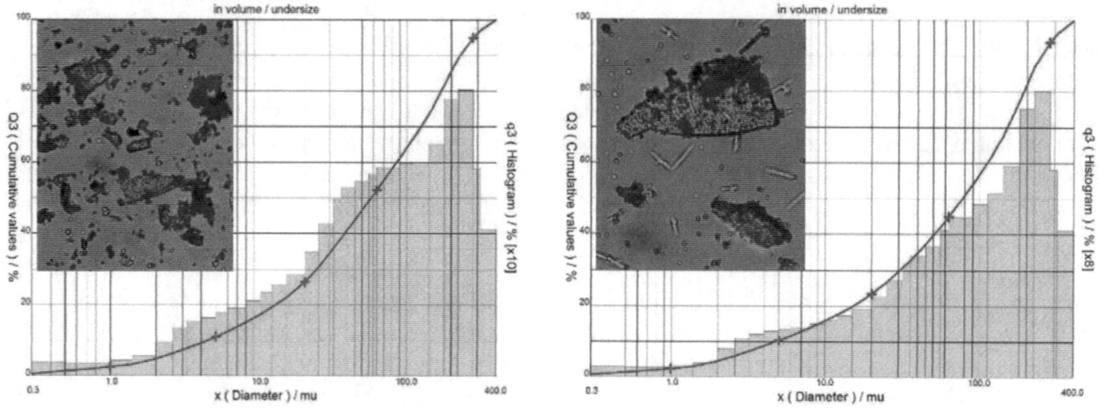

Figure 1. Optical microscopy and PSD of crushed RCA (left) and ground RCA (right).

Particle size distribution of dry mixtures was routinely carried out by a laser diffraction method. Cumulative PSD curves of the blended cement mortars were correlated with the ideal PSD curve. The effect of SCM's microsilica substitutes on cumulative particle size curves was monitored, respectively how close it is to the ideal PSD curve. Microsilica was replaced with 5, 10 and 15% wt. of L, BFS, MK and MD. The optimal amount of all SCM's was close to 10% wt., see figure 2. Compared to the ideal curve, the reference mortar TBM shows higher content of particles below 1 μm. All SCM's improve PSD below 1 μm. BFS has slightly fewer particles in the range of 1 and 10 μm.

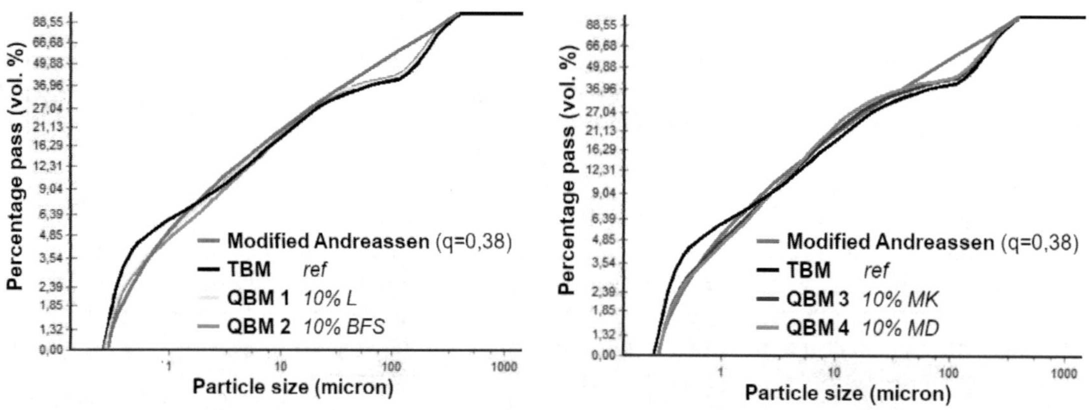

Figure 2. Ideal grading curve (red) and cumulative PSD curves for ternary (black) and quaternary blended mortars with the optimal replacement content of microsilica by SCM's.

3.2. *Rheological characterization*

The effect of RCA preparation on flow properties of reference paste is minimal. Small differences are observable at the low shear rates (figure 3), where static yield stress values can be observed.

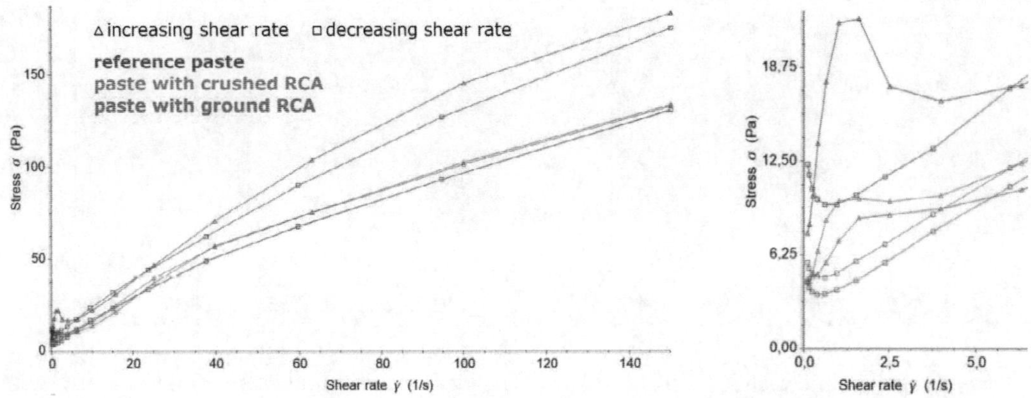

Figure 3. The effect of RCA preparation on flow properties of pastes, the low shear rates detail (right).

The samples with RCA have a lower requirement for mixing water compared to the reference paste without RCA. The measured rheological parameters also correspond to this (table 3).

Table 3. The effect of RCA preparation on rheological parameters.

Cement paste	Static yield stress (Pa)	Dynamic yield stress (Pa)	Plastic viscosity (Pa.s)	Thixotropy (Pa.s)
Reference paste	21.99	11.75	1.158	1204
Crushed RCA	9.99	7.15	0.97	566
Ground RCA	8.69	6.05	0.89	594

Fresh mortar mixture for additive manufacturing should have relatively high yield stress and a low value of plastic viscosity. The time-dependent thixotropy is important for this application in terms of gradual layering. A low value of plastic viscosity is necessary due to good pumpability. Yield stress and thixotropy in turn for shape-retention-ability during the next layering. The effect of SCM's on flow and time-dependent rheological properties after 10 and 30 minutes from the addition of water, as depicted in figure 4.

The mixture with ground limestone is liquid and does not hold its shape. The rheological parameters practically do not change during the first 30 minutes. In contrast to the reference mixture, limestone inappropriately decreases the yield stress of the fresh mixture. The mixture with MK has a significantly higher value of plastic viscosity, which indicates a possible problem with pumpability. The mixture is very plastic and holds the shape well. The fresh mixture has higher water required to achieve the same consistency as the reference mixture. The value of thixotropy decreases with time.

BFS has very similar values as the verified reference mixture, only it has lower values of thixotropy. After 30 minutes, the plastic viscosity value increases, this could lead to pumpability problems. These characteristics can be modified with chemical additives.

The mixture with MD has lower yield stress and a higher plastic viscosity in comparison to the reference mixture. Mikro-Dorsilit is an economically advantageous substitute for microsilica. Rheological properties can be further balanced using chemical additives.

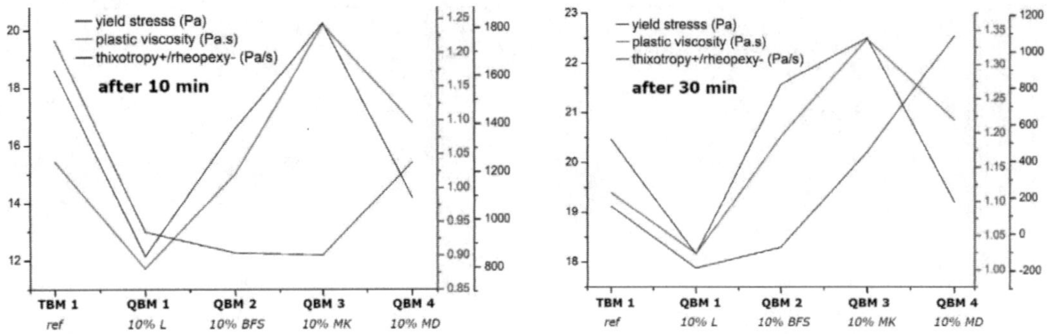

Figure 4. Flow and time-dependent rheological properties of blended cement mortars with 10% replacement of the microsilica by SCM's, 10 (left) and 30 (right) minutes after addition of water.

The effect of SCM's on the shape stability and buildability of fresh mortars was studied using an axial squeeze test, see figure 5. Mixtures with a steep increase in axial force are less plastic and therefore more stable during layering. Mixtures containing MK and MD showed the best values of shape stability. The mixture with L and the reference mixture are significantly less plastic than these mixtures. Therefore, it can be assumed that the mixtures containing MK and MD allow faster layering and the possibility of layering in height.

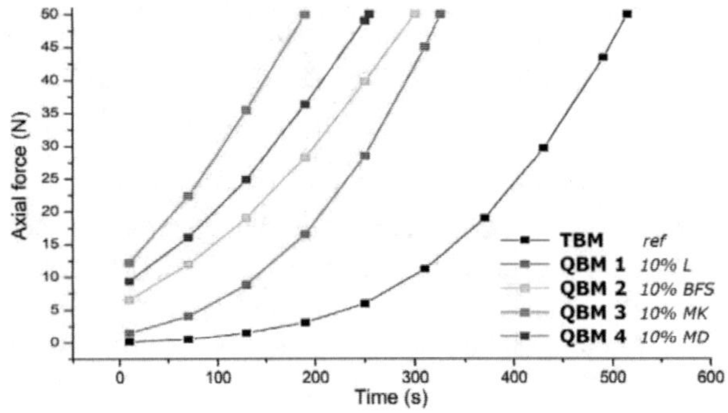

Figure 5. The effect of SCM's on shape stability and buildability of fresh mortars during layering.

The effect of SCM's on the rheological stability of fresh mortars was investigated using an amplitude sweep test. The viscoelastic behaviour of mortar mixtures was observed after 10 and 30 minutes after the addition of water. To determine the linear viscoelastic region over which the samples resist the applied deformation, the complex modulus was plotted against the strain amplitude, as depicted in figure 6. The linearity limit or critical strain is characterized as the endpoint of the LVER or also as the onset decrease in the complex modulus. The value of the upper limit of the LVER tends to proportionally correlate with rheological stability. Longer LVER suggest more physical stability of the sample.

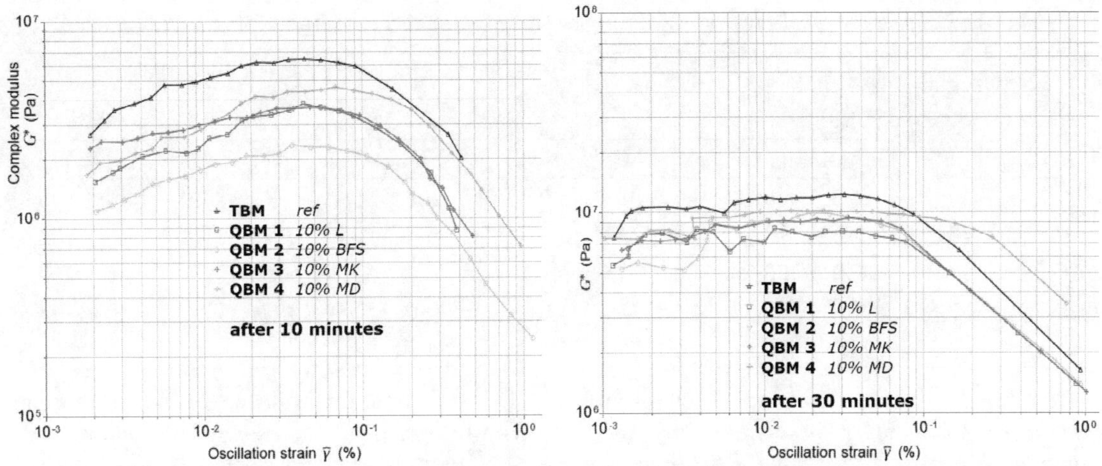

Figure 6. The effect of SCM's on rheological stability of fresh mortars.

The LVER for reference and quaternary blended mortars are very similar. Relatively greater suspension stability can be described for the mixture with MD. The critical strain values after 30 min are lower than after 10 min of hydration, see table 4. In exclusion of the mixture with BFS, a good correlation between critical strain and shape stability results can be observed.

Table 4. The effect of SCM's on critical strain (physical stability).

Critical strain γ_c (%)	**TBM** *ref*	**QBM 1** *10% L*	**QBM 2** *10% BFS*	**QBM 3** *10% MK*	**QBM 4** *10% MD*
γ_c 10 min	0.042	0.041	0.034	0.048	0.066
γ_c 30 min	0.031	0.03	0.02	0.032	0.045

3.3. Technological tests

The results of determining the technological properties of reference and quaternary blended mortars are summarized in table 5. Normal consistency and setting times were determined by penetration tests. The initial setting time of the mixed cement pastes of normal consistency did not occur earlier than 4.5 hours, the final setting time in 6.5 hours and the case of the mixture with MD in 7.5 hours. To test the consistency of the fresh mixes and to monitor the development of the strength of the cured mortars, the water coefficient was adjusted individually for each mixture so that the quaternary blended cement mortars achieved a similar consistency to the reference mix. The reference mixture had the highest water requirements (w = 0.28). The mixtures containing L, BFS, MK had a lower water-biner coefficient (w = 0.265). The least water (w = 0.255) was used for the mixture with MD. All mortars had a diameter of the slumped cone after shocking of 110 mm on the flow table before and 180 mm after shaking with 15 shocks. Individual diameter in flow table tests did not differ by more than ± 15 mm. All tested specimens showed excellent strength characteristics. SCM's as a partial replacement (10 wt.%) for microsilica does not have an adverse effect on strength development. Mixtures with L and MK show similar strengths to the reference mixture; however, MK does not have such strength development after 28 days of curing. The mixture with BFS and especially with MD achieves the highest strengths.

Table 5. Technological properties of ternary (reference) and quaternary blended mortars.

	Normal consistency (%)	Setting times (min)		Flexural and Compressive strengths (MPa)									
				1 day		2 days		7 days		28 days		90 days	
		Initial	Final	F	C	F	C	F	C	F	C	F	C
TBM	47.5	270	410	3.6	21.2	4.7	30.7	6.1	56.4	9.7	84.9	11.4	100.9
QBM 1	40.5	270	410	4.2	22.4	4.6	32.0	6.4	57.8	7.8	81.8	11.8	102.4
QBM 2	39.0	330	400	4.6	23.6	5.1	36.8	8.6	62.3	9.8	92.5	8.7	110.4
QBM 3	47.5	290	410	4.4	21.1	5.0	32.6	7.0	57.9	7.0	77.6	5.7	92.3
QBM 4	43.5	320	450	4.9	25.2	4.8	37.6	4.1	64.2	12.1	100.2	13.7	114.6

The physical properties of selected promising 3DP mixtures for additive manufacturing are summarized in figure 7. These are ternary (modified with PP fibres) and quaternary blended mortars, which have been used in pilot plant experiments. 3DP mortars showed relatively good strength characteristics. The fibre reinforced reference composition and the mixture with BFS and MD averaged compressive strengths of around 40 MPa and flexural strength of 5.5 MPa. The larger amount of fibre, as well as lime, did not have a significant effect on the strengths. The lime enriched mixtures exhibited higher water absorption and lower compressive and flexural strengths of 26 MPa and 4 MPa, respectively.

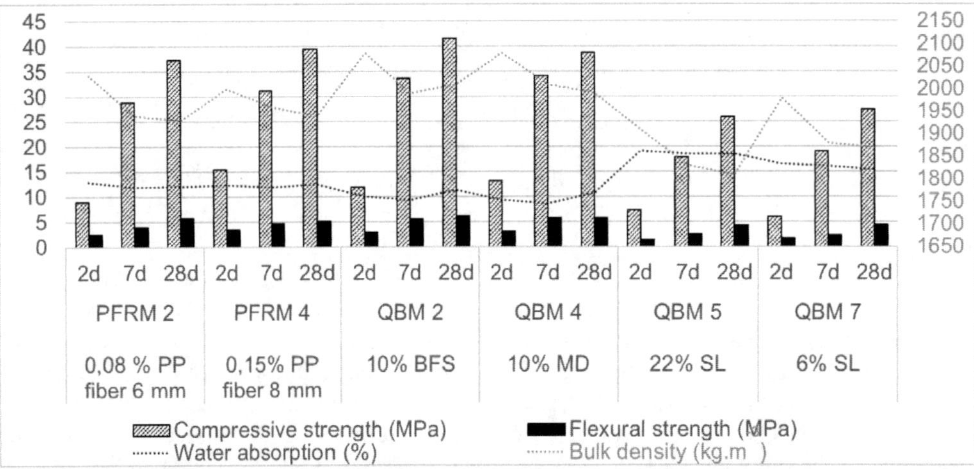

Figure 7. Physical properties of selected samples of 3DP mortars.

Durability was assessed using the frost resistance test and the accelerated durability test. Changes in the length of 3DP mortars during wet curing and after freezing cycles were continuously measured. These tests were performed on the reference and slaked lime-modified mortars. The impact of freeze-thaw cycles and also weathering changes on the physical properties of 3DP mixtures are presented in tables 6, 7 and figure 8.

During moisture curing, the longitudinal strain was very low in both samples. Comparative (unloaded) specimens of slaked lime-modified mortar had about one-third lower strengths compared to the reference mortar and also higher absorption.

The tested samples are not frost-resistant, the frost resistance coefficient decreased under the value of 0.75. The slaked lime-modified mortar had significantly lower compressive strength and especially flexural strength after freeze-thaw cycles compared to the reference mixture. The sample with SL shows enormous longitudinal strain compared to reference mortar after freeze-thaw cycles.

Table 6. The impact of freeze-thaw cycles on physical properties of selected promising 3DP mixtures.

Freeze-thaw resistance	TBM *ref* 28 days moist	50 freeze-thaw cycles	75	100	QBM 7 *6% SL* 28 days moist	50 freeze-thaw cycles	75	100
Compressive strength (MPa)	60.0	53.4	51.5	56.0	44.0	35.7	23.8	21.4
Flexural strength (MPa)	9.0	5.9	5.0	5.4	7.4	1.8	0.4	1.0
Bulk density (kg.m^3)	2060	2060	2050	2060	1980	2010	1990	2010
Water absorption (%)	12.1	12.7	12.7	12.1	17.0	18.3	19.6	19.1
Coefficient of frost resistance	-	0.65	0.56	0.60	-	0.24	0.05	0.14

Table 7. The volume changes of reference and modified 3DP mortar.

Linear strain (mm/m)	4 days moisture curing	14 days	36 days	3 days water curing	50 freeze-thaw cycles	75	100
TBM *ref*	-0.02	0.00	0.02	0.06	-0.42	-0.11	0.69
QBM 7 *6% SL*	-0.04	-0.02	-0.08	-0.06	2.93	5.63	14.69

The reference mixture had relatively good durability. After 20 years of weather changes simulation, a 4% loss in compressive strength and a 16% increase in flexural strength was observed. After 20 simulation cycles, the modified mortar had a strength of about one-fifth in both compression and flexure. The values of absorption and bulk density did not change.

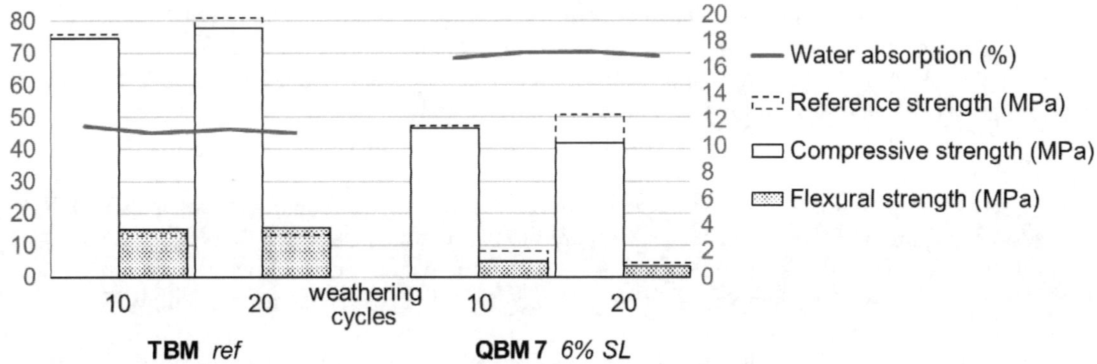

Figure 8. The impact of weathering changes on physical properties of reference and modified mortars.

3.4. Pilot plant experiments

Additive technology on ternary and quaternary blended mortars for 3D printing was verified using a 3D concrete printer. At the first stage, several printing trials of the TBM reference mixtures were carried out. To achieve a satisfactory print quality and shape stability, the water-binder ratio of 0.37 was used. The plastic shrinkage and settling resulted in the formation of cutting cracks through the profile of the printed sample.

Searched 3DP mortars (SM) and fibre-reinforced reference mixture (PFRM) were further tested to eliminate cracks or rather for practical comparison with the reference mixture. SM 1 and SM 3 formulations showed excellent print quality and dimensional stability. The fresh mortars had very good viscoelastic properties necessary for additive manufacturing. However, during the setting and hardening of the printed mortars, several cutting cracks were observed. The SM 2 mixture had relatively good print quality and shape stability, showing no surface defects or cracks. SM 4 and SM 5 mixtures with a high cement content inappropriately reduced the yield stress. These mortars had a relatively high content of plasticizing admixture, despite a very low water-binder ratio and the presence of 8 mm polypropylene

IOP Conf. Series: Materials Science and Engineering 1205 (2021) 012010 doi:10.1088/1757-899X/1205/1/012010

fibres; they were very fluid and did not keep their shape. The rheological parameters of these mixtures are incompatible with additive manufacturing. The formation of cracks on 3DP mortars was successfully eliminated by wetting the printed layers with water mist during the dormant phase of hydration.

Excellent print quality and dimensional stability can be observed in the fibre-reinforced mixtures TBFRM 1 and TBFRM2. The small addition of 3 and 6 mm polypropylene fibre (max 0.08% wt. binder) did not affect the yield stress and improved the plasticity of the mixture. Longer fibres or higher fibre content significantly increased the plasticity of TBFRM 3–5 mixtures. The material was sticking on the worm shaft and also in the funnel. The material was very difficult to push through the print head.

Based on the granulometric and rheological characterization of quaternary blended cement mortars with SCM', prospective modifications of the reference mixture were selected. Specifically, these were mixtures with BFS and MD that exhibited suitable yield stress and thixotropy values similar to the reference mixture. Different doses of slaked lime were additionally tested to improve plasticity. The mixtures QBM 2 and QBM 4 did not improve the rheological parameters of the reference 3DP mortar. These blends with BFS and MD were more fluid and did not keep the print shape. The individual layers tended to flow. A slight reduction in the water-binder ratio already resulted in a loss of print quality and surface defects. In contrast, slaked lime-modified mortar QBM 7 achieved optimum print quality and dimensional stability. SL as a partial replacement (max 6 wt. %) for microsilica has a positive effect on the plasticity and workability of the reference mortar. The mixture with SL was able to resist deformation during overlaying, the printed layers were free of surface defects, and the edges of each layer were visible and dimensionally consistent.

4. Conclusions

In this study, several blended cement mortars for additive manufacturing were experimentally verified. The effect of RCA preparation as well as various SCM's on the granulometric and rheological or physical properties was investigated. The pilot plant verification of the designed mortars was carried out by a construction 3D printer. The dimensional conformity of the printed layer with the model design as well as the ability to resist deformations during overlaying, surface defects and also cracking were monitored and evaluated. The following conclusions can be drawn:

(1) The preparation effect of the unpolluted coarse fraction of RCA on the granulometric and rheological properties was negligible. The granulometry of the crushed or ground RCA was very similar. The effect of RCA preparation on flow properties of reference paste is minimal. The samples with RCA have a lower requirement for mixing water compared to the reference paste without RCA.

(2) The effect of SCM's substitution (10 wt. %) for microsilica on the flow and time-dependent rheological parameters of fresh mixtures was significant. Quaternary blended mortar with BFS has very similar properties as the reference mixture, only it has lower values of thixotropy. The mixture with MD has lower yield stress and a higher plastic viscosity in comparison to the reference mixture. Ground limestone and metakaolin are not suitable substitutes for silica. Ground limestone inappropriately decreases the yield stress of the fresh mixture. The mixture with L is liquid and does not hold its shape. The mixture with MK has a significantly higher value of plastic viscosity, which indicates a possible problem with pumpability.

The effect of SCM's on the shape stability and buildability of fresh mortars was noticeable. Mixtures containing BFS, MD and MK were more plastic and it can be assumed that these mixtures allow faster layering and the possibility of layering in height. The mixture with L and especially the reference mixture are significantly more malleable and the possibility of layering in height is thus reduced. All SCM's had minimal influence on the rheological stability. The reference mortar and mixtures with SCM' have a very similar linear viscoelastic region in which they resist the applied deformation without destroying the structure.

(3) SCM's as a partial replacement for microsilica does not have an adverse effect on the strength development. The larger amount of fibre, as well as lime, did not have a significant effect on the strengths. Samples with BFS and MD and also fibre reinforced mortar achieved the highest strengths around 40 MPa and flexural strength of 5.5 MPa. The lime enriched mixtures exhibited higher water absorption and lower compressive and flexural strengths of 26 MPa and 4 MPa, respectively.

(4) Neither reference mortar nor slaked lime-modified mortars are frost-resistant. The slaked lime-modified mortar had significantly lower compressive strength and especially flexural strength after freeze-thaw cycles compared to the reference mixture. The longitudinal strain strength after freeze-thaw cycles was enormous.

After 20 years of weather changes simulation, the modified mortar had the strength of about one-fifth in both compression and flexure. In contrast, the reference mortar had very good durability and hardly changed strength.

(5) Reference 3DP mortar TBM by Boháč et al [9] was relatively well designed. The mortar mixture had a balanced rheological behaviour without the use of chemical additives such as viscosity modifying agents, retarders and accelerators. The printed mixture followed the contours of the circle according to the design model and resisted deformation during overlaying. Three layers were printed in 45 seconds. The printed layers exhibited minor aesthetic defects caused by shear forces that occur during extrusion when the material is stressed by compression and shear. The plastic shrinkage and settling resulted in the formation of cutting cracks through the profile of the printed sample. The formation of cracks was eliminated by wetting the printed layers with water mist during the dormant phase of hydration.

Searched 3DP mixtures SM 1 by Tay et al [10] and SM 2 and 3 by Nerella et al [11, 12] showed excellent print quality and shape stability, the printed layers were without significant surface defects. The fresh mortars had suitable rheological parameters required for additive manufacturing. The increased amount of Portland clinker in the binder (up to 55 wt. %) had a positive effect on the extrudability, shape stability and adhesion of the fresh mortar. In contrast, the SM 4–5 mixtures by Le et al and also Anell [13, 14] with high cement content were highly liquefied after printing and did not hold their shape. The low yield stress was probably due to the increased dose of superplasticizer. Le et al also used a longer 12 mm fibre and printed the mixture through a 9 mm diameter nozzle.

Fibre-reinforced reference mixtures PFRM 1 and 2 showed an exemplary print quality and shape stability similar to searched 3DP mortars. A small addition of 3 and 6 mm polypropylene fibre (max 0.08% wt. binder) improved the plasticity and shape retention ability of the reference mixture. Longer fibres or higher fibre content significantly increased the plasticity of the mixtures.

Slaked lime-modified mortar QBM 7 achieves optimum print quality and shape stability. Slaked lime as a partial replacement (max 6 wt. %) for microsilica has a positive effect on the plasticity and workability of the reference mortar. The modified mortar was able to resist deformation during overlaying, the printed layers were without surface defects, and the edges of the layers were visible and dimensionally identical.

Acknowledgements
This outcome has been achieved with the institutional support for the long-term development of research organizations by the Ministry of Industry and Trade of the Czech Republic.

References
[1] Wangler T *et al* 2016 Digital Concrete: Opportunities and Challenges *RILEM Technical Letters* **1** pp 67–75 ISSN 2518-0231 DOI 10.21809/rilemtechlett.2016.16
[2] Khoshnevis B 2004 Automated construction by contour crafting—related robotics and informationtechnologies *Autom Constr* **13** pp 5–19 DOI: 10.1016/j.autcon.2003.08.012
[3] Khoshnevis B, Hwang D, Yao K-T and Yeh Z 2006 Mega-scale fabrication by Contour Crafting *Int J Ind Syst Eng* **1** pp 301–320 DOI: 10.1504/IJISE.2006.009791
[4] Buswell R A, Soar R C, Gibb A G F and Thorpe A 2007 Freeform Construction: Mega-scale Rapid Manufacturing for construction *Autom Constr* **16** pp 224–231 DOI: 10.1016/j.autcon.2006.05.002
[5] Lim S *et al* 2012 Developments in construction-scale additive manufacturing processes *Autom Constr* **21** pp 262–268 DOI: 10.1016/j.autcon.2011.06.010
[6] Perrot A, Rangeard D and Pierre A 2016 Structural built-up of cement-based materials used for 3D-printing extrusion techniques *Mater Struct* **49** pp 1213–1220 DOI: 10.1617/s11527-015-0571-0

[7] Momeni F, Mehdi Hassani N S M, Liu X and Ni J 2017 A review of 4D printing *Materials & Design* [on-line] vol 122 pp 42–79 DOI: 10.1016/j.matdes.2017.02.068.
[8] Kovářík M, Svoboda P and Štemberk P 2018 Review of available 3d printing technologies using cement mortar *BETON TKS* 2/2018 (Praha) ISSN 1213-3116
[9] Boháč M, Nešpor B and Staněk T 2018 The role of various aggregates on rheology of fresh 3D printing concrete *SynerCrete'18* 1&2 pp 843–848 e-ISBN 978-2-35158-203-9 DOI: 10.5281/zenodo.1405563
[10] Tay Y W, Panda B, Paul S C, Tan M J, Qian S, Leong K F and Chua C K 2016 Processing and properties of construction materials for 3D printing *Materials Science Forum* vol 861 pp 177–181
[11] Nerella V N, Hempel S and Mechtcherine V 2019 Effects of layer-interface properties on mechanical performance of concrete elements produced by extrusion-based 3D-printing *Construction and Building Materials* vol 205 pp 586–601 ISSN 0950-0618 DOI: 10.1016/j.conbuildmat.2019.01.235
[12] Nerella V N, Krause M, Näther M and Mechtcherine V 2016 Studying printability of fresh concrete for formwork free concrete on-site 3D Printing technology (CONPrint3D) *Proceeding for the 25th Conference on Rheology of Building Materials, Regensburg*
[13] Le T T, Austin S A, Lim A, Buswell R A, Law R, Gibb A G F and Thorpe T T 2012 Hardened properties of high-performance printing concrete *Cement and Concrete Research* vol 42 no 3 pp 558–566
[14] Anell LH 2015 Concrete 3d printer (MSc thesis, Civil engineering, Lund University)
[15] Senthilkumar S V and Santhanam M 2003 Particle packing theories and their applicationinconcrete mixture proportioning: a review *The Indian Concrete Journal* 77(9) pp 1324–31

IOP Conf. Series: Materials Science and Engineering 1205 (2021) 012011 doi:10.1088/1757-899X/1205/1/012011

Research on the residual tensile strength of composite reinforcing bars exposed to elevated temperatures

I Rozsypalova[1], J Prokes[2], Đ Cairović[1], F Girgle[1], P Danek[1] and P Stepanek[1]

[1] Brno University of Technology, Faculty of Civil Engineering, Veveří 331/95, Brno, Czech Republic
[2] PREFA KOMPOZITY, a.s., Kulkova 10/4231, Brno, Czech Republic

Email: iva.rozsypalova@vutbr.cz; prokes@prefa.cz

Abstract. Paper describes a testing procedure for the determination of tensile strength of the composite reinforcing bars subjected to elevated temperatures. Experimentally obtained results on GFRP bars with different diameter are presented and discussed. Moreover, a brief comparison with an analytical approach was included. Almost identical temperature reduction rate of tensile strength was observed for all tested specimens, regardless diameter of the bar. Therefore, it can be expected, that different bar diameter should not significantly affect the results especially if steady state conditions were assumed.

1. Introduction

New materials and methods for improving energy performance and long-life of constructions are the main topics of the present building industry. Over the last decade, increasing attention is paid to the fiber-reinforced polymers (FRP) as a durable and corrosion resistant alternative to the traditional steel rebar.

Elevated temperatures are known to significantly affect the properties of FRP reinforcement. Still, those effects have not yet been thoroughly investigated. For that reason, this topic is the research subject of many recent experimental studies. For example, Blontrock et al. [1], presented extensive experimental results including FRP reinforcement with glass, aramid and carbon fibers. Furthermore, based on abovementioned experimental study, Saafi [2] defined temperature reduction factors of both tensile strength and modulus of elasticity.

A similar study was carried out by Hamad et al [3], showing that the FRP bars suffered significant reductions in their mechanical properties upon exposure to high temperatures up to 450°C. The percentage decrease of mechanical properties under elevated temperatures was more pronounced in specimens with FRP bars, comparing to the commonly used steel reinforcement. Also, Hajiloo et al [4] presented a comprehensive study on characteristics of glass fiber-reinforced polymers (GFRP) bars in both steady state and transient state temperature conditions. The temperature ranged from 25–500°C, while the applied tensile load varied between 20 and 70%. All tested bars exhibited loss of strength and stiffness at high temperatures. Somewhat different approach was presented by Ellis et al. [5]. Properties of GFRP bars were investigated after exposure to elevated temperatures up to 400°C and subsequent cooling to an ambient temperature. As can be expected, tensile strength generally decreases with increasing temperature exposure. However, for specimens preheated to 400°C, as much as 83% of the original tensile strength was retained.

IOP Conf. Series: Materials Science and Engineering 1205 (2021) 012011 doi:10.1088/1757-899X/1205/1/012011

Recently, Ashrafi et al. [6] presented the study that included effects of different bar diameters, type of fiber, type of resin, and fiber to matrix ratio on thermal properties of the FRP rebars. Authors also presented comparison between experimentally obtained tensile strength retention and analytical predictions available in the literature, like for ex. [2].

Within the scope of the experimental study presented in this paper, effects of exposure to elevated temperatures on tensile properties of GFRP reinforcement were assessed assuming steady state temperature conditions. The study included bars with three different diameters, exposed to five temperature levels. Finally, obtained results were compared with the analytical approach presented in [2].

2. Experimental analysis

The main aim of the experimental program was to evaluate the effects of elevated temperature on tensile properties of GFRP reinforcement. The analysis consisted of 3 distinctive phases, namely: preheating, soak-time and mechanical loading.

Firstly, specimens were preheated to desired temperature, in the range of 60 to 280°C, followed by so called "soak-time", as described in Section 2.2.1. Finally, heated specimens were continuously loaded in tension until failure (see Section 2.2.2.).

2.1. Material properties

GFRP composite reinforcing bars with various diameters were used (8, 10 and 14 mm). GFRP bars are composed of approximately E-CR glass fibers (80% by weight), and epoxide matrix. Glass transition temperature Tg of the resin is 110–120°C [7]. During a pultrusion process, surface of the reinforcing bars was sand-coated to ensure sufficient bond with concrete. The assessed bars have developed and produced in the Czech Republic, by Prefa Kompozity a.s. and further labelled as "Prefa ReBar".

Referential tensile properties were obtained experimentally at room temperature (approx. 20°C), namely tensile strength f_{fu} and modulus of elasticity E_f. Tests were performed according to the standardized test method, described in ASTM D7205 / D7205M - 06(2016) [8] or ISO 10406-1:2015 [9]. The specimens 1,000 mm in length have been loaded using displacement control at a rate of 4 mm/min, corresponding to approximately 200 MPa/min. Modulus of elasticity was determined for a load range between 15–80%, to ensure that results are not affected by the slight movement between anchors and testing grips at the beginning of loading. Obtained results were summarized in table 1.

Table 1. Material properties of GFRP reinforcement at the room temperature.

Bar Type		Prefa ReBar		
		G8	G10	G14
Fiber type 1)	[-]		E-CR Glass	
Matrix type 1)	[-]		epoxy	
Fiber content by weight 1)	[%]		80	
Nominal diameter 1)	[mm]	8	10	14
Actual diameter (with surface treatment)	[mm]	9.0	11.0	15.1
Tensile strength f_{fu}	[MPa]	1,042	1,019	871
Modulus of Elasticity E_f	[GPa]	55.0	52.5	43.3 2)

1) According to manufacturer specifications [7]

2) The G14 reinforcement was produced in a different batch (with a different ratio of the matrix components).

As defined by [8, 9], steel anchors were used, due to the low transverse strength of the FRP reinforcement. Oppose to mentioned standardized tests, the total length of the specimen was 1,600 mm, while only the middle part of 580 mm was subjected to elevated temperatures.

2.2. Methods

2.2.1. Heating. The test specimens were subjected to successive heating in the electric temperature chamber TH2700 Grip-Engineering to required temperatures of 60, 100, 200, or 280°C, referred to as preheating. The heating rate of 6°C/min on average was kept, following other similar studies. For example, [10] reported an approximate rate of 7.5°C/min, while [4] reported temperature increase rate of 5.0°C/min, simulating average conditions of the bars embedded in concrete.

After preheating phase, the temperature was maintained for an additional 10 minutes to ensure a uniform temperature throughout the rebar cross section, i.e., soak-time (see figure 2). One set of specimens was left untreated, to serve as referential.

During testing, the temperature was measured on the surface of the specimens by thermocouples, in the middle (denoted as TC2-C) and 100 mm from chamber surfaces (TC1-B at the bottom and TC3-T at the top). The temperature was also measured by a built-in sensor TC0-F near the inner surface of the chamber at one half its height to ensure required time-temperature curves.

2.2.2. Mechanical testing. Once the heating phase was completed, mechanical testing was performed while maintaining the required temperatures. Thus, the tensile stress was mechanically induced to the heated specimens. The loading test was performed using the LabTech 250 mechanical universal testing machine with hydraulic jaws.

Figure 1. Steady state testing of elevated temperature effects to GFRP rebars tensile properties.

Pure tensile stress was induced by increasing deformation at a constant displacement rate of 6 mm/min until failure. Testing the specimen using the universal tensile testing machine equipped with a temperature chamber is shown in figure 1.

3. Results and discussion

The failure of all specimens could be characterized as brittle, as it can be seen in figure 2. Besides the load-time curve (red line), temperature-time curves for four different temperatures measured using thermocouples, as described in Section 2.2.1, were also presented.

The typical failure mode of specimens after the exposure to elevated temperature and mechanical loading can be observed in figure 3.

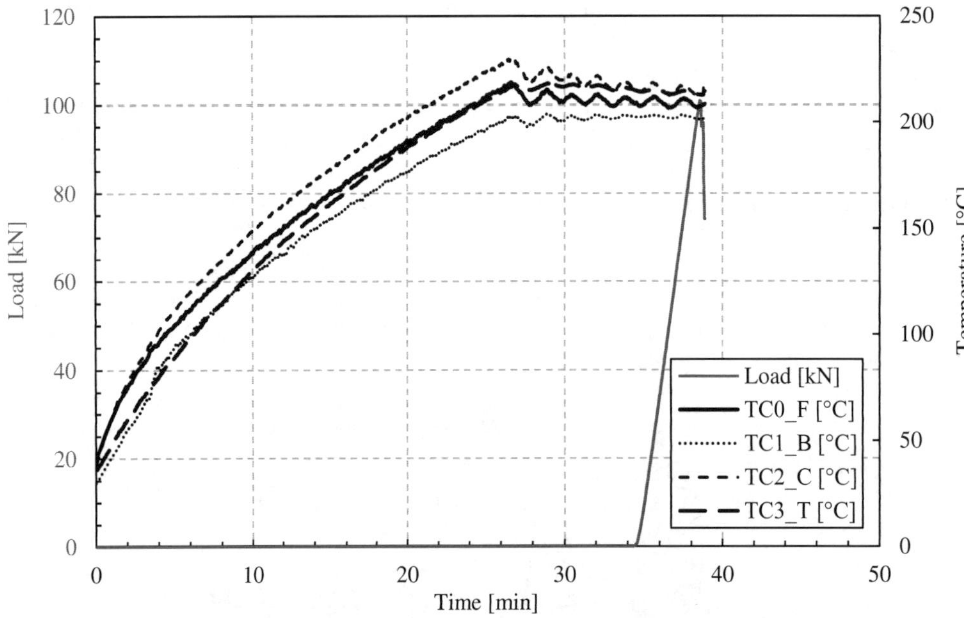

Figure 2. Dependence of load and temperature on time during a test of GFRP tensile properties – test example of specimen G14_4 (F_{max}=100.88 kN, T_{max}=212.9°C).

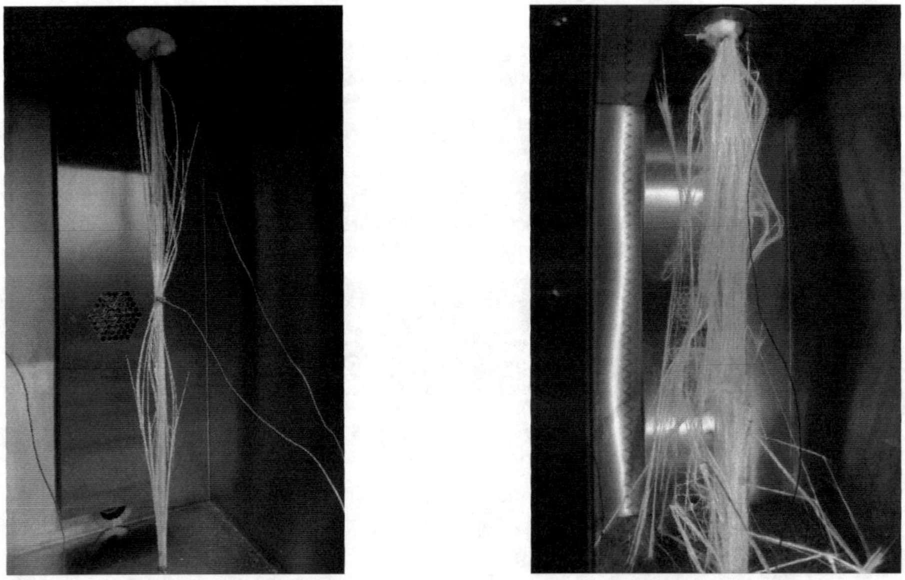

Figure 3. Typical tensile failure of the specimens subjected to moderate temperature 60°C (left) and elevated temperature 200°C (right).

Relative tensile strength was used for comparison between presented experimental results and other research studies available in literature. It was defined as the ratio between the tensile strength of the bars obtained at ambient temperature of 20°C (labeled as $f_{t,20}$) and tensile strength of the specimen subjected to elevated temperatures $f_{t,T}$.(figure 4). The obtained results, which are summarized in table 2, include nominal temperature T_{nom}, maximum temperature T_{max} and ultimate applied load F_{max}.

Maximum temperature corresponds to the ultimate applied load F_{max} at the time of the specimen failure. It is defined as a maximum value measured by the thermocouples at the surface of the specimen. Nominal temperature T_{nom} was measured inside the high-temperature chamber using built-in thermocouple probe.

Saafi [2] defined reduction of GFRP reinforcement tensile strength, based on the experimental results presented by [1] using temperature reduction factor k_f [-],

$$f_{f,T}/f_{f,20} = k_f \quad (1)$$

$$k_f = 1 - 0.0025T \quad for \quad 0 \leq T \leq 400°C \quad (2)$$

$$k_f = 0 \quad for \quad T \geq 400°C \quad (3)$$

where $f_{f,20}$ and $f_{f,T}$ are the ultimate tensile strength of FRP rebars at 20°C and T°C respectively.

Table 2. Tensile strength of GFRP bars exposed to elevated temperatures.

Type of rebar	Specimen	D [mm]	T_{nom} [°C]	T_{max} [°C]	F_{max} [kN]	$f_{t,T}$ [MPa]	$f_{t,T}/f_{t,20}$ [%]
G8	G8_1	8	20	20.0	52.38	1042.07	100.0
	G8_2	8	60	67.4	44.89	893.06	85.7
	G8_3	8	100	109.1	40.19	799.55	76.7
	G8_4	8	280	298.1	38.55	766.93	73.6
G10	G10_1	10	20	20.0	80.01	1018.76	100.0
	G10_2	10	60	66.8	64.09	816.02	80.1
	G10_3	10	100	110.1	60.13	765.60	75.2
	G10_4	10	200	211.6	57.53	732.49	71.9
	G10_5	10	280	296.3	62.40	794.50	78.0
G14	G14_1	14	20	20.0	134.10	871.16	100.0
	G14_2	14	60	66.9	116.57	757.25	86.9
	G14_3	14	100	110.6	97.96	636.36	73.0
	G14_4	14	200	212.9	100.88	655.33	75.2
	G14_5	14	280	288.9	103.45	672.02	77.1

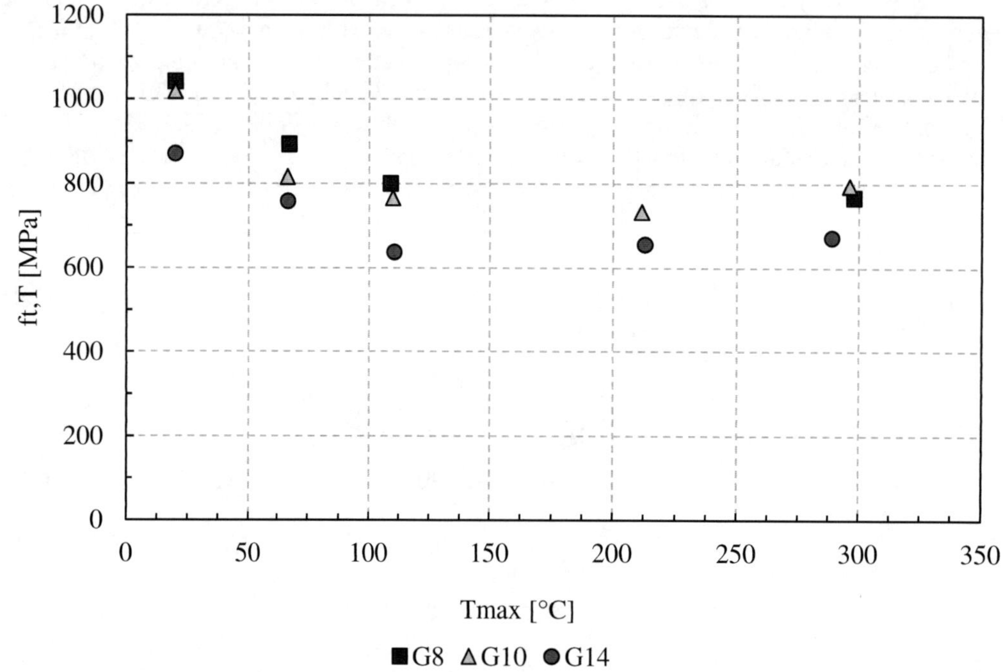

Figure 4. Residual tensile strength $f_{t,T}$ of GFRP bars exposed to elevated temperatures.

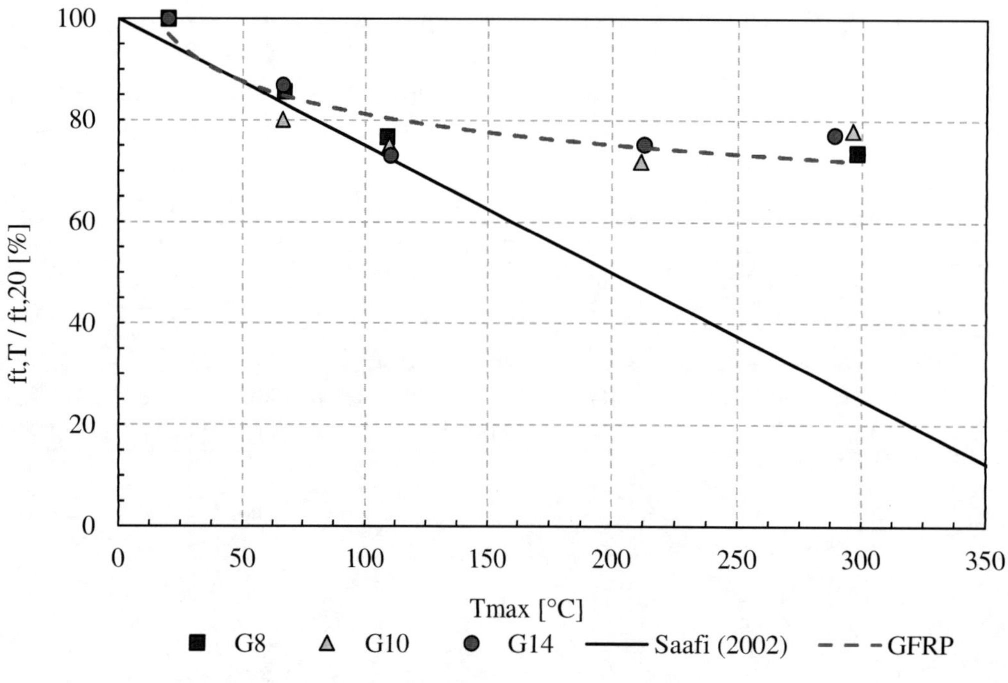

Figure 5. Comparison between relative tensile strength of GFRP bars and predicted values [2].

As it can be observed in figure 5, a good correlation between proposed analytical expression [2], defined using equations (1–3), was obtained for the temperatures up to the approximately 110°C. Due to the effects of elevated temperatures, the tensile strength of GFRP bars was reduced to about 70%. The abovementioned temperature of 110°C roughly corresponds to the glass transition temperature T_g of the used epoxy matrix of assessed GFRP bars.

On the other hand, for higher temperatures ranging up to 300°C, proposed values of the temperature reduction factor k_f seems rather conservative. Almost no further decrease in tensile strength was experimentally observed for the temperatures in this range. Similar results have been presented by Wang et al. [11], based on experimental results obtained on CFRP pultruded plates. Good correlation between experimental data and the analytical model presented by Gibson [12] was observed. Chowdhury et al [13] calibrated Gibson's [12] equations based on the tests on GFRP coupon, showing even degradation rate, comparing to the CFRP plate coupons presented in [11, 12]. This is also in accordance with CAN/CSA S806 [14], since critical temperature of 325°C was permitted for GFRP bars adequately anchored in so called "cold-zone", i.e., in the area which is not affected by elevated temperatures.

The almost identical temperature reduction rate of tensile strength was observed for all tested specimens, regardless diameter of the bar. Therefore, it can be expected that different bar diameter should not significantly affect the results, especially if steady state conditions were assumed. On the other hand, Wang et al. [11] concluded that bars with larger diameter shown higher resistance under elevated temperatures. Hence, considering the effect of a bar's diameter in further research is highly recommended.

4. Conclusion

Based on presented experimental analysis it can be concluded that:

- Most of the tensile strength of GFRP bars with epoxy matrix can be expected to be retained for temperatures up to the glass transition temperature T_g. Reaching the T_g, the tensile strength of GFRP bars was reduced only to about 70% of the reference (unheated) sample value.
- For higher temperature range, i.e., between T_g and 300°C, only slight loss of tensile strength is likely to be observed. This is also in accordance with, for example, CAN/CSA S806, since a critical temperature of 325°C was permitted for GFRP bars adequately anchored in the so-called "cold-zone", i.e., in the area that is not affected by elevated temperatures. Hence, for the temperature ranging up to 300°C, reduction of the tensile strength caused by elevated temperatures is governed mainly by the properties of the matrix, not the fibers.
- The GFRP bars can be used as reinforcement of concrete for the temperature of use below T_g. In the event of a fire or other unexpected events, the GFRP reinforcement will ensure sufficient stability of the concrete element if the bars are adequately anchored in so-called "cold-zone".

Still, further research is needed to prove the abovementioned hypothesis, including the assessment of different bar diameters, matrix types and fibers.

Acknowledgments

The paper is written under the support of the Technology Agency of the Czech Republic (grant No. TH04020431 - Extension of the application area of FRP reinforcement in concrete structures) and Brno University of Technology (grant No. FAST-S-21-7503). This support is gratefully acknowledged.

References

[1] Blontrock H, Taerwe L and Matthys S 1999 Properties of fiber reinforced plastics at elevated temperatures with regards to fire resistance of reinforced concrete members in Fourth International Symposium on Non-Metallic (FRP) *Reinforcement for Concrete Structures* **188** pp 43–54

[2] Saafi M 2002 Effect of fire on FRP reinforced concrete members *Composite Structures* **58** pp 11–20

[3] Hamad R, Megat J, Megat A and Haddad R 2017 Mechanical properties and bond characteristics of different fiber reinforced polymer rebars at elevated temperatures *Construction and Building Materials* **142** pp 521–535 DOI: 10.1016/j.conbuildmat.2017.03.113.

[4] Hajiloo H, Green M F and Gales J 2018 Mechanical properties of GFRP reinforcing bars at high temperatures *Construction and Building Materials* **162** pp 142–154 DOI: 10.1016/j.conbuildmat.2017.12.025.

[5] Ellis D, Tabatabai H and Nabizadeh A 2018 Residual Tensile Strength and Bond Properties of GFRP Bars after Exposure to Elevated Temperatures *Materials* 11(3) DOI: 10.3390/ma11030346.

[6] Ashrafi H, Bazli M , Najafabadi E P and Oskouei A V 2017 The effect of mechanical and thermal properties of FRP bars on their tensile performance under elevated temperatures *Construction and Building Materials* 157 pp 1001–1010 ISSN 09500618 DOI:10.1016/j.conbuildmat.2017.09.160.

[7] Prefa Kompozity, a.s.: Datasheet KOMPOZITNÍ VÝZTUŽE PREFA Rebar 2017 (available from: https://www.prefa-kompozity.cz/wp-content/uploads/2015/09/katalog_kompozitni_vyztuze_cze_m.pdf)

[8] ASTM D7205/D7205M-06: 2016 *Standard Test Method for Tensile Properties of Fiber Reinforced Polymer Matrix Composite Bars* (West Conshohocken, USA: ASTM International)

[9] ISO 10406-1:2015 *Fibre-reinforced polymer (FRP) reinforcement of concrete — Test methods — Part 1: FRP bars and grids*. 2. (Geneva: International Organization for Standardization)

[10] Correia J, Gomes M, Pires J and Branco F 2013 Mechanical behaviour of pultruded glass fibre reinforced polymer composites at elevated temperature: Experiments and model assessment *Composite Structures* pp 303–313 DOI:10.1016/j.compstruct.2012.10.051.

[11] Wang K, Young B and Smith S T 2011 Mechanical properties of pultruded carbon fibre-reinforced polymer (CFRP) plates at elevated temperatures *Engineering Structures* 33(7) pp 2154–2161 ISSN 01410296. DOI:10.1016/j.engstruct.2011.03.006.

[12] Gibson AG, Wu Y S, Evans J T and Mouritz A P 2006 Laminate theory analysis of composites under load in fire *Journal of Composite Materials* 40(7) pp 639–657

[13] Chowdhury E U, Eedson R, Green M F, Bisby L A and Benichou N 2009 Mechanical characterization of fibre reinforced polymers materials at high temperature *Fire Technology* 45(4) pp 1–18

[14] CAN/CSA S806-12: 2012 *Design and Construction of Building Structures with Fibre-reinforced Polymers* (Toronto, Canada: CSA Group)

IOP Conf. Series: Materials Science and Engineering 1205 (2021) 012012 doi:10.1088/1757-899X/1205/1/012012

The effect of consolidation treatment on selected mechanical properties of sandstone

Ľ Gajdos[1], M Sperl[1], Z Slizkova[1] and M Drdlova[2]

[1] Institute of Theoretical and Applied Mechanics of the Czech Academy of Sciences, Prosecká 809/76, 190 00 Prague 9, Czech Republic
[2] Research Institute for Building Materials, Hněvkovského 65, 617 00 Brno, Czech Republic

Email: sperl@itam.cas.cz

Abstract. An investigation was made into selected mechanical properties of sandstone of Božanov mining site in Czechia using both natural specimens of the sandstone and specimens impregnated with consolidants – liquid products aimed at improving strength and durability of degraded stones. Experiments in three-point bending of notched and cracked specimens made it possible to determine (i) the quasi-static notch toughness as well as the fracture toughness of specimens when they were subjected to static loading, and (ii) the impact fracture toughness and total energy of fracture when specimens were subjected to impact loading. The results of the tests are presented and are discussed with a view to the effects of consolidants.

1. Introduction

Mechanical properties of rock materials have been a subject of interest of investigators for several decades. Among those, which are paid a special attention to, belong fracture properties, namely the fracture toughness. This is a key parameter in rock fracture mechanics used to determine crack initiation and propagation. Its importance can be observed in activities such as the design of rock boring equipment, tunneling, rock cutting and bursting, assessment of stability of jointed rock masses, etc.

Besides this there is also another area where the knowledge of mechanical properties of stones is very important. This area covers conservation and restoration of historical buildings and monuments. The aim of research activities in this area is to improve the resistance of stone elements of buildings and monuments to deterioration and eventual fracture. For this purpose a number of various consolidation solutions have been developed and tested. In this study we have focused on the research of mechanical properties of the sandstone from the Božanov mining site.

2. Quasi-static loading

In order to evaluate the toughness of sandstone at various loading conditions and to consider the effect of various consolidating solutions on the toughness of sandstone we shall start with recalling the results of paper [1]. These results have been concerned with three-point bending of prismatic specimens, nominally 20 × 20 × 100 mm in dimensions with a notch 1 mm wide and 2 mm deep, the span of supports at the tests being 60 mm. The specimens were manufactured from sandstone of Božanov locality, Czech Republic. The stone material can be characterized as a quasi-brittle inelastic silicate composite [2].

Some of the specimens were given a stress cycling treatment to produce a crack at the notch root. Because it was not possible to measure the crack depth during cycling the depth was roughly estimated on the basis of the number of cycles applied. However, this led to a high scatter of crack depths. After that the most of the specimens were impregnated with liquid products designed to enhance strength and durability of degraded stone, so called "strengtheners" or "consolidants". The following products available on the market was used for the treatment: KSE 100, KSE 300, and Paraloid B 72. First two agents produced by the Remmers® company were strengtheners on a silicic acid ester (SAE) base with different silica gel deposition rate. KSE 100 had a low gel deposition rate (10%) in order to prevent excessive surface consolidation and to achieve uniform strength profiles, while KSE 300 was a solvent free product that reaches higher silica gel deposition rate (30%). The last product, Paraloid B 72 manufactured by Rohm and Haas, contained acrylic copolymer based on ethyl methacrylate with methyl acrylate (70:30) and represented a resin widely used in conservation practice for the purpose of consolidation or protective treatment of various materials. Consolidation treatments in the study were performed by partial immersion of stone specimens in liquid consolidant at room temperature for the duration of four hours. This consolidation technique is also known as cold immersion treatment and is the most used treatment in conservation practice. After finishing the immersion treatment, the treated specimens were removed from the bath and the liquid adhered on the stone surface was gently wiped off. Then impregnated stone specimens dwelled for two weeks in laboratory with an environmental condition of 20°C and 60% RH. This period was needed for evaporation of organic solvent (present in KSE 100 and Paraloid B 72) and also for finishing chemical reactions ensuring conversion of SAE to silica gel (KSE 100 and KSE 300).

Mechanical testing was then conducted in such a way that the specimens were subjected to slow monotonic loading with a deflection rate ~ 2.5 μm/s and during the loading the "force – deflection" dependences were recorded. This made it possible to obtain both the maximum force P_{max} and the absorbed energy to fracture E_f . For some specimens we have used the absorbed energy to fracture E_f , designated also as a quasi-static fracture energy, to determine the quasi-static notch toughness KCV*, and the maximum force P_{max} to determine the fracture toughness K_{IC}. The relevant results concerning the absorbed energy to fracture are presented in table 1.

Table 1. Absorbed energy to fracture in 3PB tests.

specimen	designation	h x b (mm x mm)	A (mm^2)	E_f (mJ)	KCV* (mJ/mm^2)	consolid.	note
S1	BO_N	18.94x20.66	391.3	13.27	0.034	no	notch
S2	BO_20	18.21x20.68	376.6	12.24	0.033	no	notch
S3	BO_9	18.96x21.6	409.5	8.92	0.022	no	crack
S4	BO_10	19.14x21.02	402.3	8.72	0.022	KSE100	crack
S5	BO_12	6.15x20.65	127.0	1.93	0.015	Par. B72	crack
S6	BO_14	5.4x20.56	111.0	1.24	0.011	KSE300	crack
S7	BO_16	19.32x19.18	370.6	4.24	0.011	KSE300	crack

notes: - Par. B72 in the table is a shortened form for Paraloid B 72
- E_f is a quasi-static fracture energy
- KCV* is a quasi-static notch toughness (=E_f / A)

For a better imagination of toughness properties of sandstone the quasi-static notch toughness KCV* is plotted for each specimen on figure 1.

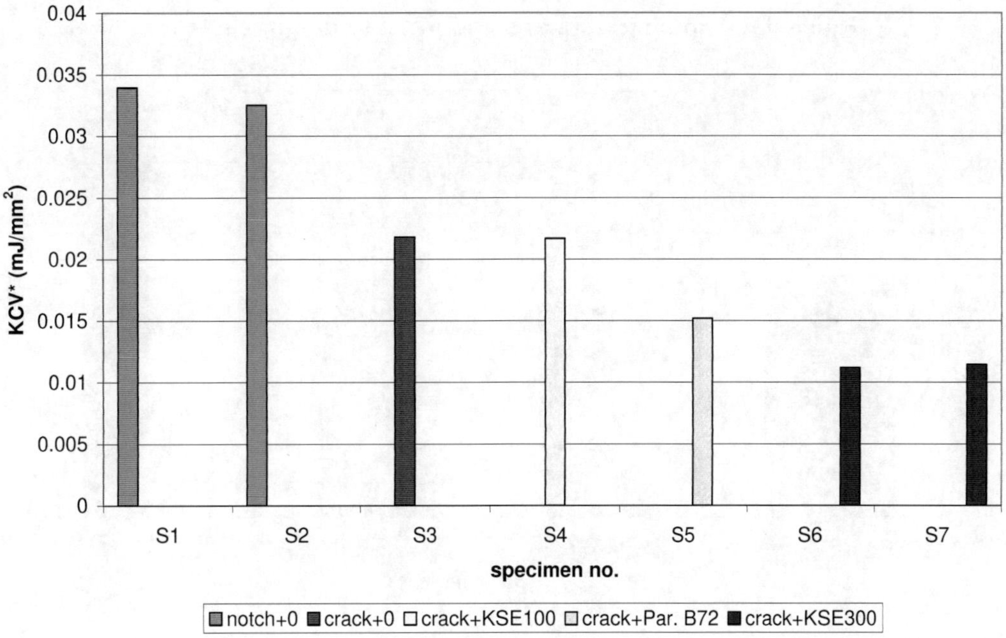

Figure 1. Quasi-static notch toughness for the specimens used.

As it is seen from table 1 and figure 1 the highest values of the quasi-static notch toughness KCV* are found for the specimens S1 and S2 representing a natural untreated material with a notch. This is rather unexpected if we consider the application of various strengtheners as a treatment to improve notch toughness of sandstone material. It is also seen in figure 1 that replacement of a notch by a crack resulted in a decrease in the notch toughness KCV* by about one third. The results also show that application of KSE 100 product for the cracked specimen S4 had practically no effect on its notch toughness. As compared to the quasi-static notch toughness of specimen S3 the application of Paraloid B 72 for the cracked specimen S5 led to lowering of its notch toughness by more than 30% and the application of KSE 300 for cracked specimens S6 and S7 resulted in a 50% reduction of their quasi-static notch toughness. These results clearly demonstrate a negligible effect of the treatments on the quasi-static notch toughness of sandstone cracked specimens. They also demonstrate a lower quasi-static notch toughness of a sandstone specimen when a notch is replaced by a crack.

It has been already mentioned that in 3PB tests the dependences of force vs. deflection were recorded. For some specimens the maximum force F_{max} at these plots was used to determine the stress intensity factor K_I characterizing the fracture toughness of a particular specimen. The resulting values of K_I are presented in table 2 together with KCV* for a comparison to see the effect of consolidators on these quantities. Graphically are the results illustrated in figure 2. Because the numerical values of both quantities differ roughly by an order when using the chosen units, the magnitudes of the fracture toughness in the figure are multiplied by 0.1.

Table 2. Comparison of quasi-static notch toughness and fracture toughness.

specimen	KCV* (mJ/mm^2)	K_{IC} (MPam$^{0.5}$)
S1	0.034	0.331
S3	0.022	0.349
S4	0.022	0.449
S7	0.011	0.561

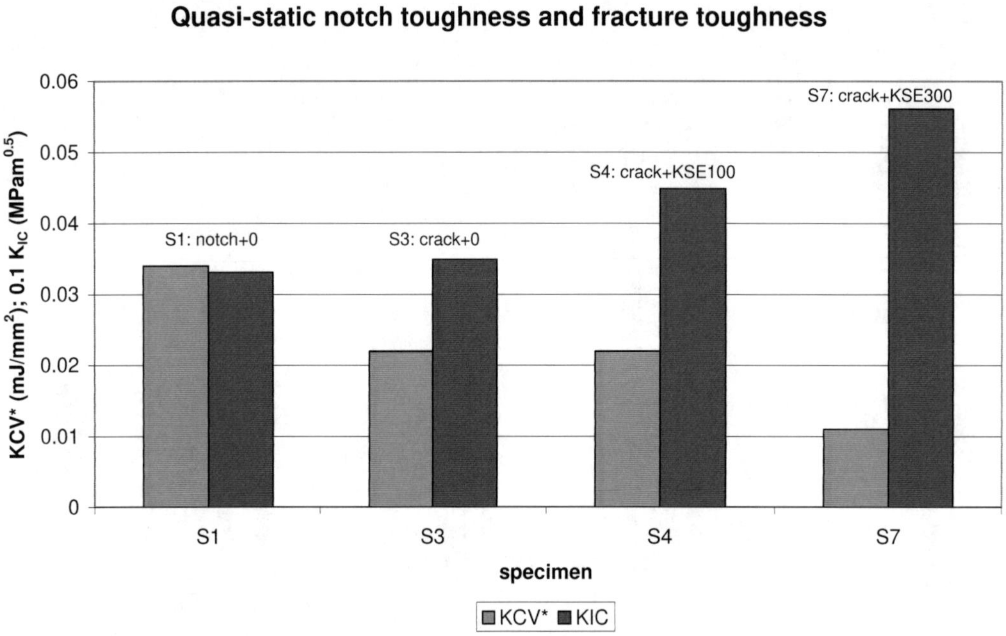

Figure 2. Comparison of quasi-static notch toughness and fracture toughness of some specimens.

As can be seen from the figure, the specimen S1 is notched and is untreated with any product; the specimen S3 is cracked and is also untreated; specimens S4 and S7 are cracked and treated with strengtheners: S4 with KSE 100 and S7 with KSE 300. In accordance with figure 1 the quasi-static notch toughness KCV* decreases when receiving a consolidation treatment. The decrease of KCV* is greater for the treatment with KSE 300. As it was already stated this decrease makes about 50% of the KCV* value for the specimen S3.

Exactly different situation is, however, for the fracture toughness K_{IC}. If we take the fracture toughness for the specimen S3 as a basis then the use of the product KSE 100 increased the fracture toughness K_{IC} by about 28% and the use of the product KSE 300 by approx. 60%.

3. Dynamic fracture toughness

It is known (see for example [3]) that dynamic rock strengths and dynamic fracture toughness are greater than quasi-static strengths and toughness, and they increase with an increasing strain rate. This means that rock materials have a stronger resistance to fracture in dynamic loading than in quasi-static loading. Because the term ‘dynamic fracture toughness’, can be used in connection with either initiation of the fast crack propagation or with the resistance against propagation of a moving crack, in this contribution we shall use the term “impact fracture toughness” to distinguish between the two. In our previous paper [4] we have reported instrumented Charpy test results as obtained on Božanov sandstone test specimens, it means on specimens of the same material as was that we have used for quasi-static notch toughness tests. Instrumented Charpy impact tester Instron Ceast 9050, used in the tests, provided a single specific loading rate given by its constructional parameters and by the use of a hammer of 1.037 kg in mass. The test configuration of the impact tester provided the striking speed of the hammer v = 3.8 m/s and the impact energy L = 7.5 J. Dimensions of the specimens were 10 x 10 x 88 mm and the initiation notch was 1.5 mm deep while its width was 1 mm. In the tests the span of supports was 60 mm, i.e. that the span to height ratio was 6. The resulting output of the tests included the total energy of fracture En as well as the “force – time” record for each specimen. Knowing the peak forces from these records it was possible to determine impact fracture toughness for the sandstone specimens. Under these conditions we have obtained the impact fracture toughness magnitudes K_{Id} . They are presented in table 3. In this study, the product KSE 300 E was used for the treatment of stone specimens. KSE 300 E, similarly to KSE 300, represented the stone strengthener on a silicic acid ester (KSE) base with silica gel deposition rate of about 30%. Compared to KSE 300, the product KSE 300 E contained an elasticized component. Elastic silicic acid esters for consolidation of natural stone were introduced by Remmers® at the end of 1990 and effects of elastic properties of gels with reduced cracking, as compared to traditional treatments without this modification, have still been an object of research [5–7]. The stone specimens to be treated are documented in figure 3. The cracked texture of silica gel created from a non-elasticized type of the strengthener KSE 300 is shown on figure 4 as the gel layer on a dish. After the treatment, thin gel layers cover the stone grains and create bridges between the grains.

Figure 3. Božanov sandstone specimens prepared for dynamic fracture toughness testing. Scale: white or black square: 1cm^2.

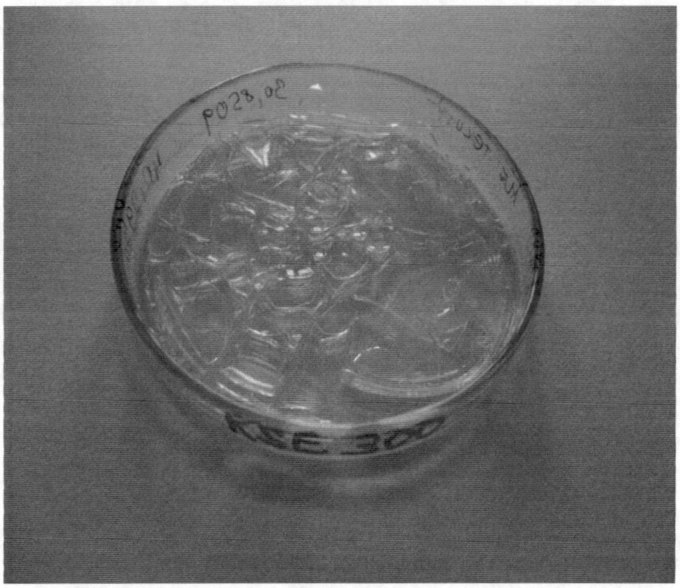

Figure 4. Cracked silica gel formed from the strengthener KSE 300 at laboratory conditions (chemical reaction of SAE with atmospheric water vapor, 1 week after pouring the product to Petri dish).

Table 3. Relevant quantities obtained from impact tests.

specimen number	P (N)	M (Nmm)	σ (MPa)	K_{Id} (MPam$^{0.5}$)	E_n (mJ)	consolidant
D1	729.7	10945	65.7	4.39	109	no
D2	1173.5	17602	105.6	7.06	112	no
D3	938.7	14080	84.5	5.65	115	no
D4	991.9	14878	89.3	5.97	111	no
D5	1037.4	15561	93.4	6.24	115	no
D6	631.0	9465	56.8	3.80	111	no
D7	658.0	9870	59.2	3.96	109	no
D8	1560.8	23412	140.5	9.39	135	KSE 300E
D9	1016.8	15252	91.5	6.12	123	KSE 300E
D10	1257.2	18858	113.1	7.56	128	KSE 300E
D11	1535.2	23028	138.2	9.24	136	KSE 300E
D12	1707.1	25606	153.6	10.27	140	KSE 300E
D13	1147.3	17209	103.3	6.90	132	KSE 300E
D14	1767.4	26511	159.1	10.63	145	KSE 300E

Notes:
P is the peak force in the "force – time" record
M is the bending moment determined as P x $s/4$, where s is the span between supports
σ is the bending stress = 6 x $M/b/h^2$ ($b = h = 10$ mm)
K_{Id} is the impact fracture toughness = $\sigma\sqrt{\pi a}\,F(a/h)$; a is the crack/notch depth ($a = 1.5$ mm)
and $F(a/h)$ is a geometric factor; for $a/h = 1.5/10 = 0.15$ its magnitude is $F =$ 0.97377
E_n is the total energy to fracture determined from the "force – time" record

Now we shall make an evaluation of the impact fracture toughness and the total energy to fracture for the group of specimens D1-D7, which were not treated, and also for the group of specimens D8-D14, which were treated with the product KSE 300E.

In order to cover in one diagram not only the absolute values of these quantities but also their statistical parameters we shall use the probability density distribution diagrams for graphical presentation. The probability density distribution diagram (PDDD) for impact fracture toughness is presented on figure 5 for unconsolidated specimens and on figure 6 for consolidated specimens. Their comparison is presented on figure 7.

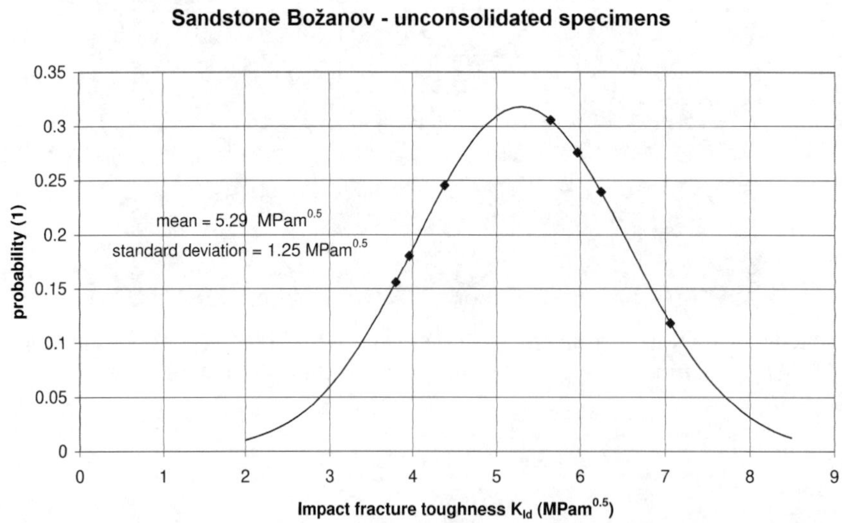

Figure 5. Probability density distribution of impact fracture toughness for unconsolidated specimens.

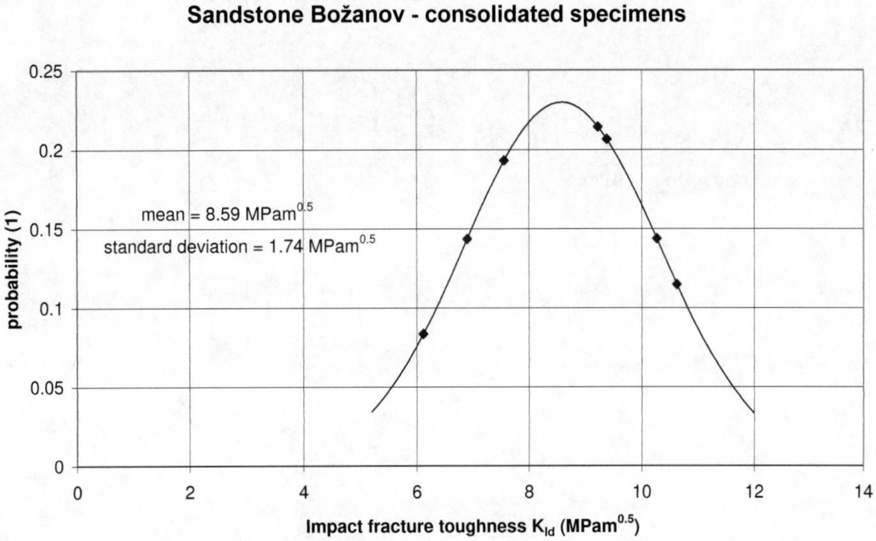

Figure 6. Probability density distribution of impact fracture toughness for consolidated specimens.

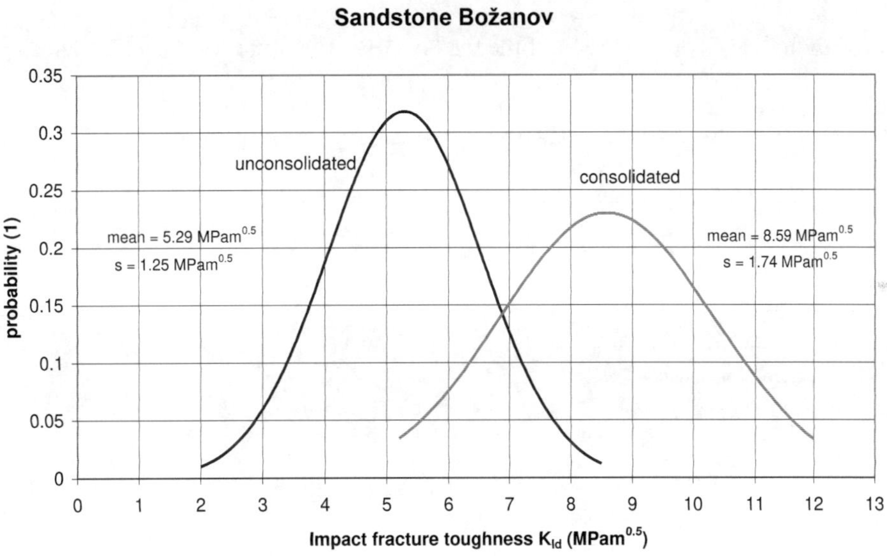

Figure 7. Comparison of probability density distribution of impact fracture toughness for consolidated and unconsolidated specimens.

It can be read from figure 5 that the mean value of the impact fracture toughness K_{Id} is 5.29 $MPam^{0.5}$ and the standard deviation s = 1.25 $MPam^{0.5}$. As follows from figure 6, application of the consolidant KSE 300E results in an increased mean value of the impact fracture toughness K_{Id} from 5.29 $MPam^{0.5}$ to 8.59 $MPam^{0.5}$. It is more than a 60% increase. This is accompanied with an increase in the standard deviation by almost 40%. The situation is digestedly illustrated on figure 7.

Now we can make a graphical representation of the total energy of fracture for consolidated and unconsolidated specimens. Alike in procedure taken for illustration of impact fracture toughness we shall plot the PDDDs for the total energy of fracture for consolidated and unconsolidated specimens. figure 8 represents the PDDD for the unconsolidated specimens and figure 9 represents that for the consolidated specimens. A comparison of these PDDDs is made on figure 10.

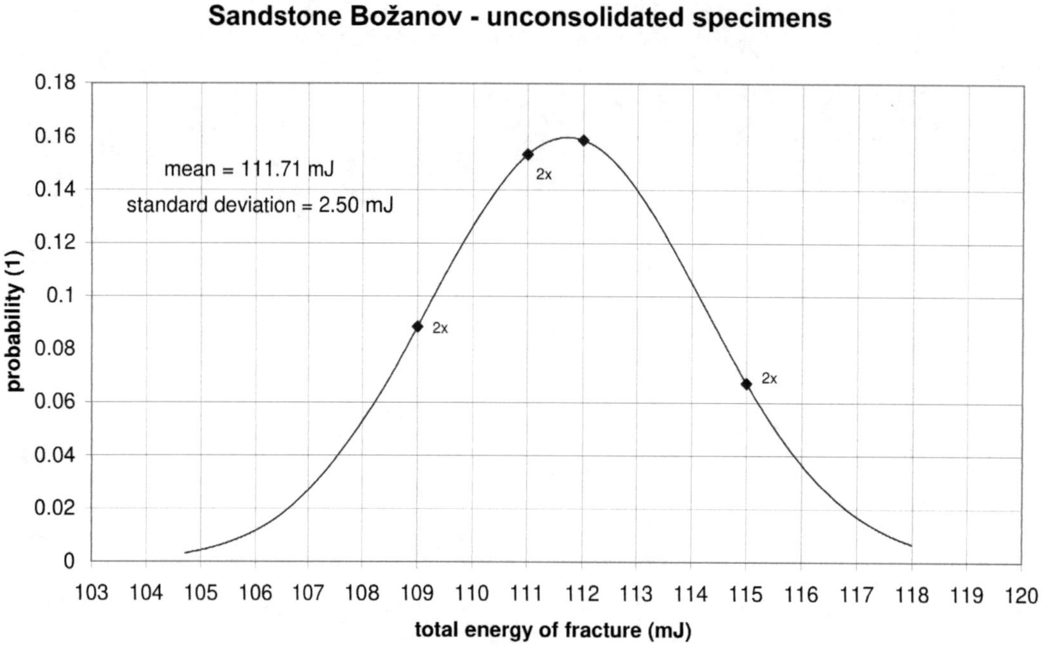

Figure 8. Probability density distribution of the total energy of fracture for unconsolidated specimens.

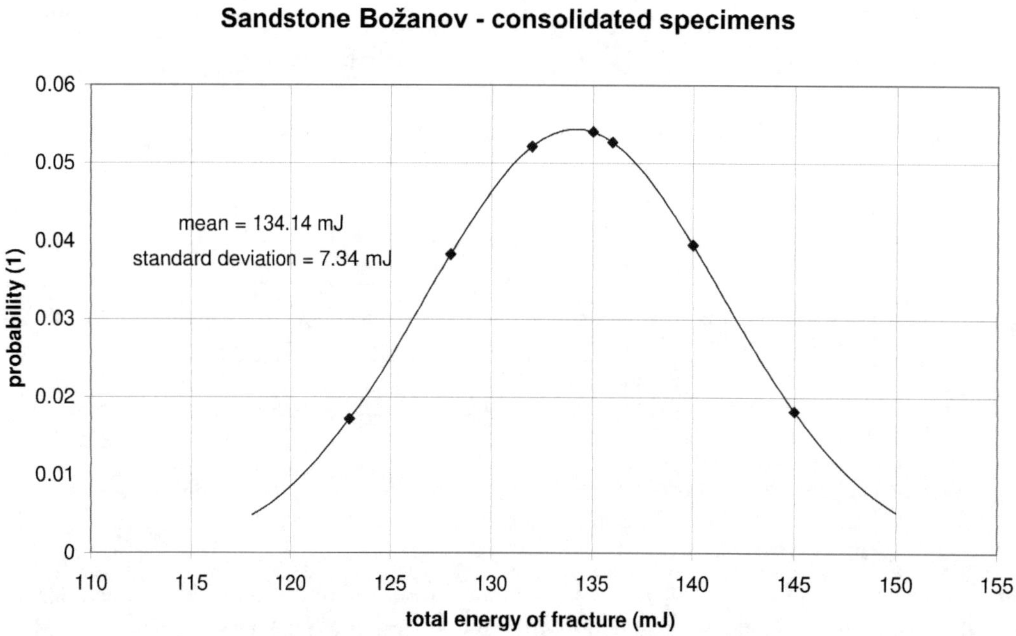

Figure 9. Probability density distribution of the total energy of fracture for consolidated specimens.

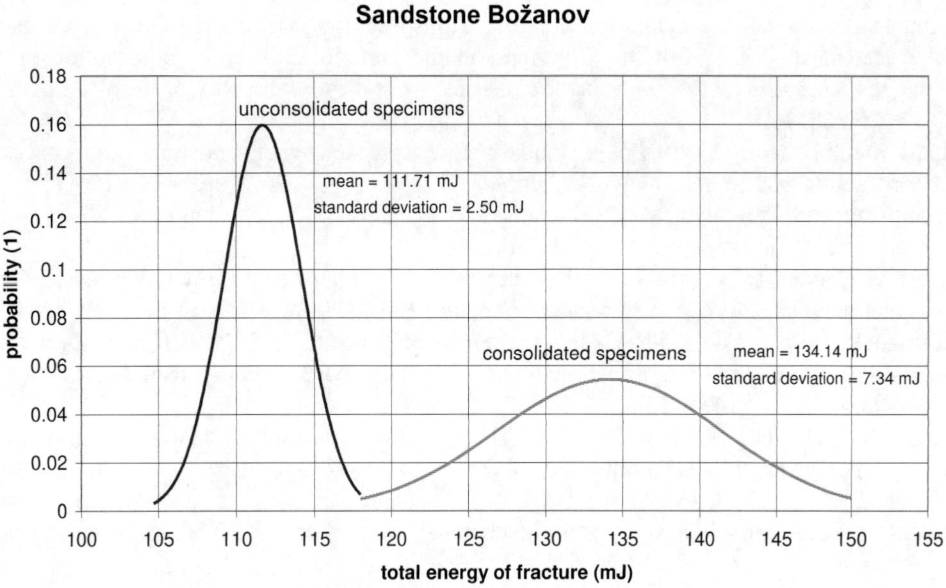

Figure 10. Comparison of probability density distribution of total energy of fracture for consolidated and unconsolidated specimens.

As it follows from table 3 and figures 8 through 10 the mean value of the total energy of fracture for unconsolidated specimens is 111.71 mJ while that for consolidated specimens is 134.14 mJ, it means it is increased by 20%. However, the standard deviation for consolidated specimens is increased from 2.5 mJ to 7.34 mJ, accordingly by a multiple of almost 3. Owing to this the PDD curve for unconsolidated specimens is rather narrow and that for consolidated specimens is wide. The peak probability of occurrence of the mean value of the total energy of fracture is 0.16 for unconsolidated specimens and 0.054 for consolidated specimens.

In fine it can be said that consolidation treatment of sandstone specimens leads to a certain improvement of both the impact fracture toughness and the total energy of fracture. On the other hand, it increases the scatter of these properties. While the increase of the scatter of impact fracture toughness is less than 40%, the scatter of total energy of fracture is almost 200%.

4. Conclusions

The results of the experimental study proved a beneficial effect of the treatments on the static and dynamic fracture toughness of sandstone specimens as well as on the total energy of fracture upon impact loading of cracked specimens. Concerning the static fracture toughness of sandstone specimens, the highest improvement was reached when KSE 300 strengthener was applied. This type of a consolidant was also used to improve the impact fracture toughness and the total energy of fracture in impact tests. However, the improvement of these quantities was accompanied by an increased scatter. The only negative effect of the consolidation treatment was connected with the quasi-static notch toughness. This result opposes the simple image of a positive effect of the treatment on mechanical properties of sandstone. There is, therefore, a need for a large-scale investigation of how exactly a consolidant works upon a quasi-static loading and why is there a different effect of this loading on the fracture toughness and the quasi-static notch toughness.

Acknowledgments

The authors acknowledge kind support from the Technological Agency of the Czech Republic (Project No. FW01010021).

References

[1] Drdácký M, Šperl M and Jandejsek I 2016 Consolidation effects on sandstone toughness In: Proceedings of the 13th Int. Congress on the Deterioration and Conservation of Stone - *SCIENCE and ART: A Future for Stone* Vol. 2 (Paisley, University of the West of Scotland, UK) pp 687–694

[2] Šperl M and Drdácký M 2018 Non-standard experimental tests of sandstone and its pre-cracking for fracture testing *IOP Conf. Ser.: Mater. Sci. Eng* 379 012029 pp 1–7

[3] Zhang QB and Zhao J 2014 Quasi-static and dynamic fracture behaviour of rock materials: phenomena and mechanisms *Int. J. Fract.* **189** pp 1–32

[4] Šperl M, Gajdoš Ľ, Slížková Z and Drdlová M 2021 Instrumented dynamic tests of building materials *IOP Conf. Ser.: Mater. Sci. Eng.* 1039 012023 pp 1–9

[5] Zárraga R, Cervantes J, Salazar-Hernandez C and Wheeler G 2010 Effect of the addition of hydroxyl-terminated polydimethylsiloxane to TEOS-based stone consolidants *J. of Cultural Heritage* 11(2) pp 138–144

[6] Leal L N P, Drdácký M and Slížková Z 2018 Flexural tests on consolidation effects on stone In: Proceedings from SAHC conference in Cusco pp 725–733

[7] Remzová M, Šašek P, Frankeová D, Slížková Z and Rathouský J. 2016 Effect of modified ethylsilicate consolidants on the mechanical properties of sandstone *Construction and Building Materials* **112** pp 674–681

Determination of ballistic resistance of cementitious composites

J Vojtisek[1], E Bystrianska[2], O Koutny[2], J Kratochvil[2] and M Drdlova[3]

[1] Brno University of Technology, Faculty of Chemistry, Purkynova 118, 612 00 Brno, Czech Republic
[2] BOGGES, spol. s r.o., Hnevkovského 30/65, 617 00 Brno, Czech Republic
[3] Research Institute for Building Materials, Hnevkovského 65, 617 00 Brno, Czech Republic

Email: xcvojtisek@fch.vut.cz

Abstract. The paper describes the possibility of using the DOP test, standardly used to evaluate the ballistic resistance of ceramics, to determine the ballistic resistance of cement composites. DOP - Depth of penetration test - is based on measurement of residual penetration of projectile in witness system after the sample perforation; this parameter is used to determine differential efficiency factor (DEF), characterizing the ballistic resistance of the particular material. To verify the method for non-ceramic materials, four variants of cementitious composites with compressive strengths in the range of 30–150 MPa have been tested. The obtained results confirm the method's suitability for determining the ballistic resistance of cement composites in terms of ease of implementation, sensitivity, and accuracy of the obtained results for both composites of common strengths and high-performance variants. The paper also discusses some aspects of the relationship between the mechanical parameters of cement composites under static loading and their ballistic resistance.

1. Introduction

Concrete has been widely used as a structural material due to its excellent and well-described static mechanical properties and their relatively low price. Nevertheless, there is the probability that the structure will be loaded in a way that has not been designed for. Beside accidental causes (car accident) or natural disasters (earthquake) the impact of a high-velocity moving object can be critical for such a structure. In other words, the protective capacity of concrete-like materials against ballistic loading has to be taken into account. The need to evaluate such properties increases nowadays in resonance with a higher risk of the terrorist attack. Due to this fact, the properties of concrete materials at high-strain rates and dynamic conditions need to be defined to determine the response of such a material on this type of loading and to estimate its protective capacity. The method for the determination of cementitious composites ballistic resistance from the material point of view is therefore needed.

The ultra-high performance fibre reinforced concrete (UHPFRC) has the greatest potential for the high ballistic resistance of cementitious composites family. The UHPFRC has superior mechanical properties and energy absorption capacity compared to ordinary construction concrete (such as C25/30, C45/55) and can resist intense dynamic loading induced by the impact of high velocity moving object. The ballistic impact resistance of concrete materials is mainly dependent on their compressive strength as well as the type, size, and hardness of coarse aggregate used. It is known that projectile penetration depth is dependent on the compressive strength of concrete materials, but when the compressive strength of the concrete material reaches a certain value (> 100–150 MPa), the penetration depth of the projectile remains more or less the same and the effect of increased compressive strength is diminished. [1]. Steel fibers used in UHPFRC improve the toughness, fracture energy and limit the initiation and propagation of cracking. However, high steel fibers volume may reduce the workability of concrete, so the optimal limit of the fiber volume ratio of 2.0% is suggested [2].

The standard and commonly used mechanical examination of cement-based materials is realized under static conditions (e.g., compressive strength and tensile strength testing). However, the behavior of cement-based materials differs under high-velocity impact loadings compared to static load [3, 4]. Therefore, a specific ballistic test is required to evaluate the response of protective materials. One of the well-known ballistic methods for characterization and evaluation of materials is the Depth of penetration test (DOP), which was initially designed for testing ceramic materials intended for add-on armour [5–7]. In the presented study, the same principle was used to verify the suitability of the DOP method to determine the ballistic resistance of the cement-based composites with various mechanical parameters. There is no available information about the method of ballistic resistance of cementitious composites. Currently, the evaluation of concrete materials' resistance is based on a simple assessment of "perforation / no perforation" at the particular projectile velocity. This method is connected with several difficulties, e.g., problematic transferability of results, inaccuracy due to projectile velocity dispersion. The proposed DOP method and the resulting DEF factor can bring the accurate output, characterizing the ballistic resistance of the cementitious materials, enabling to compare them effectively and accurately.

2. Theory

In Depth of penetration test method, the cementitious sample is backed by a semi-infinite witness material. The impacting projectile must penetrate through the sample into the backing material. By comparing the penetration of a projectile into the backing material with and without the tested sample, the cementitious sample ballistic performance may be determined.

The main goal of Depth of penetration test method is to determine the residual penetration P_{res} of the projectile in the backing witness material after perforation of testing material with defined thickness H_s and bulk density ρ_s. The witness material consists of thick homogenous reference material with a defined density ρ_{Al} (e.g., aluminum alloy). From the bulk density of the tested cementitious sample, its thickness and knowledge of the reference penetration of the projectile P_r (penetration of the projectile in reference witness material without the cementitious sample), the differential efficiency factor (DEF) can be calculated based on the equation (1):

$$DEF = \frac{\rho_{Al}(P_{Al} - P_{res})}{\rho_s H_s} \quad (1)$$

The test setup is shown in the following Figure 1:

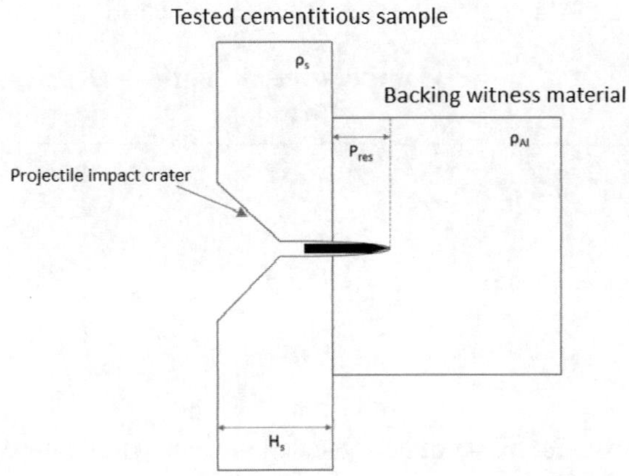

Figure 1. Principle of test method for DEF determination.

The projectile must penetrate backing witness material. Hence, the thickness of cementitious sample must be chosen so as to achieve the complete perforation of tested sample. Measuring the penetration depth in witness material (residual penetration) can be done by X-ray or cutting a witness material. Although, cutting a thick metal witness material is laborious, especially when the hard projectile AP core remains inside.

The main output of the presented test method is the dimensionless differential efficiency factor (DEF). The higher the DEF number, the higher the ballistic resistance of the tested cementitious material. If the DEF = 1, the tested material has the same ballistic resistance as the backing witness system used. By comparing the ballistic efficiencies, one may rank the ballistic performance of cementitious materials for a given projectile.

3. Experiment

Depth of penetration test provides a method for evaluating ballistic resistance of cementitious materials for specific projectiles. The test ammunition was 7.62 mm × 54R B32 API (Armor Piercing Incendiary) projectile. This projectile consists of a high hardness steel core with weight of 5.35 g, brass jacket and incendiary composition. Overall weight of the projectile is 10.4 g. The diameter and length of steel core are 6.05 mm and 23.71 mm, respectively. The standard impact projectile velocity for cementitious samples testing was set at 854 ± 20 m/s.

The test configuration was the same for each cementitious sample. The barrel-sample distance was set at 15 m from the muzzle. The sample was fixed perpendicularly to the projectile trajectory. The tested cementitious sample was backed with witness material – an aluminum cylinder. Projectile velocity was measured by LS-04 Intelligent Light Gates placed 2.5 m from tested sample.

In this study, four different cementitious samples were examined. The ballistic resistance of two ordinary concrete samples (C25/30, C45/55) and two ultra-high strength concrete samples (UHPC KŠ Prefa, UHPFRC TBG Metrostav) was compared. The Basic mechanical properties of tested samples are given in table 1. It is evident that studied cementitious samples had very different values of compressive strength. Properties of materials were evaluated according to the relevant standards valid in the Czech Republic. Compressive strength and density were determined according to standard ČSN EN 12390-3 [8] and ČSN EN 12390-7 [9], respectively and the particle size of aggregates were determined according to standard ČSN EN 12620 [10]. Before mechanical properties and ballistic resistance testing, the samples were cured for up to 28 days in water.

For DOP test, all cementitious materials were prepared into cylindrical steel mold with an inner diameter of 150 mm (20 times the projectile diameter) and 40 mm (5 times the projectile diameter) of height (H_s). From each cementitious material, 10 cylindrical-shaped samples were prepared. In total, 40 samples were subjected to ballistic testing. A single bullet was fired at the centre of each cementitious sample and the residual penetration into the backing witness material was measured.

IOP Conf. Series: Materials Science and Engineering 1205 (2021) 012013 doi:10.1088/1757-899X/1205/1/012013

Table 1. Material properties of tested cementitious samples.

	Compressive strength [MPa]	**Density [kg/m³]**	**D_{max} [mm]**
Concrete C25/30	35	2.260	16
Concrete C45/55	56	2.290	16
UHPFRC TBG Metrostav	154.5	2.360	1.5
UHPC KŠ Prefa	111	2.350	2

The backing witness system material shouldn't undergo brittle fracture after the projectile impact. Also, the backing witness material must not be subjected to strain hardening. The dimensions of backing witness material are chosen appropriately, so the deformation zone doesn't reach the edge of the material. As the backing witness material was used cylinder made from aluminium alloy AW-2030-T4 ($\rho = 2{,}809$ kg/m^3) as described in original DOP test method for ceramic materials. The diameter and thickness of aluminium witness material were 90 mm (12 times of projectile diameter) and 80 mm (11 times of projectile diameter), respectively.

For each projectile and witness material, it is necessary to determine the dependence of projectile penetration into witness material on the impact projectile velocities. This calibration curve is used to calculate the residual velocity of the projectile (projectile velocity after perforation of the cementitious sample). So firstly, the testing was conducted against monolithic witness material only (without cementitious sample). Then, the ballistic tests of cemenetititous samples backed with aluminium witness material were executed. The penetration into aluminium alloy witness material was measured after testing for both the monolithic witness material and the concrete-faced targets. X-ray technique was employed for the measurement of actual residual penetration on aluminium alloy witness material.

4. Results and Discussion

Firstly, the dependence of the projectile penetration (P_r) into aluminium witness material on projectile impact velocity (v) was determined. Obtained regression curve can be seen in Figure 2 and shows a linear dependency; therefore there was no deformation strengthening in the aluminium alloy during the projectile penetration. This confirmed selected aluminium alloy AW-2030-T4 is a suitable witness material. Aluminium alloy is the most commonly used witness system. Franzen R.R. et al. measured the depth of penetration into RHA (Rolled Homogeneous Armor) after perforation of the materials (AlN, SiC, B4C) by the projectile [11]. The aluminium alloy as a witness system is cheaper and easier for change of dimensions.

From the regression equation it is possible to calculate the residual velocity of the projectile after perforating the tested cementitious material. For the regresion equation it is desirable to obtain the dependency of projectile's pentration depth over wide range of impact velocities.

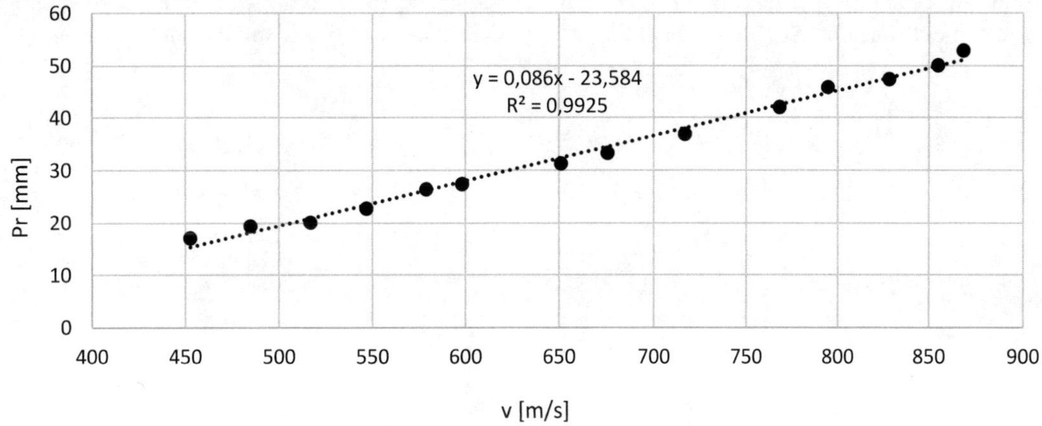

Figure 2. Calibration curve for projectile 7.62mm x 54R B32 API and witness material AW 2030-T4.

Table 2 summarized the values of measured parameters, which were obtained from 10 ballistic experiments for each cementitious material. Two ordinary concrete samples (C25/30 and C45/55) which contained large aggregates (D_{max} = 16 mm) had a similar value of obtained DEF parameter (0.528 and 0.545) with a relatively high standard deviation for DEF parameters. Deviation in DEF could be a result of the inhomogeneity of tested concrete samples. It is known that the penetration process of rigid projectile is different when it is penetrating through cementitious matrix and coarse aggregates. In most cases, aggregates have much higher strength than cementitious matrix, therefore their resistance to projectile impact is higher [12].

Unlike heterogeneous concrete materials (C25/30 and C45/55), the UHPC materials (UHPC KŠ Prefa, UHPFRC TBG Metrostav) were much more homogenous due to a smaller maximum particle size (D_{max} = 2 mm). Therefore, the DEF values for UHPC materials hadn't scattered. Furthermore, the UHPC samples reported a higher values of DEF parameter because the ballistic resistance of concrete-based materials is to some extend dependent to the compressive strength of concrete materials. It should be noted that the compressive strength of tested materials is not the only main parameter affecting the ballistic resistance of concrete-based materials. For example, the presence of high-strength aggregate could significantly increase the ballistic resistance of concrete-based materials [12].

Table 2. Summary of DOP test results for cementitious samples.

	v [m/s]	Hs [mm]	P_{res} [mm]	P_r [mm]	DEF	σ
Concrete C25/30	849.2	39.7	32.4	49.5	**0.528**	0.140
Concrete C45/55	850.5	41.8	31.0	49.6	**0.545**	0.147
UHPFRC TBG Metrostav	848.6	42.4	20.7	49.4	**0.786**	0.035
UHPC KŠ Prefa	819.4	42.6	23.2	46.9	**0.664**	0.042

Figure 3 shows a dependency of calculated DEF parameters on the compressive strength of tested cementitious materials. As shown in Figure 3, the DEF is dependent on compressive strength, although the compressive strength is not the only main factor affecting the ballistic resistance, as mentioned in the previous paragraph - a big influence on ballistic resistance has also the type and size of aggregate used in cementitious matrix (high strength aggregate will increase the ballistic resistance).

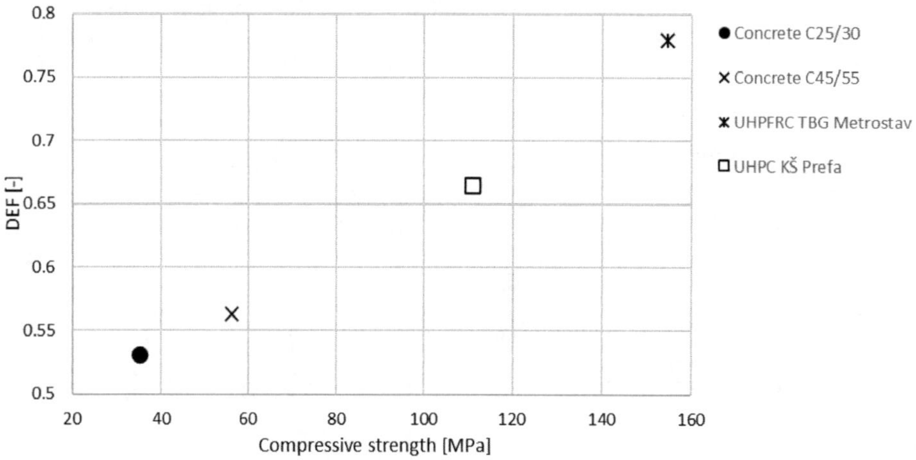

Figure 3. Dependence of DEF parameter on compressive strength of tested cementitious materials.

Results from testing concrete materials by DOP test method could be used for calculation of the ballistic limit thickness (H_{bl}) of the concrete materials by following equation (2):

$$H_{bl}(DEF) = \frac{P_{res}\rho_{Al}}{DEF\rho_s} + H_s \tag{2}$$

Ballistic limit thickness obtained from equation (2) represents the minimum thickness of the protective concrete panel, at which complete penetration doesn't occur. The dependence of ballistic limit thickness on DEF is shown in the following Figure 4.

The measured results in this article are compared with a study by Savio S.G. et al, who investigated ballistic limit thickness. Ceramics material (boron carbide) were studied on residual DOP and DEF was studied using aluminium alloy 6063-T6 as the reference witness material. Comparing the results, the DEF values have a similar character for ceramics and cementitious composites [13].

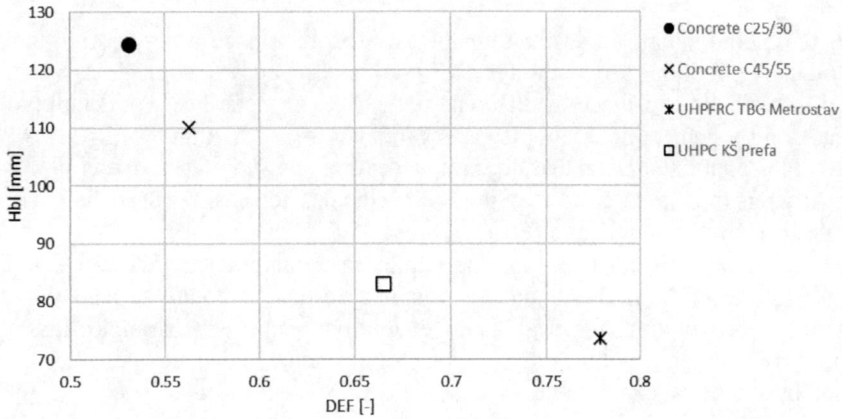

Figure 4. Dependence of H_{bl} value on DEF of cementitious composites.

Based on the results, it is possible to say that the presented ballistic test methodology could be applicable for a large spectrum of concrete-based materials with different material properties (e.g., compressive strength). Presented methodology was verified for 7.62 × 54R B32 API projectile. Moreover, it could be used for other similar-types projectiles, i.e., for projectile with rigid and non-deformable core, which is not significantly deformed during penetration of the cementitious composite. In presented test method, the core of armour piercing projectile could be considered non-deformable compared to cementitious materials (the erosion of the projectile core is not considered). For large calibre types projectiles, the following scaling for sample and witness material dimensions can be applied: the sample diameter is 20 times the projectile diameter and the sample thickness must not be less than 5 times the projectile diameter. The diameter of the witness material can be calculated as 12 times of projectile diameter and the thickness of the witness material is 11 times the projectile diameter. The dimensions of the specimens for large projectiles are calculated the same as a dimension of tested specimens in this article.

The following tables summarize the recommended dimensions for cementitious samples and witness material for different projectile calibres.

Table 3. Samples dimensions for the various AP projectiles.

Projectile	D [mm]	h [mm]
14.5 mm AP	300	80
12.7 mm AP	250	70
7.62 mm AP	150	40
5.56 – 7.62 mm Ball	115	30

Table 4. Witness system dimensions for the various AP projectiles.

Projectile	D_w [mm]	h_w [mm]
14.5 mm AP	180	200
12.7 mm AP/MP	150	160
7.62 mm AP	90	80
5.56 – 7.62 mm Ball	90	80

5. Conclusion

In this paper, the methodology for evaluating the ballistic resistance of cementitious samples was verified. The main output of this test method is a differential efficiency factor (DEF), which can be used for comparison of ballistic resistance of different cement-based materials. Applicability of presented ballistic test method for cementitious samples was validated for a 7.62 mm × 54R B32 API projectile. The advantage of this methodology is that the sample dimensions and witness material dimensions can be scaled for different projectile calibers. Suitable projectiles for scaling must have rigid and non-deformable core (e.g., armor piercing projectile).

- The methodology was verified on cementitious materials with a wide range of compressive strength (30–155 MPa). Based on the obtained results, it could be said that the ballistic resistance of cementitious materials is dependent on compressive strength. It is probable that compressive strength is not the only main factor affecting the ballistic resistance. This must be verified by additional experiments.
- Cementitious samples containing larger aggregates reported higher deviation in DEF parameter. It is known that the penetration process of a rigid projectile is different when it is penetrating through cementitious matrix and coarse aggregates, thus in inhomogeneous sample higher deviation of DEF could be observed.
- Results from Depth of penetration test could be used for ballistic limit thickness determination and prediction of cementitious protective panel thickness.
- By assessing the obtained DEF values, it is possible to conclude the good sensitivity of the method - the obtained values of the DEF parameter vary substantially for individual materials and allow a sensitive comparison of them.

Acknowledgments

This outcome has been achieved with the financial support by the project of specific research FCH-S-21-7553.

References

[1] Zhang M H, Shim V P W, Lu G and Chew C W 2005 Resistance of high-strength concrete to projectile impact *International Journal of Impact Engineering*
[2] Máca P, Sovják R, Konvalinka P and Chew C W 2014 Mix design of UHPFRC and its response to projectile impact *International Journal of Impact Engineering*
[3] Ren F, Mattus C H, Wang J J and Dipaolo B P 2013 Effect of projectile impact and penetration on the phase composition and microstructure of high performance concretes *Cement and Concrete Composites*
[4] Soufeiani L, Raman S N, Jumaat M Z B, Alengaram U J, Ghadyani G and Mendis P 2016 Influences of the volume fraction and shape of steel fibers on fiber-reinforced concrete subjected to dynamic loading – A review *Engineering Structures*
[5] Anderson C E, Morris B L, Jumaat M Z B, Alengaram U J, Ghadyani G and Mendis P 1992 The ballistic performance of confined Al2O3 ceramic tiles *International Journal of Impact Engineering*
[6] Madhu V, Ramanjaneyulu K, Balakrishna Bhat T and Gupta N K 2005 An experimental study of penetration resistance of ceramic armour subjected to projectile impact *International Journal of Impact Engineering*
[7] Gupta N, Basu B, Prasad V and Vemuri M 2012 Ballistic Studies on TiB2-Ti Functionally Graded Armor Ceramics *Defence Science Journal*
[8] ČSN EN 12390-3.: 2019 *Testing hardened concrete - Part 3: Compressive strength of test specimens*
[9] ČSN EN 12390-7: 2020 *Testing hardened concrete - Part 7: Density of hardened concrete*
[10] ČSN EN 12620+A1: 2008 *Aggregates for concrete*
[11] Franzen R R, Orphal D L and Anderson C E 1997 The influence of experimental design on depth-

of-penetration (DOP) test results and derived ballistic efficiencies *International Journal of Impact Engineering*
[12] Wu H, Fang Q, Gong J, Liu J Z, Zhang J H and Gong Z M 2015 Projectile impact resistance of corundum aggregated UHP-SFRC *International Journal of Impact Engineering*
[13] Savio S G, Ramanjaneyulu K, Madhu V and Balakrishna T 2011 An experimental study on ballistic performance of boron carbide tiles *International Journal of Impact Engineering*

IOP Conf. Series: Materials Science and Engineering 1205 (2021) 012014 doi:10.1088/1757-899X/1205/1/012014

Development of lightweight structural concrete with the use of aggregates based on foam glass

J Zach[1a], J Bubeník[1b] and M Sedlmajer[1c]

[1] Brno University of Technology, Faculty of Civil Engineering, Department of Building materials and Components, Veveri 95, 602 00 Brno, Czech Republic

Email: [a]zach.j@fce.vutbr.cz, [b]bubenik.j@fce.vutbr.cz, [c]sedlmajer.m@fce.vutbr.cz

Abstract. Lightweight concretes are increasingly being used in the construction industry, either for the overall lightweighting of the structure itself, reducing material consumption for construction and thus CO_2 emissions, or for specific reasons such as improving the thermal insulation properties of the structure or acoustic properties.
Today, lightweight concretes with lightweight expanded aggregates (expanded clay, agloporite) are most commonly used. This paper deals with the production of lightweight concretes lightweighted with foamed glass-based aggregates. Foamed glass is a lightweight material characterised by a very good ratio of thermal insulation and mechanical properties. Foamed glass is made of approximately 90% recycled glass waste (mostly mixed), which cannot be used in any other way, as well as water glass and glycerine. When concrete is lightened with foamed glass, these concretes achieve unique properties while conserving primary aggregate resources, avoiding landfilling of glass waste and efficiently using the waste material to produce lightweight concrete with higher added value. The paper discusses the possibilities of developing lightweight structural concretes using glass foam-based aggregates to achieve higher strength classes while reducing the weight and thermal conductivity of the concrete. As part of the research work, new types of lightweight concrete with a bulk density in the range of 1750–1930 kg/m^3 and a thermal conductivity from 0.699 to 0.950 W/(m·K) were developed.

1. Introduction

Ductile lightweight concretes are most similar to ordinary concrete due to their apparently compact (closed) structure. Weight reduction in these concretes is most often achieved by replacing dense (heavy) aggregates with lightweight ones. All or part of the dense aggregate may be replaced, depending on the final bulk density required. Lightweight aggregates are by definition aggregate with a bulk density of less than 2000 kg/m^3. The most commonly used aggregates for lightweight concrete include aggregates made from sintered clays (expanded clay), sintered fly ash (agloporite), vermiculite, expanded obsidian and others [1].

The use of glass foam-based aggregates for the production of lightweight concrete is not very common in practice. Foamed glass is a relatively long known material that has been used in the construction industry for the last ten years. Foam glass is made from waste glass ground into a fine powder (grain size 40–70 microns). The raw meal is then mixed with foaming and melting additive and placed on a pad, after which the mixture goes into a tunnel kiln. In the furnace, there is a preheating zone (approx. 200°C) in which the material is heated, and unwanted moisture is removed. The mixture passes through a melting zone (approx. 750–850°C) where the volume of the material increases several times, and a regular porous structure is formed. Finally, an intensive cooling process breaks the foam glass into larger pieces moved by a loader to a jaw crusher that crushes the foam glass to the desired aggregate fraction. Aggregate made of foam glass is characterized by low bulk density (120–180 kg/m^3), very low thermal conductivity (about 0.06 W/(m·K)), non-flammability (reaction to fire class A1), low water absorption, is not subject to degradation and is recyclable [2–4].

The use of foam glass-based aggregates was investigated as a partial replacement for natural quarried or crushed aggregates in concrete. Foam glass contains large amounts of amorphous SiO_2, has a regular porous structure, a rough surface and can be used in cement-based materials as a partial replacement for natural aggregate. Overall, the use of foamed glass has a minimal effect on workability. Still, over time, due to the higher water adsorption on the aggregate surface resulting from the crushing of larger units of foamed glass, workability is reduced compared to natural aggregates. At higher doses of foamed glass, there is a decrease in compressive strength and modulus of elasticity, which is related to the properties of the foamed glass. The water coefficient also has a major influence on the final properties of lightweight concrete with foamed glass; the excess water that can be adsorbed by the aggregate can be used for internal treatment of the concrete during maturation. From the point of view of durability, it seems optimal to replace 30–40% of the natural aggregate with a coarse fraction of foamed glass (fraction 4–16 mm) and up to 15% of the fine fraction of foamed glass (fraction below 4 mm); at higher doses, negative volume changes occur mainly due to the higher content of very fine fractions contained in the 0–4 mm fraction of foamed glass. Foamed glass aggregate obtained from recycled glass powder can offer many advantages in the design of lightweight concrete. The foam glass-based material is environmentally friendly by reducing landfills and increasing the use of glass recyclate, especially in the conservation of non-renewable natural resources of raw materials [5–7]. The experiment aimed to design the optimum technology for the production of lightweight concrete using glass foam aggregate and to produce lightweight concrete with the highest compressive strength.

2. Production of test specimens

Four trial formulations were produced as part of the development of lightweight concretes with foamed glass aggregate. The different recipes differed in the amount of cement and aggregates used (relation between natural and foam-based aggregate). For the fourth recipe, the amount of foam glass aggregate was significantly reduced, and the plasticizer was exchanged for superplasticizer on the base of polycarboxylate. Prior to the actual production of the test formulations, the foam glass aggregate was soaked in water for a minimum of 24 hours due to the high absorption of water on the surface of the crushed grains. The individual recipes are shown in table 1.

Crushed aggregate from the company Refaglass fraction 4–16 with a bulk density of 160 kg/m^3 and absorbency of 38% was chosen as light aggregate, cement was used CEM I 42.5R from the production Mokrá, plasticizer was used from the company MC Bauchemie based on naphthalene sulfonates, and the superplasticizer was a polycarboxylate additive from Mapei.

Two types of test specimens were produced to determine the required properties. Specifically, these were:

- Cubes 150 × 150 × 150 mm (determination of mechanical and physical-mechanical properties)
- Plate specimens 300 × 300 × 60 mm (determination of mechanical and thermal insulation properties)

Table 1. Test formulas of ultra-lightweight aggregate concrete for 1 m^3.

Entry component		Formula 1 m^3			
		LC 1	LC 2	LC 3	LC 4
Cement CEM I 42.5 R	[kg]	510	390	450	510
Water	[kg]	184	154	161	140
Natural aggregate 0-4	[kg]	730	420	420	800
Foam glass aggregate 4-16	[kg]	140	140	140	100
Natural aggregate 4-8	[kg]	380	0	0	450
Natural aggregate 8-16	[kg]	0	680	680	0
Superplasticizer	[kg]	-	-	-	2.5
Plasticizer	[kg]	3.9	3.9	3.9	-
Water-cement ratio	[-]	0.36*	0.39*	0.36*	0.27*

* the water coefficient is low because the soaked foam glass aggregate contains a considerable amount of extra water

Figure 1. Test specimen after compressive strength test.

Figure 2. Test specimen for determination of thermal conductivity.

3. Methods

Key properties were determined on the manufactured test specimens with a view to subsequent use in construction practice. The following were selected as key properties of lightweight concrete with foamed glass:

- Testing fresh concrete - Part 2: Slump-test (ČSN EN 12350-2) [8],
- The density in a fresh/hardened state (ČSN EN 12350-6, ČSN EN 12390-7) [9, 10],
- The compressive strength of the concrete (ČSN EN 12390-3) [11],
- The thermal insulating properties – coefficient of thermal conductivity (ČSN EN 12667, ISO 8301) [12, 13].

Before mixing began, test moulds were prepared to produce the required test bodies and the individual raw materials in the required quantities to produce of the lightweight concrete analysed. All raw materials were dosed by weight according to the respective recipe and then homogenized in a forced circulation mixer. After mixing, the fresh, lightweight concrete was placed in the prepared moulds. The deposition was carried out in 2 layers, each compacted by vibration for 5 seconds on a vibrating table. After 48 hours of mixing, the test specimens were de-moulded and then placed in water storage. After 28 days, the plate specimens had to be mechanically prepared to build the thermal conductivity coefficient to achieve a flat surface using a polishing grinder.

4. Results and discussion

4.1 Determination of mechanical properties

To observe the rheological properties of fresh concrete, the cone settlement method was selected to determine the consistency of fresh concrete. Furthermore, the fresh concrete was used to determine the fresh and hardened bulk density. The bulk density in the hardened state along with the compressive strength was determined at the age of 7, 28 and 90 days after mixing. The measured values are given in the following table.

Table 2. Overview of measured values of consistencies, bulk densities and compressive strengths.

Formula	Settlement of fresh concrete [mm]	Bulk density [kg/m^3]				Compressive strength [MPa]		
		Fresh state	7 days	28 days	90 days	7 days	28 days	90 days
LC 1	100	1840	1780	1750	1780	16.0	17.6	20.9
LC 2	120	1930	1850	1890	1860	20.3	23.5	20.0
LC 3	140	1960	1870	1900	1910	17.1	23.4	25.7
LC 4	180	1990	1950	1930	1950	23.2	29.3	37.5

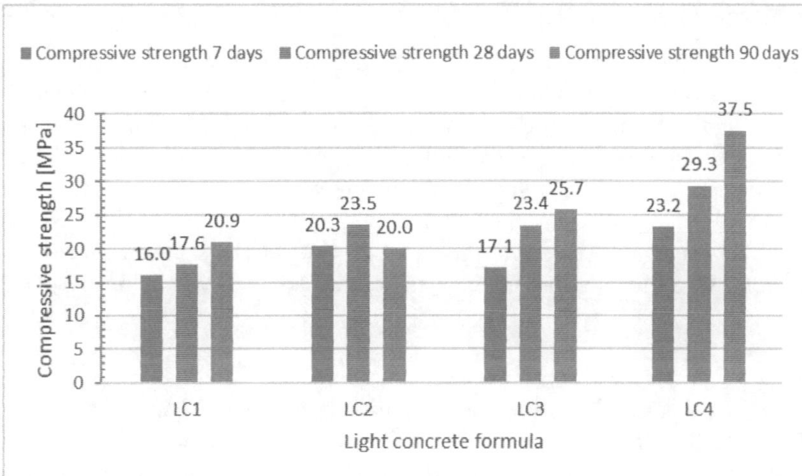

Figure 3. Compressive strengths of individual recipes after 7, 28 and 90 days from mixing.

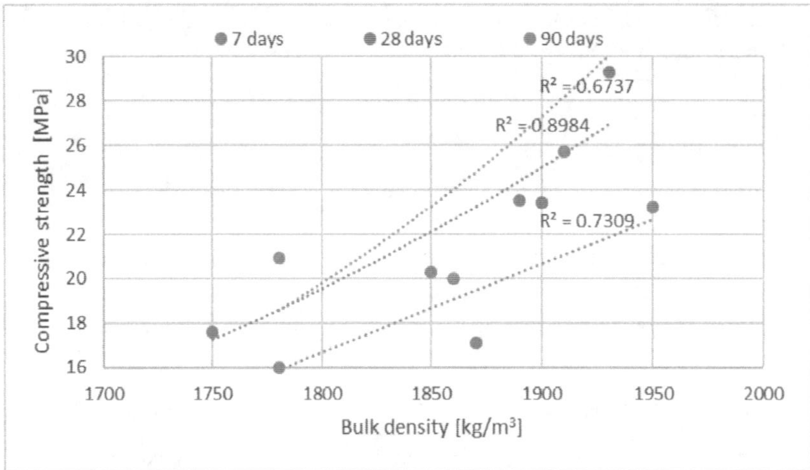

Figure 4. Dependence of compressive strength on bulk density.

The consistency of the concretes in the fresh state ranged from 100 to 180 mm cone settlement. The lowest value of cone settlement was found in the LC1 formulation (100 mm) due to the large amount of fines in the concrete compared to the other formulations. In contrast, the highest value of cone settlement was measured in the LC4 concrete (180 mm) when a more effective superplasticizer was used. The bulk density ranged from 1750 to 1950 kg/m^3 depending on the concrete formulation. Lightweight concrete LC4 (1950 kg/m^3) showed the highest bulk density due to the lower amount of glass foam aggregate in the concrete.

The compressive strengths corresponded to the bulk density of the individual concrete, ranging from 17.6 to 29.3 MPa after 28 days. The amount of foamed glass aggregate used had a significant effect on the compressive strengths, where the strength increased significantly as the amount of foamed glass in the concrete decreased, at the same time as the bulk density of the hardened concrete increased. Another factor influencing the strength of LC4 lightweight concrete was using a more effective superplasticizer, which allowed for a higher reduction of the water content. In the case of LC2, the strength decreased after 90 days. Still, this decrease was due to the variability of service samples, when in the set used to determine the compressive strength after 90 days, there were samples with a lower bulk density than in the previous sets.

4.2 Determination of thermal insulation properties

Determination of thermal insulation properties, specifically the thermal conductivity coefficient, was carried out to determine the effect of foamed glass in lightweight concrete on this physical characteristic. The thermal conductivity was determined by the stationary plate method using a Lambda 2300, Holometrix Micromet at a temperature gradient of 10 K with a mean temperature of 10°C. The thermal conductivity coefficient λ was determined on specimens of 300 × 300 × 60 mm. The fabricated plate test specimens had to be prepared using a stationary horizontal grinder with a diamond wheel. The surfaces were perfectly flat and no air gaps were created between the measuring plate and the specimen itself - figure 5. The measured values can be seen in the following table.

Table 3. Overview of values of coefficient of thermal conductivity in a dried state.

Formula	Weight [g]	Dimensions [mm]			Bulk density [kg/m³]	Average bulk density [kg/m³]	Thermal conductivity [W/(m·K)]			Average thermal conductivity [W/(m·K)]
		width	length	height						
LC 1	8920.1	303	300	57.9	1700	**1680**	0.660	0.658	0.657	**0.699**
	8943.1	302	301	59.3	1660		0.732	0.742	0.748	
LC 2	9765.4	304	301	58.3	1830	**1820**	0.768	0.770	0.784	**0.757**
	9568.4	300	301	59.0	1800		0.750	0.739	0.730	
LC 3	9748.1	300	300	58.6	1850	**1840**	0.799	0.802	0.806	**0.818**
	9718.8	300	301	58.7	1830		0.838	0.839	0.824	
LC 4	10194.1	301	300	58.5	1930	**1880**	1.073	1.064	1.062	**0.950**
	9327.4	300	304	55.8	1830		0.833	0.838	0.831	

Figure 5. Modification of plate samples for perfect placement in the measuring device.

The correlation between the thermal conductivity coefficient and the bulk density is evident from the measured thermal conductivity values. These bulk masses were determined by weighing and measuring test plate samples dried at 105°C to constant weight. The lowest bulk density for LC 1 (1680 kg/m^3) corresponds to a thermal conductivity coefficient of 0.699 W/(m·K). On the other hand, the highest bulk density for LC 4 (1880 kg/m^3) corresponds to the thermal conductivity (0.950 W/(m·K)). It can be seen that in all cases, the thermal conductivity is significantly lower than the typical values for structural concretes of classical volumetric masses according to ČSN 720540 [14].

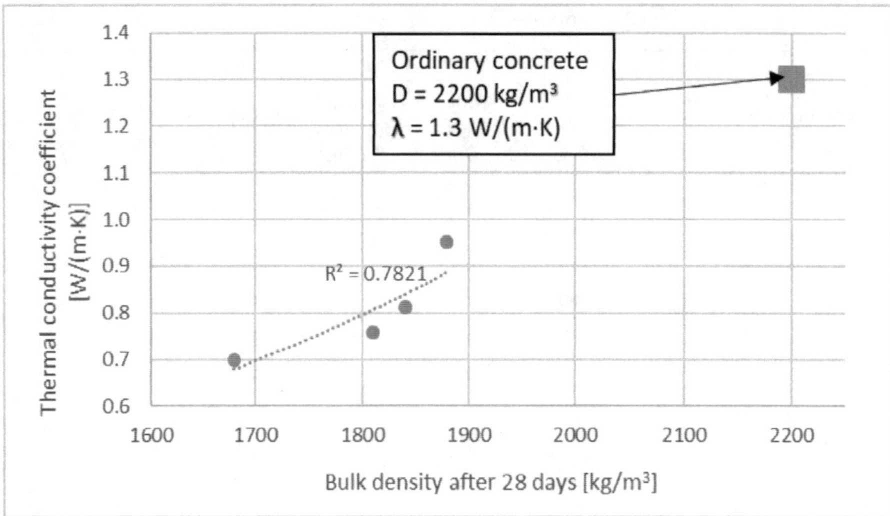

Figure 6. Dependence of the thermal conductivity on the density of lightweight concrete.

The graphical evaluation (figure 6) shows a significant relationship between the bulk density and the thermal conductivity coefficient. For comparison, the values of the bulk density and thermal conductivity of the traditional compacted unballasted concrete (bulk density 2200 kg/m^3 and thermal conductivity 1.3 W/(m·K)) are also added to the graph [14]. The graph shows that the resulting dependence for manufactured lightweight concrete with foamed glass corresponds to the value generally reported for traditional dense concrete.

As part of the research work, four test formulations of lightweight compacted concrete with aggregates based on foamed glass fraction 4–16 mm were experimentally produced in the laboratory. The foam glass aggregate is present in the concrete in quantities of 7 and 11% of the total weight of aggregate. When working with foam glass in concrete, the aggregate must first be placed in water for a minimum of 24 hours before mixing. This is to avoid any negative effects on the consistency of the fresh concrete, mainly due to the adsorption of water during the deposition of the fresh concrete on the surface of the crushed foam glass grains. It was found that a higher amount of foam glass in the formulation (11%) had a negative effect on the consistency of fresh concrete and the compressive strength of concrete. Conversely, a higher dose of foam glass significantly reduced the thermal conductivity coefficient. The compressive strength of the lightweight concrete after 28 days ranged from 17 to 29 MPa and was significantly affected by the bulk density of the concrete. The compressive strength increased with the increasing bulk density of the lightweight concrete, see figure 6. The developed lightweight concretes have very good thermal insulation properties. The thermal conductivity coefficient varied from 0.699 to 0.950 W/(m·K) depending on the bulk density. The values obtained correspond to the tabulated values of other similar types of lightweight concrete (e.g. slag/cinder concrete) and are lower than the typical thermal conductivities for some kinds of conventional lightweight aggregate concrete used, such as ceramsite concrete, which has characteristic thermal conductivity values at 1700 kg/m^3 is 1.25 W/(m·K) [14].

5. Conclusions

The use of lightweight concretes with foamed glass in building structures seems to be an interesting option that is environmentally friendly and at the same time meets the high demands on mechanical and other required properties. The developed concretes are suitable for producing prefabricated elements (wall and ceiling panels, lintels and spans) while reducing the overall load on the structure due to its weight. Due to the higher porosity of the used aggregate, the ideal use appears to be the main use in constructing buildings in structures not directly stressed by weather conditions. It is not necessary to directly address the issue of alkali-silica reaction (ASR) for this aggregate. However, common types of concrete from the ASR point of view, the glasses are also suitable for outdoor use. When comparing the developed concretes with similar concretes based on the classic light rounded (packed) clay or foam glass-based aggregate, the developed concretes show higher mechanical properties in the case of recipes with a higher bulk density [15].

The benefit of replacing part of the natural aggregate with foamed glass aggregate is not only the conservation of natural non-renewable resources, but also an increase in the performance of the elements in the structure or the whole structure. When used in wall or ceiling panels, these structures can achieve much higher thermal insulation properties than traditional concretes. It is possible to reduce the need for thermal insulation materials for these structures, which are necessary in the case of traditional concretes to meet the energy requirements for buildings.

In practice, the production of these concretes will mean the careful use of primary natural resources (aggregates, clay), a reduction in the amount of all input raw materials (less CO_2 in cement production), the use of glass recyclate and a new perspective on the production of lightweight concretes in terms of ecology and the environment.

Acknowledgements
This paper was elaborated with the financial support of the project GAČR 21-25813S "Study of the influence of organic fibers on the properties of cement composites under extreme loading".

References

[1] ČSN EN 206+A2: 2021 Beton – Specifikace, vlastnosti, výroba a shoda (Rue de Stassart 36, B-1050 Brusel) p 106 (73 2403)

[2] Arulrajah, Arul, Mahdi M. Disfani, Farshid Maghoolpilehrood, Suksun Horpibulsuk, Artit Udonchai, Monzur Imteaz and Yan-Jun DU 2015 Engineering and environmental properties of foamed recycled glass as a lightweight engineering material *Journal of Cleaner Production* **94** pp 369–375 ISSN 09596526. 10.1016/j.jclepro.2015.01.080

[3] Yassaman Y, Schneider A, Baaj H, Tighe S and Youssefi A 2016 Foam glass lightweight aggregate: The new approach p 8

[4] Bubeník J and Zach J 2019 The use of foam glass based aggregates for the production of ultra-lightweight porous concrete for the production of noise barrier wall panels *Transportation Research Procedia* **40** pp 639–646 ISSN 23521465

[5] Sedlmajer M, Zach J and Bubeník J 2019 Using secondary raw materials in lightweight open-structure concrete with good utility properties *Acta Polytechnica CTU Proceedings* **22** pp 94–98 ISSN 2336-5382 10.14311/APP.2019.22.0094

[6] Zach J, Sedlmajer M, Bubenik J and Drdlova M 2019 Development of lightweight composites based on foam glass aggregate *IOP Conference Series: Materials Science and Engineering* pp 583 ISSN 1757-899X. 10.1088/1757-899X/583/1/012016

[7] Limbachiya, Mukesh, Mohammed Seddik Meddah and Soumela Fotiadou 2012 Performance of granulated foam glass concrete *Construction and Building Materials* 28(1) pp 759–768 ISSN 09500618. 10.1016/j.conbuildmat.2011.10.052

[8] ČSN EN 12350-2: 2020 Testing fresh concrete — Part 2: Slump-test (Rue de Stassart 36, B-1050 Brusel) p 12 (731301)

[9] ČSN EN 12350-6: 2019 Testing fresh concrete – Part 6: Density (Rue de Stassart 36, B-1050 Brusel) p 16 (731301)

[10] ČSN EN 12390-7: 2019 Testing hardened concrete - Part 7: Density of hardened concrete (Rue de Stassart 36, B-1050 Brusel) p 16 (731302)

[11] ČSN EN 12390-3: 2019 Testing hardened concrete - Part 3: Compressive strength of test specimens (Rue de Stassart 36, B-1050 Brusel) p 24 (73 1302)

[12] ČSN EN 12667: 2001 Thermal performance of building materials and products - Determination

of thermal resistance by means of guarded hot plate and heat flow meter methods - Products of high and medium thermal resistance (Rue de Stassart 36, B-1050 Brusel) p 60 (73 0569)
[13] ISO 8301: 1991 Thermal insulation — Determination of steady-state thermal resistance and related properties — Heat flow meter apparatus (Rue de Stassart 36, B-1050 Brusel) p 30
[14] ČSN 73 0540-3: 2005 Thermal protection of buildings - Design values of quantities (Prague: Czech Standards Institute) p 89
[15] Adhikary, Suman Kumar, Deepankar Kumar Ashish and Žymantas Rudžionis 2021 Expanded glass as light-weight aggregate in concrete – A review *Journal of Cleaner Production* 313 ISSN 09596526. 10.1016/j.jclepro.2021.127848

Decalcification resistance of various alkali-activated materials

P Hruby[1], V Bilek[1], L Topolar[2], L Kalina[1], M Marko[1], F Soukal[1], R Dvorak[2] and T Hercik[3]

[1] Brno University of Technology, Faculty of Chemistry, Purkynova 118, 612 00 Brno, Czech Republic
[2] Brno University of Technology, Faculty of Civil Engineering, Veveri 331/95, 602 00 Brno, Czech Republic
[3] Brno University of Technology, Faculty of Mechanical Engineering, Technicka 2896/2, 616 69 Brno, Czech Republic

Email: xchrubyp@fch.vut.cz

Abstract. The resistance of alkali-activated materials (AAMs) to degradation processes, particularly the decalcification, was studied in this paper. The ground granulated blast furnace slag was alkali-activated using various activators with the same activator dosage 6% Na_2O by slag weight (sodium hydroxide, sodium waterglass and sodium carbonate) and subjected to testing of decalcification resistance (immersion in 6M NH_4NO_3) for 84 days. The reference samples were stored in water. The progress of degradation was studied using the phenolphthalein technique, mechanical properties testing (compressive and flexural strength), and dilatometry analysis or weight measurements. The results obtained were compared to the CEM III/A 32.5R. The significant loss of mass along with the deterioration of mechanical properties were observed for all binder types, still some of the AAMs showed better durability than the cementitious one.

1. Introduction

The production of Ordinary Portland Cement (OPC) increases every year due to the needs of the building industry. However, the production of OPC represents a highly energetically and economically demanding process. Furthermore, the environmental aspects should be considered as well since the production of one tone of OPC is accompanied by the formation of one tone CO_2 [1]. Alkali-activated materials represent a suitable alternative from the economic and environmental point of view if properly composed (selection of suitable alkaline activator and aluminosilicate precursor). The current state of art remarks higher resistance against the action of aggressive media (high temperature [2], sulphates [3], chlorides [4] or acidic solutions [5] for the AAMs contrary to OPC. Nevertheless, the real applications of the AAMs in the building industry are limited due to the absence of relevant standards and insufficient verification of long-term properties when exposed to degradation processes during the service life as well.

Degradation processes can be realized due to several reaction mechanisms: formation of new reaction products (e. g. carbonation, sulphate corrosion, alkali-silica reaction, …); processes of ionic exchange and regarding the topic of this paper mainly by processes of leaching of components of hydration products, thus through decalcification of binder gel or minor reaction products. The decalcification processes are based on the simultaneous leaching of Ca^{2+} from the binder gel together with the formation of corresponding calcium salts. The decalcification is related to simultaneous reduction of C/S ratio in binder gel to form only the residual S-H gel with limited utility value instead of the C-S-H or the C-A-S-H [6].

The alkali-activated blast furnace slag (AABFS) shows better resistance against the decalcification processes than the Ordinary Portland Cement, as stated in studies like [6–9] because of several aspects. Firstly, the AABFS binder gel has more intensely cross-linked silicate chains due to the incorporation of alumina units in C-A-S-H gel compared to the main hydration product of cement hydration (C-S-H gel). Besides this, the A-S-H can form a relatively stable passivation layer. Another benefit arises from the absence of portlandite in AABFS, since it represents a potential "first" target for decalcification. This causes opening of the porosity (an increase of permeability) after the complete leaching which is accompanied by the increase of degradation rate. Other like lower initial C/S ratio or lower porosity also explain the greater stability of the AAMs in acidic media.

The experimental study of decalcification resistance can be realized using the accelerated test – immersion in a solution of 6M NH_4NO_3 [10]. The benefit of this method contrary to other acidic leaching is that there are no changes in silicon concentration. Thus, the changes of microstructural and psychico-mechanical properties are related only to decalcification itself.

The present study analyses the influence of the type of alkaline activator used for the alkali-activation of blast furnace slag to the resistance against the degradation by decalcification reactions using the immersion in 6M solution of ammonium nitrate. A comparative study of CEM III/A-32.5R was carried out too.

2. Experimental procedure

2.1. Raw materials

2.1.1. Blast furnace slag (BFS). The slag was from Kotouč Štramberk, s.r.o. (Blaine specific surface area of 400 $m^2 \cdot kg^{-1}$). The chemical composition is shown in Table 1. The amorphous content was 84%, while the rest was crystalline (akermanite – 9.5%, calcite – 3.7%, merwinite – 2.3% and quartz – 0.5%) according to the X-ray diffraction (XRD) using Rietveld method.

Table 1. The chemical composition of the BFS determined by X-ray fluorescence (XRF) analysis.

Chemical composition in wt. %									
CaO	SiO_2	MgO	Al_2O_3	SO_3	TiO_2	K_2O	MnO	Na_2O	Fe_2O_3
41.1	34.7	10.5	9.1	1.4	1.0	0.9	0.6	0.4	0.3

2.1.2. Alkaline activators. Liquid sodium water glass (NaWG) obtained from Vodní sklo, a.s. – Silicate modulus (M_s) = 1.89; Sodium hydroxide (SH) 50% solution – PENTA s.r.o. (pure); Sodium carbonate (SC) – INCHEMA s.r.o (pure).

2.1.3. Other materials. Standard siliceous sand according to ČSN EN 196-1 was used for the mortar preparation. The cement for universal use – Optimcem – CEM III/A 32.5R (Cemmac, s.r.o.) served as comparison binder to the AAMs. The pure ammonium nitrate of obtained by Lachner s.r.o. was used as degradation medium in concentration of 6M.

2.2. Tests and methods

The samples of individual binders had dimensions of 40 × 40 × 160 mm and were prepared according to the ČSN EN 196-1. The samples were stored in sealed moulds for 24 h hours, and then they were demoulded and stored in water up to 28 days (to reach a certain degree of hydration before the degradation tests took place). The leaching method consisted of submerging the 28-day water cured specimens (time zero in all of the figures below) in 6M NH_4NO_3 and water.

The testing temperature was kept constant (25°C) during the experiment (28, 56 and 84 days of exposition to the media). The sampling and analysis of the samples with selected techniques took place in the same period. The immersion solutions were renewed every 28 days.

The binder systems were mortars with sand to binder (slag or cement) ratio of 3:1. The water to binder ratio was the same for all alkali-activated materials (0.45), same as the activator dosage (6% Na_2O by the slag weight). Water to binder ratio was equal to 0.45 in case of the cementitious system as well, but there was no alkaline activator added.

2.2.1. Mechanical properties testing. The determination of mechanical properties (compressive and flexural strength) was done as prescribed in ČSN EN 196-1 using the Desttest 4310 Compact A (Beton System, s.r.o.). The flexural strength was measured in the three-point arrangement. The flexural strength was determined on three samples of each binder type in each interval, meanwhile the compressive strength was tested on six samples. The final values below were calculated as the mean values.

2.2.2. Phenolphtalein technique. The phenolphthalein technique was used for the determination of the progress of the degradation reactions since they are accompanied by the decrease of pH of matrices. The phenolphthalein is an acid-base indicator, which changes colour with dependence on the pH. It changes colour from purple-red (alkaline pH) to colourless. The fracture surfaces of the samples after the three-point bending test were treated with a 1% solution of phenolphthalein in ethanol. The purple-red colour can be seen in case of the non-degraded sample. On the other hand, the damaged samples show no colour change (since the decrease of pH took place) and remain colourless. The treated fracture surfaces were then photographed, and the dimension of the colourless area was measured and evaluated using the optical analysis method.

2.2.3. Dilatometry analysis, weight measurement. The samples with dimensions of 40 × 40 × 160 mm with implemented brass contacts were used for dilatometry analysis, same as the weight measurement. The samples were measured before and during the degradation tests in a defined period and the data obtained were then evaluated.

3. Results and discussion

According to the literature that for the decalcification in 6M NH_4NO_3 to take place pH below the 9.25 is necessary [9]. Thus, the pH was monitored for various systems and reached values of 8.10; 8.35; 8.58 and 8.66 after 84 days for NaOH; NaWG, Na_2CO_3 and CEM, respectively. The most noticeable changes in pH were observed during the initial 14 days period since the pH increased to 9.01; 9.22; 9.31, and 9.48 in the same order. Thus the need of solution exchange during the degradation reactions was confirmed.

3.1. Mechanical properties

The development of mechanical properties in time as a function of activator type and the storage medium (water/6M NH_4NO_3) is depicted in Figure 1 and Figure 2. The lowest absolute values of compressive and flexural strength were obtained for the NaOH activated blast furnace slag. Nevertheless, this system showed the lowest drop of compressive strength when exposed to 6M ammonium nitrate compared to water. The residual compressive strength (calculated as percentage difference compared to water) follows: NaOH>NaWG>Na_2CO_3>CEM with values of 80.2; 77.7; 58.3, and 27.3% after 84 days of deposition to the ammonium nitrate solution. Secondly, residual flexural strength compared to water storage about 82.5; 69.6; 54.6 and 21.6% was observed in the following alkaline activator order NaWG; NaOH; Na_2CO_3 and ending with the cementitious system after 84 days of deposition to ammonium nitrate solution. The whole development of residual compressive and flexural strength over the testing period is shown in Figure 3.

The cementitious system showed the lowest residual compressive and flexural strength, same as the highest loss of mass as specified below. Both trends have simultaneous character and increase with prolonging time of deposition to NH_4NO_3. This is related to the decalcification reactions taking place and simultaneous leaching of Ca^{2+} from the gel structure. The highest resistance to the decalcification was observed for the NaOH and NaWG activated systems. This can be related to factors like higher pH of pore solution or character of hydration products. Quite similar trends were reported in Komljenović et al. [7] study, where residual compressive strength about 41; 19, and 18% after 30; 60, and 90 days of the exposition were determined. Meanwhile, the residual strength for AABFS system activated using the mixture of NaWG+NaOH (M_s=0.6) was 94; 91 and 85%. These results are comparable with the ones in the present study and confirm higher resistance of the AAMs, especially when water glass is used. In contrary to study [7] no signs of retardation of degradation rate for the cementitious system were not detected in the present paper. Significant deterioration of mechanical properties of OPC and AABFS (NaWG+NaOH; M_s=1.2) due to the decalcification in 6M NH_4NO_3 was also presented in the study of Varga et al. [8]. They stated a decrease of compressive strength about 34% and 77% for AABFS and OPC after 21 days of immersion. Both research groups attribute better resistance of AABFS against decalcification to more cross-linked C-A-S-H gel structure along with the absence of portlandite, lower porosity and permeability and last but not least positive effect of alumina units forming protective barrier retarding the further progress of decalcification. The degradation in a solution of ammonium nitrate is accompanied by a decrease in the C/S ratio and deterioration of mechanical properties.

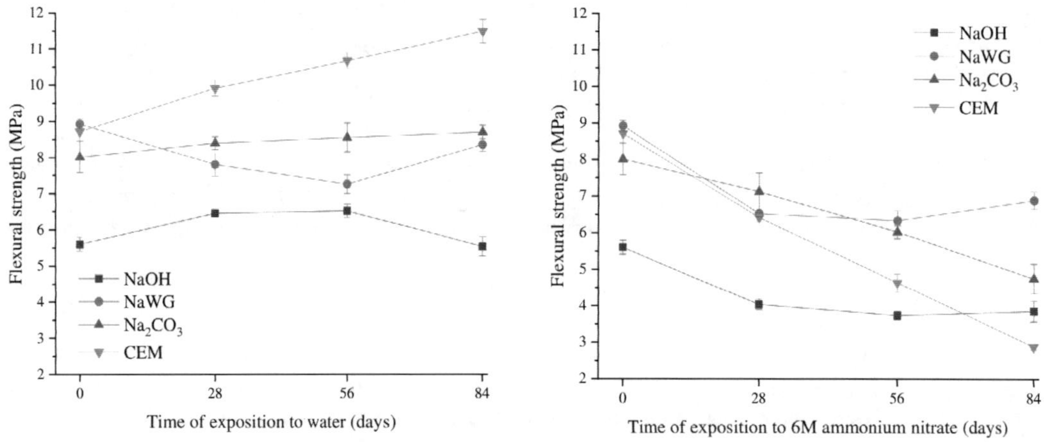

Figure 1. Dependence of flexural strength of various binder types on the time of exposition to water and 6M ammonium nitrate.

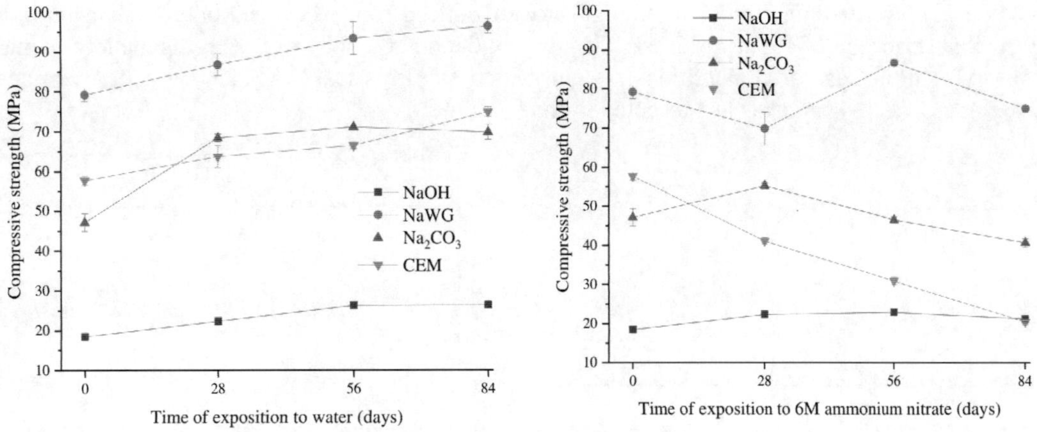

Figure 2. Dependence of compressive strength of various binder types on the time of exposition to water and 6M ammonium nitrate.

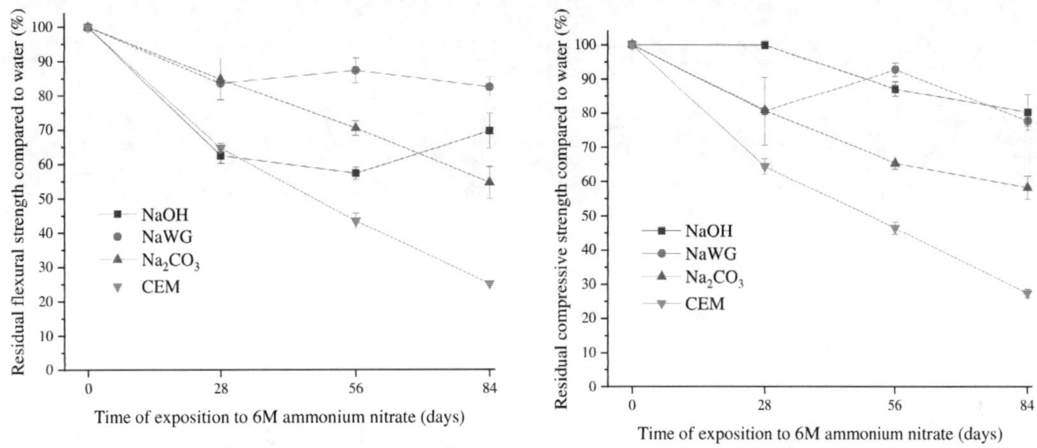

Figure 3. Residual flexural and compressive strength of various binders stored in 6M NH_4NO_3 compared to water as a function of time.

3.2. Dimensional and weight changes

None of the tested binder systems showed a remarkable loss of mass due to the water storage. The weight changes were between -0.2% (NaWG) and +0.5% (NaOH) after 84, respectively total of 112 days of exposition to water. Those changes are not related to degradation processes but only different sorption of water and leaching of activator in time. On the other hand, the significant loss of mass was observed for the cementitious system stored in 6M NH_4NO_3, since the simultaneous drop of the weight of the samples about 1.1; 1.7; 2.6 and 3.3% after 14; 28; 56 and 84 days was observed. The loss of mass of the cementitious system is noticeably higher than for any of AAMs. The loss of mass of the AAMs decreased in the following alkaline activator order NaOH>NaWG>Na_2CO_3. Still, the values of weight change are only about -1.1; -0.7 and -0.4%, thus not such crucial compared to the CEM. Overall comparison is presented in Figure 4.

Secondly, Figure 5 presents the development of length changes of the samples stored in water and the one exposed to 6M NH_4NO_3. It can be concluded that storage in water causes no changes in the dimension of the samples since it lies between -0.0005 and 0.009% after the 84days of water storage. The exposition of the NaOH and NaWG activated samples to ammonium nitrate did not cause any significant length changes (0.0015 and 0.004% after the 84 days), contrary to the Na_2CO_3 activated system where simultaneous shrinkage of the samples was observed. It reached the value of -0.1249%

after 84 days. Nevertheless, similar trend was as well noticed for the cementitious system but with a higher final shrinkage of about -0.1776%. To sum up, the highest shrinkage same as the loss of mass was observed in the case of cementitious system, which confirms that AAMs show better resistance against aggressive environments as presented in studies [6–9].

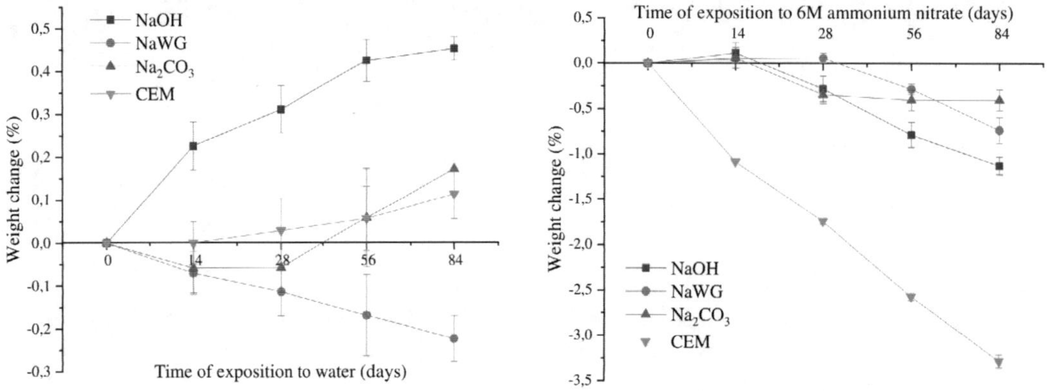

Figure 4. Time development of weight loss of the various binders due to the immersion in water and 6M NH_4NO_3.

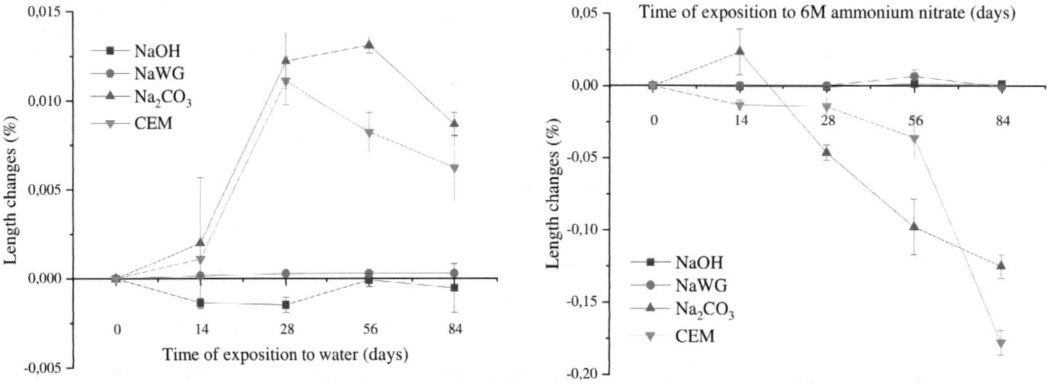

Figure 5. Time development of length changes of the various binders exposed to water and 6M NH_4NO_3.

3.3. Phenolphthalein technique

There were no colour changes (samples remained purple-red after the treatment using the phenolphthalein solution) for any of the tested binder systems stored in water during the whole testing interval (0–84 days), as shown in Table 2. Only minor changes in colour intensity were observed between the various AAMs and the CEM. These minor changes correspond to the alkalinity of the pH of the pore solution. Thus, the system activated using the NaOH had the most intense-purple red colour contrary to the less intense purple-red colour of the cementitious system. The intensity of colouration decreases as follows NaOH>NaWG>Na_2CO_3>CEM. Overall, any of the tested AAMs same as the CEM did not show any signs of decrease of pH of matrices. Thus it indicates that no degradation took place in water.

Table 2. The pictures of phenolphthalein treated fracture areas of various AAMs and CEM binders.

	H_2O			$6M\ NH_4NO_3$		
	28 days	56 days	84 days	28 days	56 days	84 days
NaOH						
NaWG						
Na_2CO_3						
CEM						

On the other hand, immersion of the samples to the 6M NH_4NO_3 was related to significant changes (corresponded with the presence of colourless area – degraded parts). The area of non-affected (non-degraded) layer decreases with prolonging exposition time to the ammonium nitrate for all binder types (Table 2). The depth of degraded area calculated using the optical analysis method (the value corresponds to the depth from one side of the sample to the middle) is in Figure 6. This graph shows that the depth of the degraded layer increases in time in the following order of binder systems NaOH>Na_2CO_3>NaWG>CEM. Still, the differences between the activators for the AAMs are not convincing. The cementitious systems showed the lowest degraded layer. However, the loss of mass or deterioration of mechanical properties was the highest.

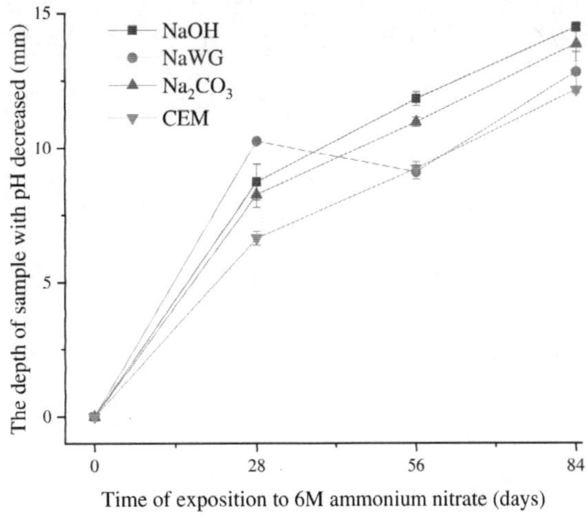

Figure 6. The depth of sample with pH decreased for various binder types as function of time.

4. Conclusion

This paper investigated the effect of alkaline activator type used for the alkali activation of blast furnace slag on the decalcification resistance. Decalcification was realized under the effect of 6M solution of ammonium nitrate for 28, 56 and 84 days. This resistance was characterized with a meaning of mechanical properties, weight and length changes and phenolphthalein technique. The CEM III/A 32.5R was used as comparison material. Based on the obtained results, the following conclusions can be drawn:

- AABFS showed superior durability compared to the cementitious system.
- Simultaneous decrease of residual strength was observed for all types of the binder. However, the cementitious system showed the highest deterioration (residual compressive and flexural strength about 30% after 84 days).
- Significant loss of mass was not detected for any of the AAMs tested (<1%), but CEM system showed a loss of about 3.3%.

Acknowledgments

This outcome has been achieved with the financial support by the following projects „The use of non-destructive methods for testing of the condition of degraded alkali-activated concrete“ GAČR 19-04703S supported by Grant Agency of Czech Republic and others like FCH/FSI-J-21-7402, FCH-K-21-6944; FCH-S-21-7553.

References

[1] Pacheco-Torgal F, Abdollahnejad Z, Camoes M, Jamshidi M and Ding Y 2012 Durability of alkali-activated binders: A clear advantage over Portland cement or an unproven issue? *Construction and Building Materials*. **30** pp 400–405
[2] Zhao R and Sanjayan J 2011 Geopolymer and Portland cement concretes in simulated fire *Magazine of Concrete Research*. **63** pp 163–173
[3] Bakharev T, Sanjayan J G and Cheng Y B 2002 Sulfate attack on alkali-activated slag concrete *Cement and Concrete Research*. **32** pp 211–216
[4] Shi C 2003 Corrosion resistance of alkali-activated slag cement *Advances in Cement Research*. **15** pp 77–81
[5] Bakharev T, Sanjayan J G and Cheng Y B 2003 Resistance of alkali-activated slag concrete to acid attack *Cement and Concrete Research*. **33** pp 1607–1611
[6] Pacheco-Torgal F, Labrincha J A, Leonelli C, Palomo A and Chindaprasirt P 2015 *Handbook of Alkali-Activated Cements, Mortars and Concretes* (Woodhead Publishing)
[7] Komljenović, Miroslav M, Baščarević Z, Marjanović N, Nikolić V and Illikainen M 2012 Decalcification resistance of alkali-activated slag: A review *Journal of Hazardous Materials*. pp 112–121
[8] Varga C, Alonso M M, Mejía Gutierrez R and Puertas F 2015 Decalcification of alkali-activated slag pastes. Effect of the chemical composition of the slag: A review *Materials and Structures*. **48**(3) pp 541–555
[9] Baščarević Z, Komljenović M, Miladinović Z, Nikolić V, Marjanović, Žujović Z and Petrović R 2013 Effects of the concentrated NH4NO3 solution on mechanical properties and structure of the fly ash based geopolymers: A review *Construction and Building Materials*. **41** pp 570–579
[10] Provis J L and Van Deventer J S J 2014 *Alkali Activated Materials: State-of-the-Art Report*, RILEM TC 224-AAM

Effect of alkali cation type on the plasticizing effect of ligno-sulfonate in alkali-activated systems

V Iliushchenko[1], V Bilek Jr.[1], L Kalina[1], P Hruby[1], T Opravil[1] and J Bojanovsky[2]

[1] Brno University of Technology, Faculty of Chemistry, Purkyňova 118, 612 00 Brno, Czech Republic

[2] Brno University of Technology, Faculty of Mechanical Engineering, Technická 2896/2, 616 69 Brno, Czech Republic

Email: xciliushchenko@vutbr.cz

Abstract. The rheological properties of alkali-activated systems are significantly affected by the nature of the alkaline activator. Hydroxide-activated systems' workability is typically lower than that of alkali-activated systems but can be improved by lignosulfonate plasticizer. However, the lignosulfonate plasticizer's effectivity depends on the dosage of lignosulfonate, the nature of hydroxide and pH of their solutions. Therefore, in this study, the effectiveness of lignosulfonate plasticizer with respect to alkali ion type (Na^+, K^+, Li^+) in alkali hydroxide-activated systems based on ground granulated blast furnace slag was evaluated. The concentration of the alkaline activator (NaOH, KOH and LiOH) was the same in all cases of 4M. The superplasticizer dosage was 0%, 0.5% and 1.0% of dry matter of lignosulfonate plasticizer to the slag weight. Rheological properties were determined using a rotational rheometer equipped with vane in-cup geometry working in oscillation amplitude sweep mode, from which critical strain and corresponding viscoelastic variables were obtained.

1. Introduction

Over time, humankind has become more trying to protect the planet against global warming and is looking for innovative ways to reduce the negative impact of the rapidly industrializing world. Ordinary Portland cement (OPC) with regard to the scale of its production around the world (annual production is around 3Gt) is considered as the dominant construction binder. One of the crucial problems with the use of OPC is the relatively high CO_2 footprint related to the production of its main component - Portland clinker. Approximately one tonne of CO_2 is formed per tonne of produce Portland clinker. Given the current position in relation to the ecological and economic situation, alternative solutions are being sought to replace (completely or partially) the high-emission binder. Aluminosiliceous by-products are gaining attention as the viable alternative to OPC. The reaction processes, which are taking part during the setting and hardening of the alkali-activated systems (AAS), are quite well-described and understood [1–4]. Meanwhile, there are only a few publications that clarify the rheological aspect of AAS.

An appropriate design of AAS could be leading the attainment of earlier compressive strength, better chemical resistance, lower permeability over the OPC. Although it cannot be expected that AAS will replace OPC on wide scale, these advantages can be used in some applications [1]. Nevertheless, the corresponding rheological behaviour of AAS is indispensable for its utilization. The rheological models used to interpret the behaviour of cement-based systems such as the Bingham model, modified

Bingham model as well as Hershel-Bulkley are also applied to characterize AAS [5]. Puertas et al. [6] found out that the nature and the dosage of the alkaline activator has a significant influence on the rheological behaviour of AAS. Behaviour of AAS activated by NaOH followed Bingham model, while AAS activated with sodium silicate rather fits Herschel-Bulkley model [6].

The rheological behaviour of cement-based binders are well controlled by using organic admixtures like superplasticizers, which greatly improve their workability by reducing yield stress and plastic viscosity. The mechanism of the efficiency of superplasticizers in cement-based systems is well known, where due to the adsorption on the cement particles, the electrostatic or steric forces arise, resulting in to the repulsive effect of these particles. However, not all superplasticizers are in AAS as effective as in cement-based binders [7]. Palacios et.al. [8] found that polycarboxylate or polyacrylated-based superplasticizers are non-stable in highly alkaline environment (pH > 13). Lukkonen et al. [9] reported that among the tested superplasticizers, lignosulfonates showed the best efficiency due to their steric repulsion effect, but it was emphasized that type and concentration of the activator strongly affect the plasticizer suitability. Significant efficiency of lignosulfonate plasticizer in NaOH-activated slag in contrast to analogous AAS pastes with Na_2CO_3, and sodium silicate was proven in our previous study [10].

Consequently, this paper deals with the efficiency of different dosages of lignosulfonate (0, 0.5, and 1.0%) in AAS with various alkali hydroxides (NaOH, KOH, LiOH) and its comparison to water- and cement-based pastes. In contrast to commonly used rotational tests, non-traditional approach based on oscillatory rheology in terms of the strain sweep tests was used in this study. Such test enables the determination of viscoelastic parameters of the arising paste's structure including stress needed to its destruction.

2. Experimental Procedure

2.1. Raw materials

2.1.1. Blast furnace slag (BFS). In this study, the slag was obtained from Kotouč Štramberk, s.r.o. (Blaine specific surface area is 400 $m^2 \cdot kg^{-1}$) was used as an aluminosilicate precursor. Its density was determined by pycnometer method and makes 2.88 $g \cdot cm^{-3}$. The phase composition was determined by using X-ray diffraction analysis (XRD, on an Empyrea X-ray diffractometer from Panalytical with Bragg-Brentano goniometer; Cu anode, with parameters 40 kV and 30 mA), Rietveld method, where the amorphous part was about 84% whilst the rest was in crystalline forms (akermanite – 9.5%, calcite – 2.3%, merwinite – 2.3% and quartz – 0.5%.

2.1.2. Ordinary Portland cement (OPC). Then, ordinary Portland cement (OPC) CEM I 42.5 R from Mokrá-Českomoravský cement, a.s. Heidelberg Cement, Mokrá-Horákov, Czech Republic with the specific surface area 377 $m^2 \cdot kg^{-1}$. The chemical compositions of BFS and OPC (XRF, using Olympus Delta professional) are presented in table 1.

Table 1. Chemical composition of BFS and OPC.

Blast Furnace Slag

CaO	SiO_2	MgO	Al_2O_3	SO_3	TiO_2	K_2O	Mn_2O_3	Fe_2O_3	SrO	ZrO_2	BaO
37.35	39.80	8.64	8.15	1.37	0.30	1.19	0.89	0.74	0.06	0.02	0.12

Ordinary Portland Cement CEM I 42.5 R

CaO	SiO_2	MgO	Al_2O_3	SO_3	K_2O	Na_2O	Fe_2O_3
65.0	19.0	1.0	4.0	3.0	0.77	0.17	3.0

2.1.3. Alkaline activators. The BFS was activated by three different alkaline activators, more precisely sodium hydroxide, potassium hydroxide (from Vodní sklo a.s.), lithium hydroxide (from Schem a.s.), and also mixed only with water. The suspension of OPC with water was used as well. The stock activating solutions were prepared somewhat more concentrated (4.23 mol·dm^{-3}) than target concentrations of 4.00 mol·dm^{-3} due to the dilution by addition of plasticizer and water to simultaneously reach slag volume fraction in the final suspension of 0.50. The same volume fraction was set for the paste of slag in water as well as for OPC paste to ensure the comparability of all mixtures.

Suspensions were modified by the commercial plasticizer, namely lignosulfonate (LS). Three different doses of LS were tested, more specifically, 0%, 0.5% and 1.0% of the dry matter to the weight of slag.

2.1.4. Rheological experiment. The measurement was performed on hybride rheometer Discovery HR-2 from TA-Instruments, using the vane-cup geometry. The experiment itself was started from the preparation of the suspension using the polypropylene beaker and hand mixer. The whole duration of the mixing process was 3.0 minutes, of which during the first minute the mixture was mixing at the slow rate, for the following 30 s the speed was increased to the maximum, while the following 30 s the mixture was stirred manually using spatula and the stick residuals were homogenized with the suspension. In the last minute, the suspension was mixed using the maximum rate.

The rheological parameters were studied with the stainless-steel vane (diameter 20.3 mm and length 19.5 mm) in cup (diameter 30.4 mm) geometry. The rheological experiment started 4.5 minutes after the mixing with the 30 s pre-shear at the shear rate of 100 s^{-1}, followed by 30 s rest before the amplitude sweep measurement. The angular frequency was set to 10 rad/s, while the strain was logarithmically increased from the $5{\cdot}10^{-4}$% to 500%.

The domain of linear viscoelastic (LVE) behaviour of the samples was obtained from the graphical dependency of storage modulus G' on oscillation strain. According to the literature, the critical strain of LVE is usually evaluated as the deviation of about 5–10% from the linear part of the curve [11]. In this study, the smaller deviation of 2% was selected, especially due to the evolution of viscoelastic parameters in time. The stress corresponding to the critical strain (yield point) as well as the stress corresponding to the destruction of the pastes' structure shown as a sharp drop of viscoelastic moduli (flow point) was evaluated.

3. Results and discussion

One representative rheological response of each paste is given in figure 1 as the evolution of complex modulus plotted depending on the oscillation stress. In most cases, there is an increase in the complex modulus for low stresses corresponding to LVE region due to the simultaneous structural build-up. At certain stress, the complex modulus starts to decrease. Further increase in the oscillation stress results in a drop of complex modulus by 2–3 orders of magnitude. In this region, the paste starts to flow, which is indicated by higher loss modulus than storage modulus (not shown here). As was already mentioned, the stress needed to start breaking of the evolved structure was termed as the yield point (green on the graph), while the latter as the flow point (purple on the graph). Their clearer comparison depending on the activator type and LS dose is given in figure 2.

Firstly, the influence of the type of alkaline activator on the rheological behaviour of BFS is commented. The complex modulus provides the information about rigidity of the paste, whereas the yield and flow points characterize the solidity.

Generally, in some literature [12, 13] it is shown that suspensions activated by NaOH usually exhibit higher yield stress than the suspensions containing KOH. The reason why it can be performed is that K^+ has lower charge density, which causes lower ion-dipole forces. Zeta potential of slags activated by KOH is higher than NaOH-based suspensions due to the lower absorption of Na^+ on the BFS grains surface. Higher adsorption of K^+ on the particles leads to the increasing of repulsive force of double-layer between these particles, which contributes to the decreasing of the yield point. In contrast with Na^+

cations have bigger solvation shell. Related to the better adsorption ability of KOH on the particles, negatively charged lignosulfonates groups can have higher affinity to bond with them [13–15].

The yield point was evaluated from the LVE area on graphical dependence of complex modulus on oscillation stress. In the case of NaOH and KOH-based activated systems, it can be observed that with the increasing dosage of LS, the oscillation stress gradually decreases. The course of the suspensions responding to the LS adding is similar, and it can be assumed that even 1% of LS is still effective. For the clear arrangement, the yield point was plotted to the graph with the LS dosage. From figure 2, it can be seen that the addition of LS (0.5 and 1%) in BFS activated by KOH leads to the lower values of yield point (about 1.20 and 1.05 times respectively), with regard to the fact that reference values (0% of LS) of KOH-activated systems were higher than NaOH-activated (1.08 times). On the other hand, these differences can be accepted with regard to the standard deviation. In any case, with the adding LS, the smooth decrease of yield point can be observed.

Different rheological behaviour with respect to the LS dose can be observed in the case of LiOH and BFS-water-based systems. From figure 2, it can be observed that the 0.5 and 1.0% of LS contribute to the similar yield point decreasing. This kind of behaviour can indicate that the limit value of the LS, which is effective in LiOH and water-based systems, is between 0 and 0.5%.

The behaviour of BFS was compared with the rheological response of the traditional binder on LS adding. The rheological behaviour of OPC is noticeably different from the BFS-activated materials. In OPC pastes, 0.5% of LS had only limited effect on yield point and flow point, while they both were greatly decreased by 1% of LS and the same stands for complex modulus. It should be noted that the results of the paste with 0.50% of LS is affected by the rate of its structural build-up prior to reaching the yield point as the evolution complex modulus starts from much lower values compared to the reference paste without LS. Therefore, it can be expected that if the structure is destroyed (e.g. by mixing), the flowability of the paste with 0.5% LS will be much better than that of the reference paste. Hence the effectiveness of LS at intermediate dose is time and shear dependent.

The flow point in all cases was evaluated from the loss factor, where the destruction of substantial amount of bonds is performed, and the suspension loses its properties. The rheological behaviour of tradition binder is remarkable different from the measured alternative binders. In OPC, the flow point is 42 times decreased by 1.0% LS, whereas its use in AAS offers about six times reduction at most (for NaOH).

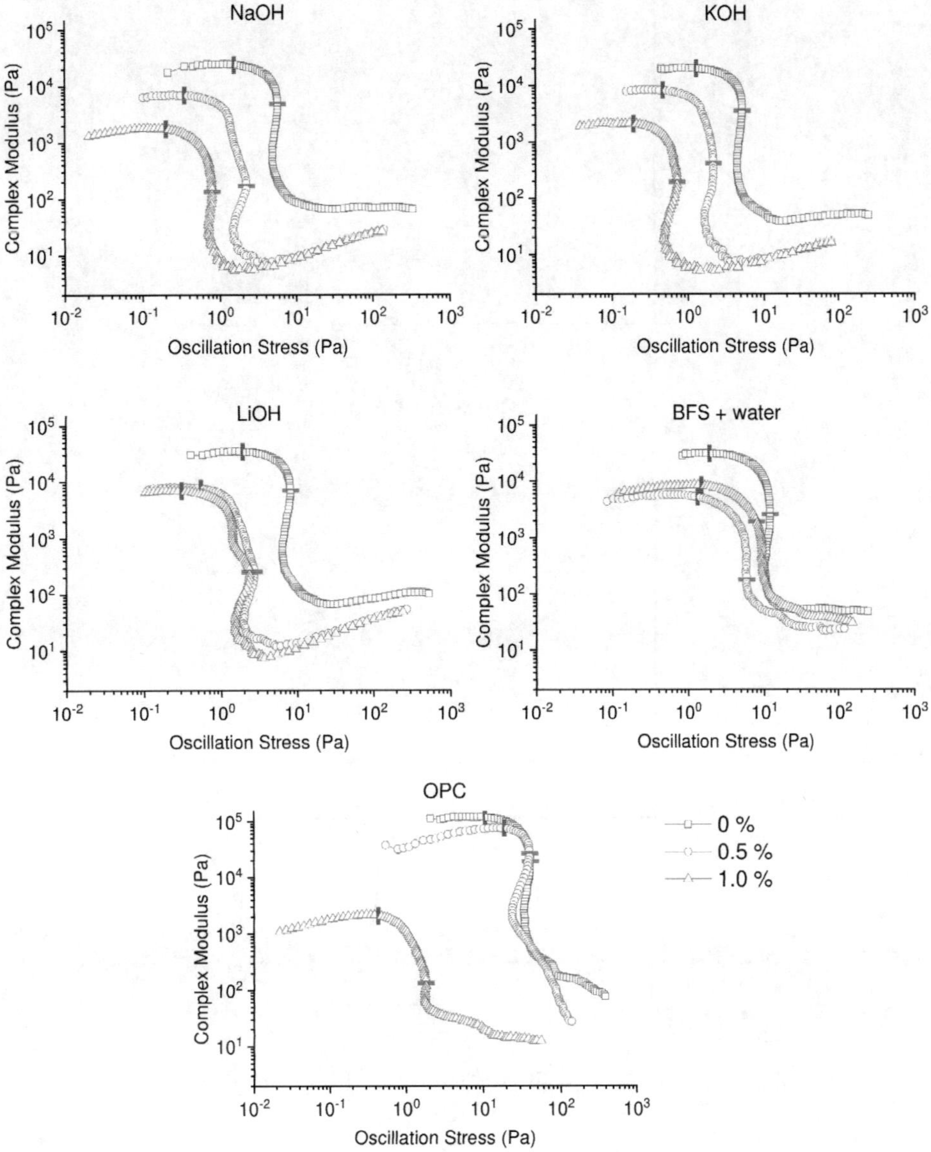

Figure 1. Complex modulus on oscillation stress of BFS activated by various activators, different dosages of LS, with the increasing strain from 5·10–4 to 500% (yield point marked by olive green colour, flow point by pink colour).

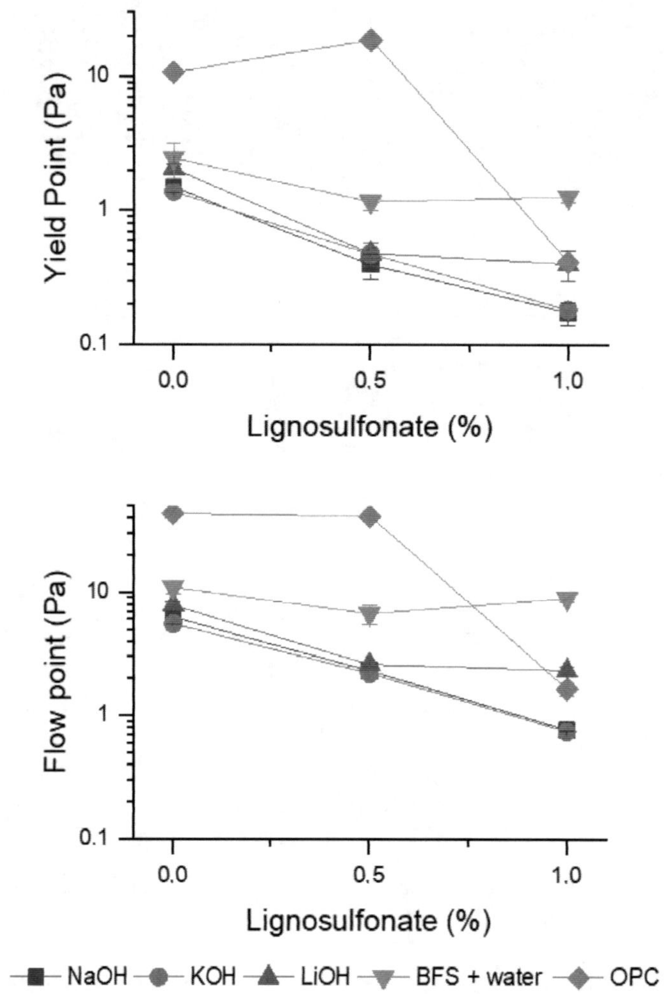

Figure 2. Yield/flow point on the dosage of LS in various pastes.

4. Conclusion

Within this paper, the effectiveness of the plasticizing effect of lignosulfonate on the type of alkali cation in alkali-activated systems was investigated from the rheological point of view. The gained results have shown that the response of alkali-activated systems depends not only on the kind of alkaline activator but on the dosage of admixture (LS) as well. Sodium and potassium hydroxide demonstrated the similar behaviour in terms of gradual weakening of the paste with increasing LS. In contrast, when the lithium hydroxide or water was used, a higher dosage than 0.5% of LS was not accompanied with improvement of workability, so the limit dosage of LS, which can be still effective, is between 0 and 0.5%. Plasticizing efficiency of LS in OPC paste was by order of magnitude higher than that in AAS, as illustrated by flow point reduction by 42 and 6 times, respectively.

These results indicate that the nature of alkali cation and paste composition in general affect the plasticizer efficiency. It can be expected that it is related to the different surface chemistry of slag and plasticizer performance in various environments, but further study is needed to shed more light on this issue.

Acknowledgement:
This outcome was obtained thanks to the contribution of the following projects: GA20-26896S "Towards the organic admixtures tailoring for alkali-activated slag-based systems" founded by the Czech science foundation, and the project FCH/FSI-J-21-7387.

References

[1] Provis J L and van Deventer S J J 2014 Alkali Activated Materials: State-of-the-Art Report RILEM TC 224-AAM (Brooklyn Publishing)

[2] Nedeljković M, Zhenming L and Guang Y 2020 Setting, Strength, and Autogenous Shrinkage of Alkali-Activated Fly Ash and Slag Pastes: Effect of Slag Content *Materials* **11** pp 1–20

[3] Hewlett P C 2004 *Lea's Chemistry of Cement and Concrete* (Butterworth-Heinemann: Elsevier Science & Technology Books)

[4] Fernández-Jiménez A et al 2020 Structure of Calcium Silicate Hydrates Formed in Alkaline-Activated Slag: Influence of the Type of Alkaline Activator *Journal of the American Ceramic Society* **86** pp 1389–1394

[5] Li L, Lu J-X, Zhang B and Poon C-S 2020 Rheology behavior of one-part alkali activated slag/glass powder (AASG) pastes *Construction and Building Materials* **258** pp 1–13

[6] Puertas F, Varga C and Alonso M M 2014 Rheology of alkali-activated slag pastes. Effect of the nature and concentration of the activating solution *Cement and Concrete Composites* **53** pp 279–288

[7] Lu C, Zhang Z, Shi C, Li N, Jiao D and Yuan Q 2021 Rheology of alkali-activated materials: A review. *Cement and Concrete Composites* **121** pp 40–48

[8] Palacios M, Banfill P F G and Puertas F 2008. Rheology and setting of alkali-activated slag pastes and mortars: Effect of organic admixture *ACI Materials Journal* **105(2)** pp 140–148

[9] Luukkonen T, Abdollahnejad Z, Ohenoja K, Kinnunen P and Illikainen M 2019 Suitability of commercial superplasticizers for one-part alkali-activated blast-furnace slag mortar *Journal of Sustainable Cement-Based Materials* **8** pp 244–257

[10] Bilek V Jr, Iliushchenko V, Hruby P and Kalina L Effectiveness of common water-reducing admixtures in alkali-activated slag pastes with different types of activator (article submitted to same volume of IOP Conf. Ser.: Mater. Sci. Eng.)

[11] Mezger T G 2014 *Applied Rheology: With Joe Flow on Rheology Road* (Anton Paar GmbH)

[12] Vance K, Dakhane A, Sant G and Neithalath 2014 Observations on the rheological response of alkali activated fly ash suspensions: the role of activator type and concentration *Rheologica Acta* **53** pp 843–855

[13] Kashani A, Provis J L, Qiao G G and van Deventer J S J 2014 The interrelationship between surface chemistry and rheology in alkali activated slag paste *Construction and Building Materials* **65** pp 583–591

[14] Poulesquen A, Frizon F and Lambertin D 2011 Rheological behavior of alkali-activated metakaolin during geopolymerization *Journal of Non-Crystalline Solids* **357** pp 3565–3571

[15] Ngele E and Schneider U 1989 The zeta-potential of blast furnace slag and fly ash, *Cement Concr. Res.* **19** pp 811–820

IOP Conf. Series: Materials Science and Engineering 1205 (2021) 012017 doi:10.1088/1757-899X/1205/1/012017

Reuse of waste tire textile fibers from tires in plaster mixtures

I Khongova[1], I Chromkova[1] and V Prachar[1]

[1] Research Institute for Building Materials, Hněvkovského 30/65, 617 00 Brno, Czech Republic

Email: khongova@vustah.cz

Abstract. The waste tire textile fibers from tires are generated as a by-product and are classified as special waste. To avoid landfilling and incineration of this waste, it is necessary to focus on its further use. This article describes an experimental study of the use of waste tire textile fibers for plaster mixtures. Waste fibers were added in different amounts of 2.9%, 3.6%, and 4.3% related to the weight of cement, the effect of fibers on selected properties was monitored. The results of the experimental work show that the lower addition of waste fibers (2.9% and 3.6%) had a predominantly positive effect on mechanical properties. By exceeding of the optimal dose of fibers, the tested properties of the plasters got worse.

1. Introduction

The increasing population is also associated with the increasing need for vehicles thus producing tires. Tires are an essential part of the economy of every country; about 17 million tons of end-of-life tires (ELTs) are produced worldwide each year. ELTs are solid wastes generated in large amounts, and the decomposition of these materials takes approximately 600 years [1–3].

Generally, there are four ways to handle ELT: recycling (3–15)%, reuse (5–23)%, landfilling (20–30)%, and energy recovery (25–60)% [4]. Up to 60% of ELTs are used for energy recovery. The calorific value of ELT is around 26 MJ/kg, which is very similar to black coal. However, burning of ETLs is not environmentally friendly, it can emit hazardous gases that can pollute the environment. Burying ELTs in landfills or sending them to landfills is also very unecological as landfills take up a lot of space that could be used in other ways [5, 6].

The material composition of tires varies by category, i. e. passenger cars or trucks. All tires contain natural and synthetic rubbers, reinforcing materials (steel wires and textile fibers), and facilitators. Tire life can be extended by retreading, but this process cannot be repeated all the time. The ELTs are processed by cryogenic grinding or shredding in dedicated mills. The output of the process is rubber material of various sizes (cuts, shred, chips, granulate, powders, and fine powders), steel reinforcement (5–30)%, and textile fibers (10%) [7, 8].

Rubber and steel are currently widely used in construction. Steel wires are being separated and used by factories to produce new steel [1]. Musci et al [9] used steel wires from waste tires as reinforcement for geopolymers. Another application of steel was investigated in the study [10], the mechanical properties of concrete with steel fibers from tire waste were tested. The reuse of steel fibers from tires as reinforcement in concrete is the most common topic in experimental studies. In civil engineering, waste rubber is used to improve the properties of concrete (increase the durability of concrete) as an additive or substitute for sand in concrete and asphalt. Rubber chips can be used to produce athletic tracks or as flexible coverage options in playgrounds and sporting surfaces [11]. Another less popular product obtained from the recycling process of ELTs is waste tire textile fibers (WTTFs).

WTTFs represent about 10% by weight of ELTs and are classified as special wastes for disposal. Usually, WTTFs are landfilled, which may pose a risk in the future, such as the accumulation of rodents, soil and underground water pollution, and the potential risk of a wildfire [7, 12]. Some authors have already dealt with the issue of WTTFs reuse, in the study [13], the influence of textile fibers on the mechanical and thermal properties of concrete was determined. Many studies [1, 5, 12] have investigated the application of fibers as a reinforcing material to strengthen sandy and clayey soils. Marconi et al [14] focused on the reuse of WTTF in plastics compounds. In a recently published article [15], WTTFs are used to produce a rubber aerogel that has super-hydrophobic properties and is suitable for oil spill-cleaning applications.

This paper aims to verify another possible application of WTTFs in plaster mixtures. The influence of the addition of WTTFs on the physical and mechanical properties of plaster and assessment of the suitability of the use of waste material in plaster mixtures are studied.

2. Experimental part

2.1. Materials and methods

The materials used to produce plaster mixtures were Portland cement 42.5 R, lime hydrate Cemix CL 90-S, quartz sand ST 56 Střeleč, Experlit EP 150 OM, Walocel MKX 15000 PP 25 (modified hydroxyethyl methylcellulose improves workability and water retention), VINNAPAS 5010 N (polymer powder increases adhesion and flexibility), water and waste tire textile fibers. A sample of WTTFs was supplied by RPG-Recycling, which deals with tire recycling and the sale of outputs from the recycling process (steel fibers, rubber granulate, and WTTF). The obtained waste material was adjusted to the desired form before use in the plaster mixture because the WTTFs were very inhomogeneous and contained residues of rubber and steel fibers. For application to the plaster mixture, the WTTFs were cleaned, as shown in figure 1.

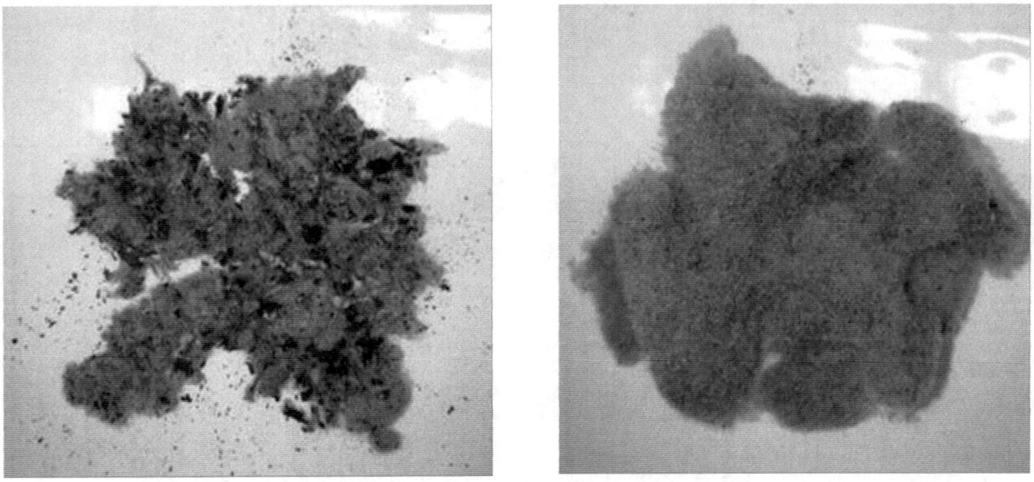

Figure 1. WTTFs before cleaning (left) and clean WTTFs (right).

IOP Conf. Series: Materials Science and Engineering 1205 (2021) 012017 doi:10.1088/1757-899X/1205/1/012017

WTTFs tend to exist in different lengths and widths (figure 2) and are composed of synthetic fibers, predominantly nylon. Fiber width was determined from 150 measurements from images taken with an Eclipse LV100ND optical microscope. Figure 3a shows the range of fiber widths and their expression in percent. The range of fiber widths from 10 to 31 μm was determined from the obtained data, the average width of WTTFs was approximately 20 μm. The length of the fibers was made by hand extended and measured. The measured values of the fiber length are shown in figure 3b. The length of the fibers ranged from 3 to 20 mm, the average length of the fibers was approximately 8 mm.

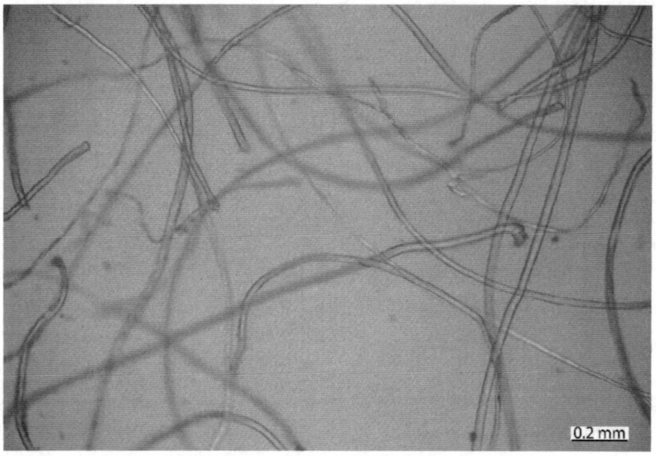

Figure 2. Detail of WTTFs.

a)

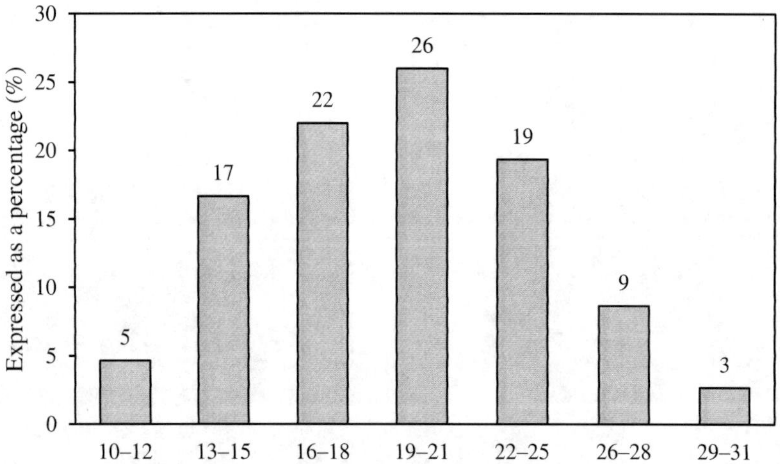

b)

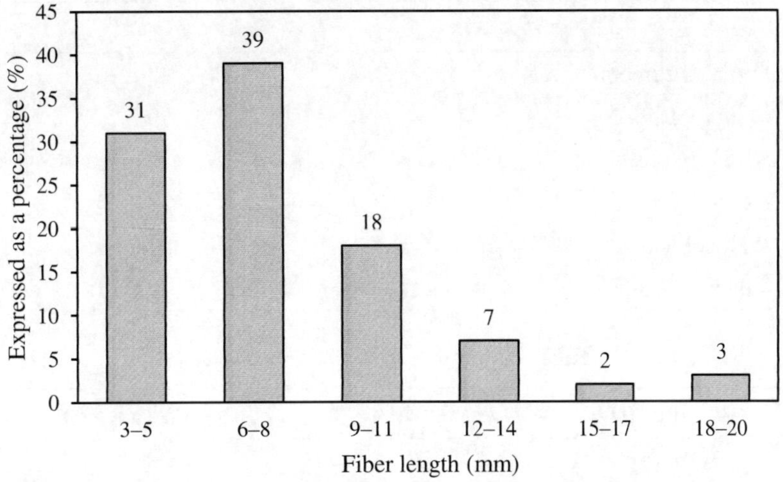

Figure 3. Fiber size.

The following properties were monitored for the proposed plasters. The consistency of fresh plaster was determined by flow table (ČSN EN 1015-3). After 28 days, the plasters were tested for compressive strength, flexural strength (ČSN EN 1015-11), and adhesive strength on substrates (ČSN EN 1015-12). Another test method was chosen to determine the coefficient of water absorption due to the capillary action of hardened plasters (ČSN EN 1015-18), the test specimens were covered with a sealing material, broken into two parts, and dried to constant weight before testing.

Determination of the water vapor permeability of hardened plasters (ČSN EN 1015-19) was tested on circular specimens that were attached to a circular vessel with solution. Water vapor pressure was created using a saturated LiCl solution. The test specimens were tested in the Feutron climate chamber, where the required humidity and temperature were determined.

2.2. Sample preparation

A total of 4 test plaster mixtures were prepared, one reference and three with different addition of WTTFs. The composition of the basic mixture is shown in table 1, the amounts of water and fibers are listed in table 2. The fibers were dosed by weight. The amount of water was adjusted according to the required consistency. The waste fibers were dosed as the last component of the plaster, the mixing and homogenization of the fibers with the plaster mixture took place in a forced circulation mixer. The chosen dosing method did not allow a higher addition of fibers because there was nonuniform dispersion and agglomeration in the mixture. A suitable application for higher fiber dosing would be dispersion in water. The test specimens were stored for 7 days in controlled environment with relative humidity of (95 ± 5)% and temperature of (20 ± 2)°C, for another 21 days, the samples were in relative humidity of (65 ± 5)% and a temperature of (20 ± 2)°C.

Table 1. Composition of compounds.

Dry component	OM ref (%)
Portland cement 42.5 R	26.9
Cemix CL 90-S	9.2
ST 56 Střeleč	57.5
Experlit EP 150 OM	5.4
Walocel MKX 15000 PP 25	0.2
VINNAPAS 5010 N	0.8

Table 2. Amount of water and fibers.

Mixture ID	w (%)	WTTF (%)[a]
OM ref	1.57	-
OM P1	1.61	2.9
OM P2	1.62	3.6
OM P3	1.64	4.3

[a]the amount of fibers is related to the weight of the cement

3. Results and discussion

3.1. Workability

The addition of WTTF was expected to affect the mixture's workability, therefore, the plaster mixture's consistency was adjusted to the required plaster flow diameter (175 ± 10) mm. It is clearly illustrated that the increasing addition of WTTFs increases the water-cement ratio (w/c). The consistency is shown in figure 4, where the values are measured immediately after mixing the plaster mixture and then after 90 minutes. Plaster mixtures with the highest addition of waste fibers met the requirement initial consistency even after 90 minutes.

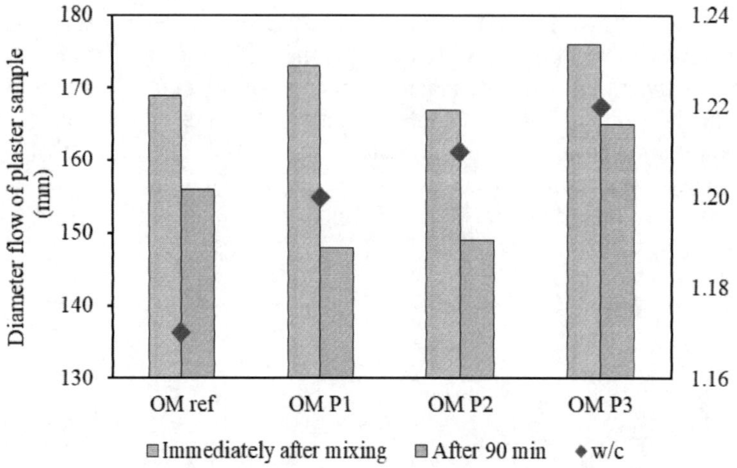

Figure 4. Results of a consistency determination.

3.2. Compressive and flexural strength

The compressive strength and flexural strength of the plaster mixtures were tested after 28 days, the results of strength are depicted in figure 5. The flexural strength of the reference sample (OM ref) was 1.6 N/mm^2. According to the results, it is clear that a certain amount of fibers positively affects flexural strength. The highest flexural strength was 2.0 N/mm^2 (OM P2), further increase of the fibers already resulted in a decrease of flexural strengths.

A similar trend can be observed when monitoring compressive strength. The highest compressive strength of 5.9 N/mm^2 was recorded for the OM P1 sample, which contained the lowest fiber addition. The OM P2 mixture also had higher compressive strength than the reference sample. On the other hand, the lowest value of compressive strength was 4.6 N/mm^2 for OM P3, which contained the highest fiber content. From the results, it can be concluded that a low dose of fibers improves strength; after exceeding the optimal dose, the effect is the opposite.

In the study [16], the same decreases in strengths were observed. Compressive and flexural strengths also decreased with increasing fiber addition. The optimal ratio between the matrix and the reinforcement is achieved with a lower addition of fibers, the matrix coats the fibers well and transfers the load. Higher addition of fibers disrupts the optimal ratio and thus leads to poor load transfer and lower resulting strengths.

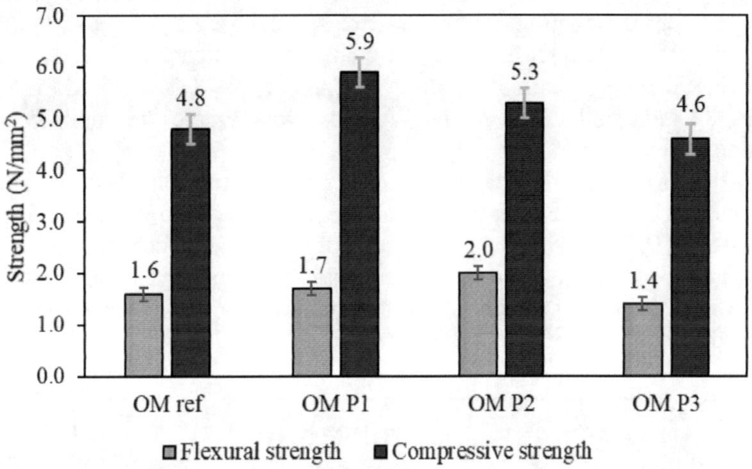

Figure 5. Development of compressive strength and flexural strength after 28 days.

3.3. Adhesive strength on the substrate

The results of testing the adhesion to the substrate are shown in table 3. A concrete slab was used as a substrate for testing the plaster, on which a layer of plaster was applied. During testing, the adhesive strength and type of rupture are recorded. The adhesive strengths were very similar, the highest adhesive strength was 1.05 N/mm^2 in the plaster with the highest addition of WTTFs (OM P3), the standard deviation of all measurements was 0.10. Another monitored parameter was the type of rupture. The results show that all the proposed mixtures were damaged in the plaster so that the adhesion of the plaster is in fact higher than the measured values. The specific value of material adhesion would be determined if there were damage between the plaster and the substrate. However, based on the obtained results, it is clear that the addition of fibers does not reduce the adhesion to the substrate. The adhesion of the plaster can also be affected by the selected type of substrate, an important factor is especially the surface roughness and porosity of the material [17].

Table 3. Pull-off test.

Mixture ID	Adhesive strength on the substrate (N/mm^2)	Type of rupture
OM ref	1.01	Damage in plaster
OM P1	0.90	Damage in plaster
OM P2	0.92	Damage in plaster
OM P3	1.05	Damage in plaster

3.4. Water absorption coefficient

The capillary water absorption coefficient is an important parameter that characterizes the water absorption of porous materials and is necessary for moisture transfer simulations. The results of capillary absorption are illustrated in figure 6, the lowest value of capillary absorption was 0.3 $kg/(m^2 \cdot min^{0,5})$ for the reference plaster sample (OM ref). Further addition of fibers increased capillary absorption, plaster mixtures OM P1 and OM P2 had a capillary absorption value the same 0.4 $kg/(m^2 \cdot min^{0.5})$. The highest capillary absorption of 0.5 $kg/(m^2 \cdot min^{0.5})$ was recorded for the mixture with the highest fiber addition.

Generally, plasters for interior and exterior use are classified in the category W_c 0 to W_c 2. The results of mixtures OM ref, OM P1, and OM P2 may be included in the category W_c 1, where the capillary water absorption coefficient is ≤ 0.40 $kg/(m^2 \cdot min^{0.5})$. Category W_c 1 is required for thermal insulation plasters. The OM P3 mixture belongs to the W_c 0 category where the capillary water absorption coefficient is not prescribed.

The same trend as for capillary absorption was for water absorption. With the increasing addition of fibers, the water absorption increased; only for OM P1, the measured value was lower.

Higher addition of WTTFs also required higher amounts of water to achieve the desired consistency. Water evaporates during the setting process, forming pores and capillary channels. Therefore, the water ratio is one of the most important factors that affect material's porosity, as stated in the study [18]. The results obtained correspond to the theory mentioned above.

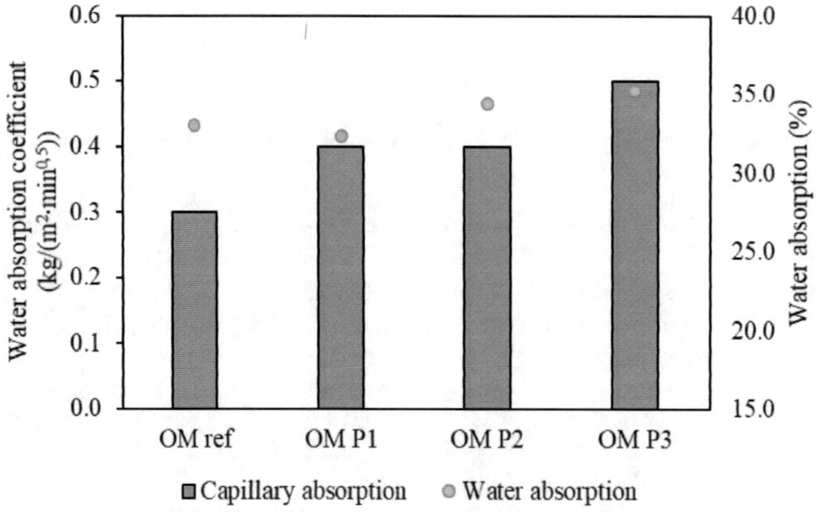

Figure 6. Absorption measurement results.

3.5. Water vapor permeability

Table 4 presents the results of the proposed plasters. For the reference plaster, the water vapor resistance factor was 12.2. The highest value of the water vapor resistance factor 12.4 was for the OM P1 plaster, which contained the lowest addition of waste fibers. Other plasters that contained a higher addition of fibers achieved lower values of water vapor resistance factor than OM P1 plaster. According to ČSN EN 998-1, the maximum value of the water vapor resistance factor of 15 is determined for renovation and thermal insulation plasters. All proposed plasters meet this requirement, the values obtained are favorable for ensuring the passage of water vapor out of the structure.

This test is used to determine the breathability of plasters. The lower the value of the diffusion resistance factor, the easier the movement of water vapor in the structure. Permeability is very much related to the porous structure of the material, and according to some studies, permeability is responsible for the long-term durability of materials [19, 20].

Table 4. The water vapor resistance factor of the studied samples.

Mixture ID	μ (-)
OM ref	12.2
OM P1	12.4
OM P2	11.6
OM P3	12.0

4. Conclusion

This experimental work aimed to verify the use of waste fibers in the proposed plaster mixtures. The morphology of the waste fibers was determined which is one of the critical factors for the material's mechanical properties. The effect of the addition of fibers on the selected properties of plaster mixtures was monitored. The lowest addition of fibers had a positive effect on the compressive strength; with a further increase in fibers, there was a slight decrease. The application of the WTTFs also had a positive effect on the flexural strength, the mixtures containing 0.6% and 0.7% of the WTTFs had higher strengths than the reference mixture. Adhesion to the substrate is one of the most important properties of plasters. No negative effect on adhesion to the substrate was observed for all proposed plasters with waste fibers. The results of capillary absorption of plaster mixtures showed a dependence on the water coefficient. Higher water consumption and WTTF dose led to the formation of pores and capillary channels, which resulted in an increased value of the capillary water absorption coefficient. According to the water vapor resistance factor, the permeability of the plasters did not deteriorate either, all the proposed plasters even met the criterion for thermal insulation plasters.

The application of waste fibers to plaster mixtures is possible, and in particular, the low addition of WTTFs improves the mechanical properties. The disadvantage of using waste fibers is the need for purification before applying the fibers and also poorer incorporation of the fibers into the mixture. Nevertheless, all experimentally produced plasters with waste fibers achieve comparable technical parameters as commercially available products.

Acknowledgments

The present study is part of long-term conceptual development of a research organization Research Institute of Building Materials 2018-2022.

References

[1] Abbaspour M, Aflaki E and Nejad F M 2019 Reuse of waste tire textile fibers as soil reinforcement *Journal of Cleaner Production* **207** pp 1059–1071

[2] Mucsi G, Szenczi A and Nagy S 2018 Fiber reinforced geopolymer from synergetic utilization of fly ash and waste tire *Journal of Cleaner Production* **178** pp 429–440

[3] Júnior A B, Battistelle R A, Bezerra B S and Castro R 2012 Use of scrap tire rubber in place of SBS in modified asphalt as an environmentally correct alternative for Brazil *Journal of Cleaner Production* **33** pp 236–238

[4] Forrest M 2014 Recycling and Reuse of Waste Rubber

[5] Narani S S, Abbaspour M, Mir Mohammad Hosseini S M, Aflaki E and Moghadas Nejad F 2020 Sustainable reuse of Waste Tire Textile Fibers (WTTFs) as reinforcement materials for expansive soils: With a special focus on landfill liners/covers *Journal of Cleaner Production* **247**

[6] Gheni A A, Alghazali H H, ElGawady M A, Myers J J and Feys D 2019 Durability properties of cleaner cement mortar with by-products of tire recycling *Journal of Cleaner Production* **213** pp 1135–1146

[7] Shulman V L 2011 Chapter 21 - Tyre Recycling *Waste* Academic Press pp 297–320

[8] Landi D, Marconi M, Meo I and Germani M 2018 Reuse scenarios of tires textile fibers: an environmental evaluation *Procedia Manufacturing* **21** pp 329–336

[9] Mucsi G, Szenczi A and Nagy S 2018 Fiber reinforced geopolymer from synergetic utilization of fly ash and waste tire *Journal of Cleaner Production* **178** pp 429–440

[10] Samindi S M, Samarakoon M K, Ruben P, Pedersen J and Evangelista L 2019 Mechanical performance of concrete made of steel fibers from tire waste *Case Studies in Construction Materials* **11** e00259

[11] Mohajerani A, Burnett L, Smith J V, Markovski S, Rodwell G, Rahman M T, Kurmus H, Mirzababaei M, Arulrajah A, Horpibulsuk S and Maghool F 2020 Recycling waste rubber tyres in construction materials and associated environmental considerations: A review *Resources, Conservation and Recycling* **155** 104679

[12] Yadav J S and Tiwari S K 2017 Effect of waste rubber fibres on the geotechnical properties of clay stabilized with cement *Applied Clay Science* **149** pp 97–110

[13] Medina N F, Medina D F, Hernández-Olivares F and Navacerrada M A 2017 Mechanical and thermal properties of concrete incorporating rubber and fibres from tyre recycling *Construction and Building Materials* **144** pp 563–573

[14] Marconi M, Landi D, Meo I and Germani M 2018 Reuse of Tires Textile Fibers in Plastic Compounds: Is this Scenario Environmentally Sustainable? *Procedia CIRP* **69** pp 944–949

[15] Thai Q B, Duyen K L, Nga H N D, Phung K L, Nhan P-T, Chien Y W and Hai M D 2020 Advanced aerogels from waste tire fibers for oil spill-cleaning applications *Journal of Environmental Chemical Engineering* **8** 104016

[16] Parres F, Crespo-Amorós J E and Nadal-Gisbert A 2009 Mechanical properties analysis of plaster reinforced with fiber and microfiber obtained from shredded tires *Construction and Building Materials* **23** pp 3182–3188

[17] Faria P, Lima J, Nabais J and Silva V 2019 Assessment od adhesive strength of an eart plaster on different substrates through different methods *5th Historic Mortars Conference* ed J. I Alvarez, J. M Fernandez, I Navarro, A Duran and R Sirera (Pamplona: Universidad de Naverra) pp 51–64

[18] Lopez-Zaldivar O, Lozano-Diez R, Herrero del Cura S, Mayor-Lobo P and Hernandez-Olivares F 2017 Effects of water absorption on the microstructure of plaster with end-of-life tire rubber mortars *Construction and Building Materials* **150** pp 558–567

[19] Casnedi L, Cappai M, Cincotti A , Delogu F and Pia G 2020 Porosity effects on water vapour permeability in earthen materials: experimental evidence and modelling description. *J. Build. Eng.* **27** 100987

[20] Westgate P, Paine K and Richard J B 2018 Physical and mechanical properties of plasters incorporating aerogel granules and polypropylene monofilament fibres *Construction and Building Materials* **158** pp 472–480

IOP Conf. Series: Materials Science and Engineering 1205 (2021) 012018 doi:10.1088/1757-899X/1205/1/012018

Decomposition of asbestos materials using fluoride wastes

K Necasova[1], P Buchta[1], I Chromkova[1], T Stanek[1] and T Simbera[1]

[1] Research Institute for Building Materials, Hnevkovskeho 65, 617 00 Brno, Czech Republic

Email: necasova@vustah.cz

Abstract. Asbestos is one of the materials causing ecological stress. Due to its health harmfulness, an effective, ecological, and economic decomposition is highly desirable. One of the decomposition possibilities is a chemical decomposition, which could compete with commonly used thermal decomposition. The chemical decomposition can be accomplished both with the use of pure chemicals and waste chemicals from production technologies. This work deals with the use of technological wastes containing hydrofluoric acid or fluorides. Fluorides release hydrofluoric acid in the acid medium, which acts as the main decomposition medium. The source of fluorides was waste from the glass and metallic material industry. The efficiency of degradation processes was studied by mass analysis. Materials and decomposition products were characterized by X-ray powder diffraction.

1. Introduction

Asbestos belongs to natural silicate minerals with a fibrous structure. It can be found in two mineral subgroups in nature – amphiboles and serpentines. The serpentinites are the largest asbestos group, where it is mainly found as chrysotile, which occurs in approximately 95% of industrial products. Materials with asbestos fibres were massively used thanks to their thermal and chemical resistance. These materials found use mainly in civil engineering, where they were used for the production of gaskets, ropes, pipes, boards, and roofing [1, 2].

The positive attributes of these materials are accompanied by big drawbacks in the form of negative influences on living beings and carcinogenicity. Asbestos is rated as a first-class carcinogen in the present day. Asbestos fibres are the main danger that can cause serious illnesses when inhaled. For this reason, the usage of these materials is forbidden. Manipulation and disposal of materials with asbestos contents is governed by "The law about chemical compounds and mixtures", Act no. 350/2011 Sb. [3, 4].

A large amount of waste with these harmful fibres causes ecological stress. Effective disposal or at least conversion of asbestos fibres to harmless and useful materials is the aim of many researchers. Various disposal methods for these materials are already in use.

One of these methods is the thermal disposal of asbestos fibres, which is done in high temperatures (approx. 1000°C). Thanks to the high temperature, the bound water is liberated, and asbestos forms non-fibrous structures [5, 6, 7]. Another quick and effective method is microwave decomposition, where thanks to the higher temperature generated inside the material, the fibrous structures degrade into harmless and further useful magnesium trisilicate [8]. High energy grinding is one of the possible mechanical disposals [9].

Significant opportunities lie in the chemical disposal of these materials. Alkaline, organic, and inorganic acids or waste solutions from various technologies can be used as decomposition mediums.

IOP Conf. Series: Materials Science and Engineering 1205 (2021) 012018 doi:10.1088/1757-899X/1205/1/012018

In research papers of foreign authors, mixtures of sulphur and calcium hydroxide in combination with ball grinder were used [10]. Solutions of phosphoric acid [11], oxalic acid [12], or mixtures of HCl and HF from warm freon decomposition were also used [13]. Some other acids, such as acetic, hydrochloric, sulphuric, or hydrofluoric acids, can be used as decomposition mediums. The use of these acids as decomposition medium has been studied in the work Necasova K. et al. [14]

Hydrofluoric acid and its salts are very prospective mediums for the decomposition of asbestos materials because the acid very readily reacts with the silicon in the asbestos structure and thus disrupts the fibrous nature of this material [15]. Waste with these acid precursors can be used for these applications. Some of the sources of this waste are glass, fluoride pesticides or metallic materials industry [16]. The city and industrial wastewater which are contaminated by this compound can also be the sources. According to the research of Jiang K. et al [17], we can get 70% pure calcium fluoride in the form of sand from this water by multistep crystallization. Fluorides have a wide range of uses. They are commonly used in the manufactory of grease, colours, ceramics, and polymers. For example, sodium fluoride is used for drinking water treatment. Other applications of NaF may include glass and porcelain enamel production or glue, insecticide, and timber protective varnish manufacture. It has its place in metallurgy, steel, and aluminium manufacture, where it is used as a fluxing agent. Fluorides are further used in the cosmetic and pharmaceutic industry. They are found in dental paste and mouth wash as protection against dental cavities. It is also added to drugs combating cancer or skin conditions. Fluorides in small quantities are essential for humans. In higher quantities, they can cause serious illnesses, for example fluorosis. Even the hydrofluoric acid has a range of uses, mainly in the glassmaking industry, in aluminum fluoride making, as an additive for swimming pools, freons, and chemical syntheses [18]. Hydrofluoric acid and fluorides are generally marked as dangerous substances, except for water dissolved calcium fluoride [19].

2. Experimental

2.1. Materials and methods

We used samples of four different types of waste for this series of asbestos decomposition experiments: asbestos rope, wool, disc and roofing (figure 1). These samples were initially used as a gasket material and thermal insulation. The roofing sample comes from the roof of a holiday cottage.

A) B)

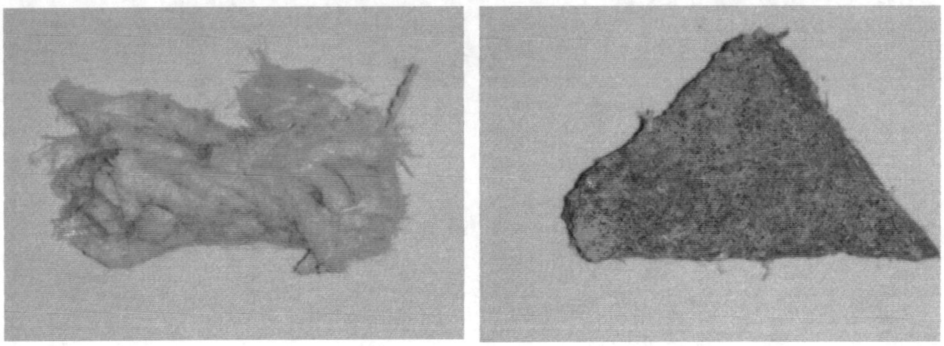

IOP Conf. Series: Materials Science and Engineering 1205 (2021) 012018 doi:10.1088/1757-899X/1205/1/012018

C) D)

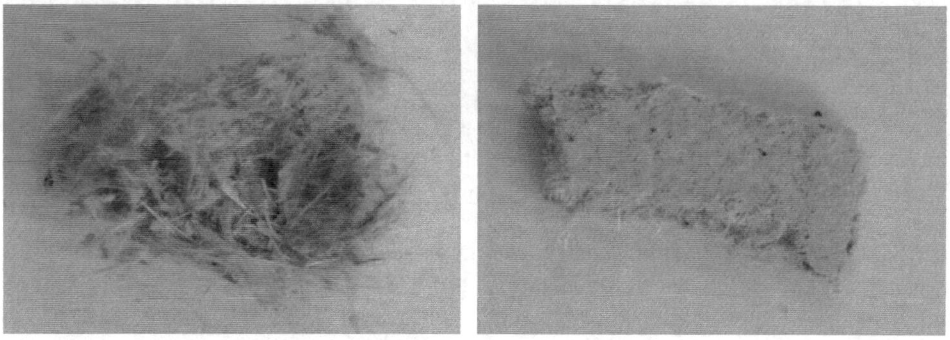

Figure 1. A) Asbestos rope sample; B) Asbestos roofing sample; C) Asbestos wool sample; D) Asbestos disc sample.

All materials were characterized by X-ray powder diffraction analysis (XRD) which was carried out on Bruker D8 Advance apparatus with Cu anode ($\lambda K\alpha$ = 1.54184 Å) and variable divergence slits at Θ-Θ Bragg-Brentano reflective geometry. For the measurement, we used original asbestos rope and asbestos disc samples. Samples of asbestos roofing and wool were dissolved in 35% hydrochloric acid. All samples were analyzed on zero background sample holder in an alcohol medium. The measurement results of asbestos rope, roofing, wool and disc are shown in figure. 2.

A) B)

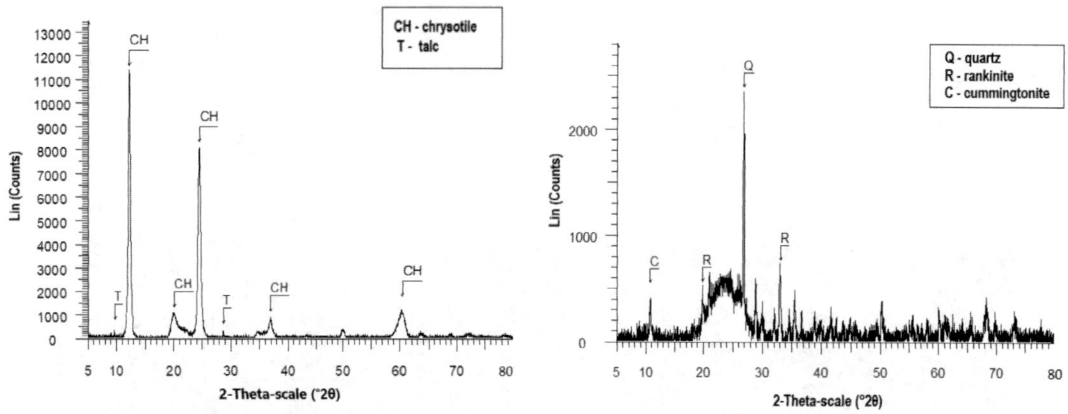

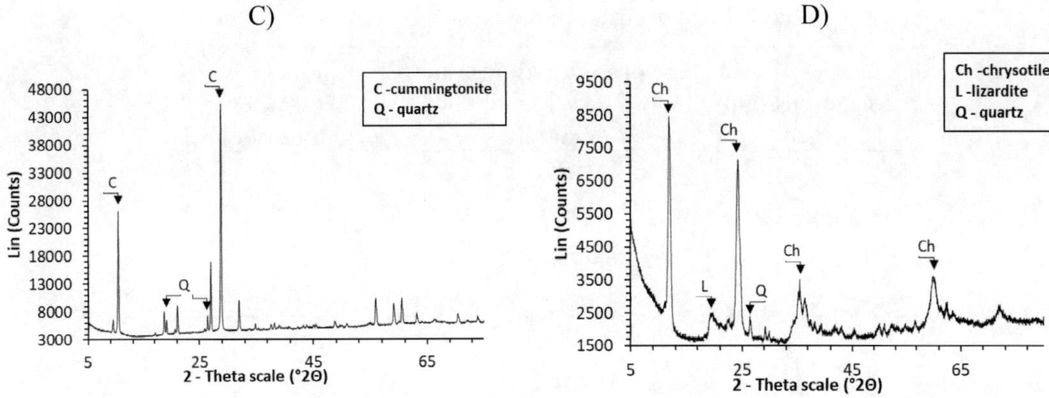

Figure 2. X-ray powder diffraction analysis of asbestos A) rope; B) roofing; C) wool; D) disc.

The results of asbestos rope analysis show the material is composed of chrysotile and talc. Chrysotile is an asbestos material ($Mg_3Si_2O_5(OH)_4$), and talc is a hydrated magnesium silicate ($Mg_3Si_4O_{10}(OH)_2$). Asbestos wool is composed of minerals cummingtonite and quartz. The asbestos disc is composed of minerals chrysotile, lizardite and other identified mineral in the sample is quartz. According to the results, the asbestos roofing is composed of cummingtonite, quartz and rankinite. Cummingtonite is a material consisting of magnesium silicate hydroxide ($Mg_7Si_8O_{22}(OH)_2$), and quartz is silicon dioxide (SiO_2). Rankinite is a calcium silicate ($Ca_3Si_2O_7$), and lizardite is a hydroxide magnesium silicate $Mg_3(Si_2O_5)(OH)_4$.

The decomposition experiments were carried out at laboratory temperature (approx. 22°C) in one month. The decomposition was done in plastic closable containers, in which we placed a precisely weighted sample and a decomposing agent. The first used decomposing agent was a waste etching solution (ES) from the glassmaking industry, which contains residues of hydrofluoric acid and its salts. The second decomposing agent was a degreasing solution (DS), used for degreasing and cleaning of components before surfacing, mainly plating. We can also find residues of hydrofluoric acid and its salts in this solution. To increase the efficiency, small amounts of inorganic acids were added to the decomposing agents: sulphuric, phosphoric, hydrochloric and nitric acid. The decomposing agent was added to the sample in 1:7 ratio. The supportive inorganic acids were added in 1:5 ratio to the decomposing agent. Decomposition products were characterized by X-ray powder diffraction analysis too. The presence of asbestos minerals in the decomposition product was verified.

3. Results and discussion

3.1. Decomposition in the etching solution and its mixtures

Results of asbestos material decomposition in an etching solution (ES) and mixtures of ES and inorganic acids are shown in table 1 and discussed in the following chapters.

Table 1. Mass decrease of the decomposed material.

Decomposition Agent	Laboratory temperature Sample	Mass after decomposition[%]
Etching solution ES	*Rope*	17,7
	Asbestos wool	*-35,3*
	Asbestos disc	*-0,4*
	Roof tile	*-0,7*
Sulfuric acid ES + H_2SO_4	*Rope*	*79,4*
	Asbestos wool	*95,0*
	Asbestos disc	*100,0*
	Roof tile	*-49,7*
Phosphoric acid ES + H_3PO_4	*Rope*	*19,4*
	Asbestos wool	*74,3*
	Asbestos disc	*53,2*
	Roof tile	*-3,2*
Hydrochloric acid ES + HCl	*Rope*	*73,3*
	Asbestos wool	*92,0*
	Asbestos disc	*75,0*
	Roof tile	*48,5*
Nitric acid ES + HNO_3	*Rope*	*80,8*
	Asbestos wool	*95,0*
	Asbestos disc	*82,2*
	Roof tile	*86,7*

The ES disrupted the rope structure, released the end fibres and decomposed a small amount of asbestos fibres. The reaction of a decomposing agent with the asbestos wool caused most of the fibres to decompose and it created grey solid matter with thin translucent layer (calcium sulphate – $CaSO_4{\cdot}2H_2O$ and quartz SiO_2 by results XRD). The asbestos disc sample shows a disruption of the surface and an overall increase of porosity with a fibre release. The asbestos roofing sample shows only a change in colour. The results of the decrease of the mass of the samples are shown in table 1.

The decomposition mixture of ES with sulfuric acid disrupted the woven structure of the asbestos rope and partially decomposed the end of the fibres. The mass loss was 79.4%. (figure 3) The asbestos wool shows a nearly complete decomposition of the fibres confirmed by the mass decrease of 95%. The disc shows a complete decomposition of the outer shell and partial disruption of the fibres (mass decrease of 100%). The asbestos roofing shows disruption and swelling of the whole structure and creation of white solid phase. The solid phase was gypsum – $CaSO_4{\cdot}2H_2O$, calcium sulfide anhydrite - $CaSO_4$ and quartz SiO_2 by results XRD, and it distorts the results of the mass analysis (see table 1).

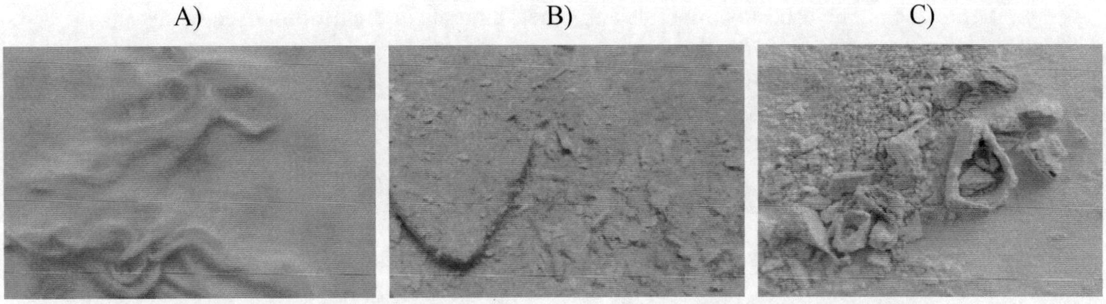

Figure 3. Decomposed sample of asbestos A) rope; B) disc; C) roofing.

The decomposition mixture of ES with phosphoric acid caused a complete disruption of the woven structure, thinning and a decomposition of a part of the asbestos fibres (mass decrease of 19.4%). One-tenth of fibres in the asbestos wool withstood the decomposition agent, mass decrease was 74.3%. The mixture caused the asbestos disc to change colour, a strong disruption of the surface and a mass decrease of 53.2% were also observed (figure 3). No longer fibres were seen in the sample. The asbestos roofing shows disruption of surface and crystallisation of white solid phase. The solid phase was probably a mix of gypsum – $CaSO_4{\cdot}2H_2O$, clinochlore – $Mg_5Al(AlSiO_3O_{10})(OH)_8$ and quartz – SiO_2 by results XRD.

The decomposition mixture of ES with hydrochloric acid had a strong degradation effect on all of the tested samples. It completely disrupted the woven structure and thinned the fibres of the asbestos rope sample. The mass decrease was 73.3%. In the case of the asbestos wool, it almost completely decomposed all the fibres (mass decrease of 92%). It decomposed the filling of the asbestos disc, thinned the fibres and decomposed small amount of them, the mass decrease was 75%. The whole structure of the asbestos roofing was disrupted by the agent and there are no visible fibres in the rest (mass decrease of 48.5%) (figure 3).

The decomposition mixture of ES with nitric acid caused thinning and release of the fibres in the woven structure of the asbestos rope (mass decrease of 80.8%). In case of the asbestos wool, it caused a complete decomposition of the fibres, which is confirmed by 95% decrease in the mass of the sample. The sample of asbestos disc showed a change of colour to grey, complete disruption of surface with a small change in overall size (mass decrease of 82.2%). The agent caused a complete disintegration of the asbestos roofing sample (mass decrease of 86.7%). The remaining solids contained no fibres.

Samples were characterized by X-ray powder diffraction analysis to evaluate the hazardous properties of the decomposed material residues. Some samples could not be measured due to a very small amount of a representative sample. The presence of asbestos minerals in the samples was recorded in table 2.

Table 2. Contain asbestos minerals and other minerals in the decomposed material.

Decomposition agent	Sample	Other minerals	Asbestos mineral
Etching solution ES	*Rope*	*orthoclase*	*no*
	Wool	*jakobsite, Fe- compound*	*no*
	Disc	*ankerite, NH_4MgAlF_6*	*no*
	Roof tile	-	-
Sulfuric acid ES + H_2SO_4	*Rope*	-	-
	Wool	-	-
	Disc	*quartz*	*no*
	Roof tile	*gypsum, rankinite, orthoclase, quartz*	*no*
Phosphoric acid ES + H_3PO_4	*Rope*	*bayerite, orthoklase*	*no*
	Wool	-	-
	Disc	*quartz, ankerite, NH_4MgAlF_6*	*lizardite*
	Roof tile	*quartz, sylvite, magnetite*	*no*
Hydrochloric acid ES + HCl	*Rope*	*no*	*chrysotile*
	Wool	-	-
	Disc	*quartz, ankerite, NH_4MgAlF_6*	*no*
	Roof tile	-	-
Nitric acid ES + HNO_3	*Rope*	*no*	*chrysotile*
	Wool	*quartz*	*no*
	Disc	*illite, muskovite, quartz*	*no*
	Roof tile	*meniaylovite, quartz*	*no*

In table 2, we can see that only 3 of 13 samples contained any asbestos mineral after the decomposition. Disc in decomposition agent ES + H_3PO_4, rope in agent ES + HCl and rope in agent ES + HNO_3. The residues of these three samples remain a hazardous waste. Other samples in decomposition reagent ES (and its combinations with acids) don´t contain asbestos minerals. These decomposition residues are not dangerous and may be reusable. For example, in the building industry. Because decomposition products as an accompanying mineral most often contain quartz, other aluminosilicates and minerals (see table 2).

3.2. Decomposition in degreasing solution and its mixtures

Results of asbestos material decomposition in degreasing solution (DS) and mixture of DS and inorganic acids are presented in table 3 and discussed in following chapters.

Table 3. Mass decrease of the decomposed material.

Laboratory temperature		
Decomposition Agent	**Sample**	**Mass after decomposition [%]**
Degreasing solution DS	*Rope*	17,1
	Asbestos wool	*4,2*
	Asbestos disc	*1,5*
	Roof tile	*-0,1*
Sulfuric acid DS + H_2SO_4	*Rope*	*31,3*
	Asbestos wool	*42,0*
	Asbestos disc	*15,2*
	Roof tile	*16,4*
Phosphoric acid DS + H_3PO_4	*Rope*	*41,3*
	Asbestos wool	*62,1*
	Asbestos disc	*21,1*
	Roof tile	*-66,3*
Hydrochloric acid DS + HCl	*Rope*	*34,0*
	Asbestos wool	*14,3*
	Asbestos disc	*72,7*
	Roof tile	*49,9*
Nitric acid DS + HNO_3	*Rope*	*38,5*
	Asbestos wool	*42,6*
	Asbestos disc	*50,5*
	Roof tile	*-25,7*

Thanks to DS, we can see only a release of end fibres in the case of the asbestos rope sample. The reaction of the decomposing agent with the asbestos wool caused caking of the material and the decomposition of very small amount of fibres. After the experiment, the asbestos disc shows a disruption of the surface and release of a small amount of fibres. The sample of asbestos roofing shows a change in colour, a partial surface decomposition, and a fibre release. All of the mass decrease data are shown in table 3.

The mixture of DS and sulphuric acid caused the asbestos rope to harden, and disruption in the end fibres was observed (mass decrease of 31.3%). The asbestos wool sample was partially decomposed, which was shown by the mass reduction of 42%. But there were still some long visible fibres resisting the rest of the material, which turned tangled and dark grey. The disc sample disintegrated and formed a white solid phase with fibres. The mass decrease was only 15.2%. Concerning the asbestos roofing sample, there was discolouration and a surface disruption (mass decrease of 16.4%).

When using a mixture of DS and phosphoric acid, the asbestos rope shows partial disruption of the woven structure. Further, there was a decomposition of a small amount of fibres (mass decrease of 41.3%) and creation of fine firm matter, formed by bayerite – $Al(OH)_3$ and orthoclase – $KAlSi_3O_8$ according to XRD results. The asbestos wool shows a mass decrease of 62.1%, but the remains still contain long fibres. The asbestos disc sample was disintegrated into a grey solid matter with fragments of fibres, as shown in figure 4 (mass decrease of 21.1%). The asbestos roofing sample shows a change of colour and a disruption of structure. There was a white decomposition by-product which distorts the mass reduction results (the mass increase of 66.3%). This by-product was combination quartz – SiO_2, sylvite – KCl and magnetite – Fe_3O_4 (according to XRD analysis), which caused the increase in the weight of the tested sample.

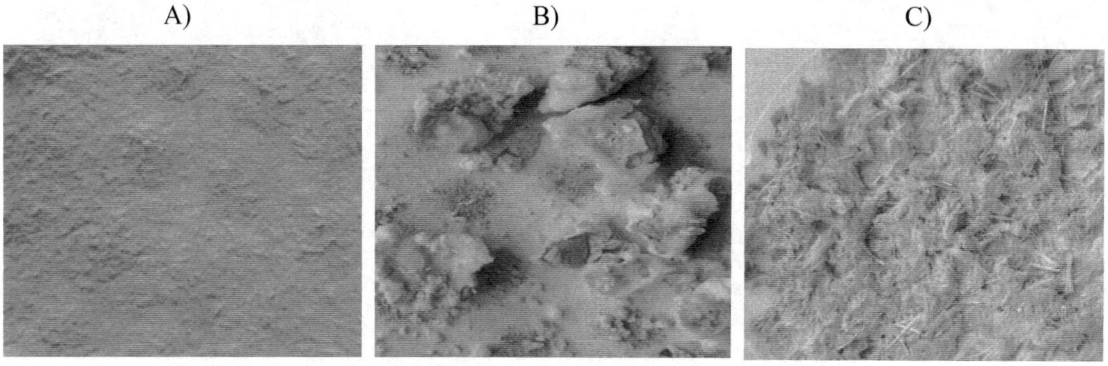

Figure 4. Decomposed sample of asbestos A) disc; B) roofing; C) wool.

The mixture of DS and hydrochloric acid freed up the end fibres of the asbestos rope and partially decomposed them, which is shown by the change of their thickness and a mass decrease of 34%. The remaining woven structure hardened. In a reaction with the asbestos wool, there was a minimal visual decrease in the amount of fibres and only a small amount of fibres were decomposed (mass decrease of 14.3%). The rest was composed of grey airy matter with fibres. The decomposing agent strongly disrupted the surface of the asbestos disc and decomposed some fibres (mass decrease of 72.7%). Concerning the sample of the asbestos roofing, there is a change in size, surface and colour. There is also a large amount of white solid matter which doesn`t contain any fibres and devalues the mass decrease result (mass decrease of 49.9%, see figure 4).

In the asbestos rope sample, there was a disruption of the surface and the woven structure. Further, there was a thinning and a release of a small amount of fibres. The mass decrease of this sample was 38.5%. The asbestos wool sample shows a mass reduction of 42.6%, but there are long visible fibres in the rest comprised of grey airy matter. (figure 4) The asbestos disc sample was disintegrated into grey solid matter with a small amount of thin fibres (mass decrease of 50.5%). There was a change of colour and a disruption of structure in the roof tile. There was a white decomposition by-product on the surface again. This by-product could be a combination meniaylovite – $Ca_4[(SO_4)(SiF_6)(AlF_6)]F \cdot 12H_2O$ and quartz – SiO_2 (by XRD results). The results of the mass analysis and mass decreases are shown in table 3.

Samples were characterized by X-ray powder diffraction analysis to evaluate the hazardous properties of the decomposed material residues too. Some samples could not be measured due to a minimal amount of a representative sample. The presence of asbestos minerals in the samples was recorded in table 4.

Table 4. Contain asbestos minerals and other minerals in the decomposed material.

Reagent	Sample	Other minerals	Asbestos mineral
Degreasing solution DS	*Rope*	*gypsum, talc*	*no*
	Wool	*gypsum, quartz*	*cummingtonite*
	Disc	*gypsum, anhydrite*	*no*
	Roof tile	*gypsum, quartz, rankinite*	*riebeckite*
Sulfuric acid DS + H_2SO_4	*Rope*	*gypsum, talc*	*chrysotile*
	Wool	*quartz*	*cummingtonite*
	Disc	-	-
	Roof tile	*gypsum, anhydrite, quartz*	*no*
Phosphoric acid DS + H_3PO_4	*Rope*	-	-
	Wool	*quartz*	*cummingtonite*
	Disc	*gypsum, talc*	*no*
	Roof tile	*gypsum, quartz, clinochlore*	*no*
Hydrochloric acid DS + HCl	*Rope*	*talc, clinochlore*	*no*
	Wool	*quartz*	*cummingtonite*
	Disc	*quartz*	*chrysotile*
	Roof tile	*gypsum, quartz, clinochlore*	*no*
Nitric acid DS + HNO_3	*Rope*	*talc*	*chrysotile*
	Wool	*quartz*	*cummingtonite*
	Disc	*quartz*	*no*
	Roof tile	*gypsum, quartz*	*riebeckite*

In table 4, we can see that 10 of 18 samples contained some asbestos minerals after the decomposition. All asbestos wool samples contain asbestos after decomposition, specifically cummingtonite. Further, the residues of decomposition of disc in agent ES + HCl, rope in decomposition agent DS + HNO_3 and agent ES + H_2SO_4. Roof tile in agent ES + HNO_3 and in agent DS samples contained asbestos.

The residues of these 10 samples remain a hazardous waste. The decomposition medium used was not sufficient to break up the structure of asbestos materials. Other samples in decomposition reagent ES (and its combinations with acids) don´t contain asbestos minerals. These decomposition residues were not dangerous and may be reusable in the building industry as decomposition products ES media. Other minerals contained in the decomposition products are listed in the table 4.

Generally, it can be said that each material needs a different decomposing agent. For the description of the decomposition rope sample process efficiency, it can be said that the surface was disrupted, and the fibres were released. Concerning the asbestos wool, it can be said that in some cases we can see complete decomposition of the fibres. The asbestos disc sample shows strong disruption of the surface and an absence of long fibres. In the case of the samples of asbestos roofing, there was a change of colour, swelling and partial disruption of the surface. In some experiments, there was a crystalline substance forming on the surface of the roofing. In the first series of experiments, the most suitable agent for the decomposition of asbestos rope is the mixture of ES and nitric acid, where the mass decrease was 80.8%. For comparison, in the second round of experiments, the highest mass decrease for the same sample was 41.3% in the case of the mixture of DS and phosphoric acid. The most suitable agent for asbestos wool was the mixture of ES with sulphuric acid and nitric acid, where both mixtures reached a mass decrease of 95%. In the second series, the highest mass decrease was noted with mixture of DS and phosphoric acid, which reached 62.1%. In the first series of experiments, the most suitable candidate for decomposition of the asbestos disc is the mixture of ES and sulphuric acid with the mass decrease of 100%. The best candidate from the second round was the mixture of DS and hydrochloric acid with the mass decrease of 72.7%. Concerning the asbestos roofing sample, the best result was a mass decrease of 86.7% with the mixture of ES and nitric acid. In the second series, the most effective mixture was the one of DS and hydrochloric acid with the mass decrease of 49.9%.

Table 5. Comparison of degradation efficiency.

Sample	Reagent	Mass after decomposition [%]
Rope	H_2SO_4	*-56.1*
	H_3PO_4	*-132.7*
	HCl	*42.4*
	$ES + H_2SO_4$	*79.4*
	$ES + H_3PO_4$	*19.4*
	ES + HCl	*73.3*
	$DS + H_2SO_4$	*31,3*
	$DS + H_3PO_4$	*41.3*
	DS + HCl	*34*
Roof tile	H_2SO_4	*-4.3*
	H_3PO_4	*12.9*
	HCl	*62.4*
	$ES + H_2SO_4$	*-49.7*
	$ES + H_3PO_4$	*-3.2*
	ES + HCl	*48.5*
	$DS + H_2SO_4$	*16.4*
	$DS + H_3PO_4$	*-66.3*
	DS + HCl	*49.9*

In an overall comparison of the decomposition results of all the materials used, ES is a more effective decomposing agent than DS. Also, the XRD results show that the decomposing agent ES is more effective in decomposing asbestos minerals than the agent DS.

The work [13] studied the decomposition of asbestos rope and asbestos roof tile by acids. For comparison with the results of this work, the critical decompositions of the two materials in sulfuric acid, phosphoric acid and hydrochloric acid. For a sample of asbestos rope, a mixture of acids and used waste solutions of ES and DS is more effective than pure acids. In the case of asbestos roof tile, on the other hand, pure acids achieve greater decomposition efficiency. Table 5 shows the comparison of results.

4. Conclusion

This work aimed to study the chemical decomposition of asbestos waste using industrial waste with the content of fluorides and residues of hydrofluoric acid in laboratory conditions. Industrial wastes used in the experiments were glassworking etching solution (ES) and metallurgic degreasing solution (DS). Sulphuric, phosphoric, hydrochloric and nitric supporting acids were used for the decomposition. The results show that the etching solution and its mixtures with acids are better decomposing agents than the degreasing solution and its mixtures. A different decomposing mixture is more suitable for each type of material. The results of the XRD analysis of the decomposed samples confirm the better degradation efficiency of the ES reagent. The research will continue with the improvement of already tested decomposition agents and with new suitable waste solutions for the decomposition of asbestos materials. Decomposition products, their safety and their use will be studied more.

Acknowledgments

This paper was elaborated with the Institutional support for the long-term development of research organizations by the Ministry of Industry and Trade of the Czech Republic.

References

[1] Petranek J *et al Encyklopedie geologie* (Ceska geologicka sluzba) ISBN: 978-80-7075-901-1
[2] Kusiorowski R *et al* 2015 Study on the thermal decomposition of crocidolite asbestos *Journal of thermal analysis and calorimetry* **120** pp 1585–1595
[3] Tovarkova J and Adamkova M 2001 *Rizika spojena s likvidaci materialu obsahujici azbest* Prirucka pro nakladani s chemickymi latkami a pripravky vcetne nebezpecnych
[4] SZU 2012 Problematika azbestu ve vnitrnim prostredi skol – prednasky
[5] Kusiorowski R *et al* 2013 Thermal decomposition of asbestos-containing materials *Journal of Thermal Analysis & Calorimetry* **113** pp179–188
[6] Alessandro F G *et al* 2008 In situ ESEM study of the thermal decomposition of chrysotile asbestos in view of safe recycling of the transformation product *Journal of Hazardous Materials* **156** pp 260–266
[7] Stanek T, Bohac M, Zezulova A and Rybova A Hydraulic Binder from Hazardous Waste *Solid State Phenomena* **276** pp 3–8
[8] Leonelli C *et al* 2006 Microwave thermal inertisation of asbestos containing waste and its recycling in traditional ceramics *Journal of Hazardous Materials* **135** pp149–155
[9] Iwaszko J *et al* 2018 Influence of high-energy milling on structure and microstructure of asbestos-cement materials *Journal of Molecular Structure* **1155** pp 51–57
[10] Tabata M *et al* 2016 Decomposition of asbestos by a supernatant used for immobilization of heavy metals in fly ash *The Journal of Material Cycles and Waste Management* **3** pp 483–492
[11] Pawełczyk A 2017 Chemical elimination of the harmful properties of asbestos from military facilities *Waste Management* **61** pp 377–385
[12] Valouma A *et al* 2017 Magnesium oxide production from chrysotile asbestos detoxification with oxalic acid treatment *Journal of Hazardous Materials* **336** pp 93–100
[13] Kazumichi Y *et al* 2009 A novel decomposition technique of friable asbestos by CHClF2-decomposed acidic gas *Journal of Hazardous Materials* **163** pp 593–599
[14] Necasova K *et al* 2019 Chemical elimination of asbestos material *IOP Conf. Ser.: Mater. Sci. Eng.* **583** 012031
[15] Klikorka J, Hájek B and Votinský J 1982 Obecna a anorganicka chemie I a II (Praha: Statní nakladatelství technické literatury)
[16] Yang Li *et al* 2015 Treatment and resource recovery from inorganic fluoride-containing waste produced by the pesticide industry *Journal of Environmental Sciences* **311** pp 21–29
[17] Jiang K *et al* 2017 Recovery and removal of fluoride from fluorine industrial wastewater by crystallization process *Technologies and Environmental Policy* **9** pp 2335–2340
[18] *Fluoridy* (*Online* http://www.arnika.org)

[19] *Zakon č. 541/2020 Sb. o odpadech* (*Online* https://www.mzp.cz)

IOP Conf. Series: Materials Science and Engineering 1205 (2021) 012019 doi:10.1088/1757-899X/1205/1/012019

Fracture parameters of fly ash geopolymer mortars with carbon black and graphite filler

H Simonova[1], C Mizerova[1], P Rovnanik[1], M Lipowczan[1] and P Schmid[1]

[1] Brno University of Technology, Faculty of Civil Engineering, Veveří 331/95, 602 00 Brno, Czech Republic

Email: simonova.h@vutbr.cz

Abstract. In this study, the effect of carbon black and graphite filler on the crack initiation and fracture parameters of fly ash geopolymer mortar is investigated. The carbon black was added in the amount of 0.5 and 1.0% and graphite powder in the amount of 5 and 10% relative to the fly ash mass. The reference mixture without any filler was also prepared. The fracture characteristics were determined based on the results of the three-point bending test of prismatic specimens provided with an initial central edge notch. The fracture experiments were conducted at the age of 48 days. The vertical force (*F*), the displacement measured in the middle of the span length (*d*), and the crack mouth opening displacement (*CMOD*) were continuously recorded during the test. The records of fracture tests were subsequently evaluated using the effective crack model, work-of-fracture method, and double-*K* fracture model. The addition of both fine fillers led to a decrease in monitored mechanical fracture parameters in comparison with reference mortar.

1. Introduction

The incorporation of novel additives and functional fillers into cement-based composites has been frequently studied in recent decades and especially materials with increased electrical conductivity represent a potential for advanced smart applications. Since the variation of external ambience conditions (temperature, humidity, stress due to mechanical loading etc.) can be detected via alteration of electrical resistivity, the development of self-sensing or self-heating structures is now in the scope of investigations [1]. Hence, the traditional cement-based composites are considered an electrical insulator and the applicability of plain binder for such purposes is low, various admixtures are used to establish a functional conductive network within the composite and improve its self-sensing performance. The most common conductive fillers include carbon or metal materials, e.g. carbon fibres, carbon nanotubes, graphite powder, carbon black, steel fibres or nickel powder [1–3]. Graphite powder and carbon black are fine carbon materials that have been widely used in asphalt or polymeric conductive composites, carbon black as a colloidal form of carbon with a high specific surface is also an inexpensive filler [4].

Geopolymers are a group of alternative alkaline activated binders manufactured from solid aluminosilicate precursors and alkaline activator [5]. Apart from favourable mechanical and durability performance, research of their electrical properties has shown that increased availability of mobile hydrated cations in the pores of the geopolymeric structure contributes to improved piezoelectric behaviour in comparison to cementitious binders [6, 7]. This phenomenon makes alkaline activated matrix a promising material for the above mentioned self-sensing structural components. Since

IOP Conf. Series: Materials Science and Engineering 1205 (2021) 012019 doi:10.1088/1757-899X/1205/1/012019

the addition of alternative and not commonly used fillers can alter the mechanical properties of these composites, it is necessary to study their mechanical performance. Therefore, the determination of mechanical fracture parameters is an important issue for the practical application of such conductive composite materials. In this study, the effect of carbon black and graphite filler on the crack initiation and fracture parameters of fly ash geopolymer mortar is investigated.

2. Experimental part

2.1. Materials

The basic aluminosilicate precursor for the preparation of the geopolymer mortars was fly ash from the high-temperature combustion of coal (Dětmarovice power plant, ČEZ). Sodium silicate solution with the SiO_2/Na_2O ratio of 1.6 was used as an alkaline activator and quartz sand with the maximum grain size of 2.5 mm was used to prepare the reference geopolymer. The chemical composition of fly ash and activator solution is given in table 1. The geopolymer mortar was modified by graphite powder or carbon black as fine conductive fillers. Graphite (PMM 11) was supplied by Graphite Týn (CZ) having a particle mean size of 11 μm. Carbon black (Vulcan 7H) was supplied by CS Cabot (CZ). Since carbon particles are hydrophobic, Triton X-100 (Sigma-Aldrich) in the form of a 2% aqueous solution was used as a dispersing agent and 1% solution of Lukosan S (Lučební závody, CZ) was added as an air detraining agent to avoid foaming and increased porosity.

Table 1. Chemical composition of raw materials.

	Composition (%)								
Material	SiO_2	Al_2O_3	Fe_2O_3	CaO	MgO	K_2O	Na_2O	MnO	TiO_2
Fly ash	51.67	23.31	7.08	4.45	0.36	2.95	0.77	1.14	1.00
Sodium silicate	26.43	–	–	–	–	–	16.61	–	–

2.2. Specimens preparation and curing

The composition of geopolymer mixtures is presented in table 2. The carbon black was added in the amount of 0.5 and 1.0% (CB0.5 and CB1.0, respectively) and graphite powder in the amount of 5 and 10% (G5 and G10, respectively) relative to the fly ash mass. The reference mixture (REF) without any filler was also prepared.

Table 2. Mixing proportions of geopolymer mortars.

	Mortar				
Component	REF	CB0.5	CB1.0	G5	G10
Fly ash (g)	350	350	350	350	350
Sodium silicate (g)	280	280	280	280	280
Graphite (g)	–	–	–	17.5	35
Carbon black (g)	–	1.75	3.5	–	–
Sand (g)	1050	1050	1050	1050	1050
Dispersing agent (ml)	–	1.75	3.5	17.5	35
Air detraining agent (ml)	–	–	–	7.5	7.5
Water (ml)	35	32	32	35	35

IOP Conf. Series: Materials Science and Engineering 1205 (2021) 012019 doi:10.1088/1757-899X/1205/1/012019

Geopolymer mortars were prepared in the Hobart planetary mixer according to the following procedure. At first, graphite powder/carbon black was dispersed in the sodium silicate and Triton X-100 solution. Then, fly ash and quartz sand together with additional mixing water were added and stirred for at least 5 minutes to prepare a fresh slurry. Finally, an air detraining agent was added and the prepared geopolymer mixture was cast into prismatic moulds with dimension 40 × 40 × 160 mm. After 2 hours at laboratory temperature, the moulds with fresh mixtures were cured at 40°C for 24 hours to accelerate the geopolymerization reaction. Hardened specimens were wrapped in plastic bags and stored under laboratory conditions until fracture testing. Three specimens made of each mortar were used for the fracture tests.

2.3. *Fracture tests*

The fracture experiments in the three-point bending configuration were conducted at the age of 48 days. The above-mentioned prismatic specimens were provided with an initial central edge notch one day before testing. The notch length was approximately one-third of the specimen's height. The span length was set to 120 mm. The stiff multi-purpose mechanical testing machine LabTest 6.250 with a load range of 0–250 kN was used for fracture tests. The loading process was controlled by a constant increment of displacement of 0.02 mm/min in the whole course of the experiment.

The vertical force (F), the vertical displacement (d) measured in the middle of the span length using the inductive sensor, and the crack mouth opening displacement ($CMOD$) measured using a strain gauge mounted between blades fixed on the bottom surface of the specimen were recorded during the experiment. In this way, the continual records of the fracture tests in the form of F–d and F–$CMOD$ diagrams were obtained. The raw data were further processed to obtain the correct input values for diagrams evaluation using the selected fracture models described below. The modification of diagrams consisted of the removal of duplicate points, the shifting of the origin of the coordinate system, the smoothing of the diagram, and the decrease of the number of points. The adjustment of raw data was performed in GTDiPS software [8] which is based on advanced transformation methods used for the processing of extensive point sequences.

2.4. *Microstructural characterization*

Pore distribution was evaluated utilizing mercury intrusion porosimetry analysis that was conducted on samples using Micromeritics Poresizer 9310 porosimeter that can generate a maximum pressure of 207 MPa and can evaluate a theoretical pore diameter of 0.006 μm. Micrographs of the geopolymer composites were taken on a TESCAN MIRA3 XMU scanning electron microscope in secondary electron imaging mode. The micrographs were taken on samples that were sputtered with gold.

3. Assessment of fracture parameters

The mechanical fracture parameters were determined based on the direct evaluation of the F–d and F–$CMOD$ diagrams using the chosen fracture models. The modulus of elasticity was assessed from the initial parts of F–d diagrams according to Karihaloo [9]. The effective fracture toughness was determined from F–d diagrams using the effective crack model introduced by Karihaloo [9]. This approximative model combines the linear elastic fracture mechanics and crack length approaches.

The F–d diagrams were further used for the calculation of work of fracture according to the RILEM method [10]. Then, the specific fracture energy is assessed by dividing the work of fracture by the ligament area (i.e. the projection of the original uncracked area).

The double-K fracture model [11] was chosen for the evaluation of F–$CMOD$ diagrams. The double-K fracture model works on the combination of the concept of the cohesive forces on the fictitious crack with a criterion based on the stress intensity factor. The different stages of the fracture process in materials with the brittle matrix can be predicted using this model. The two parameters are used for this purpose, namely initial cracking toughness and unstable fracture toughness which are given in terms of stress intensity factor. The unstable fracture toughness is defined as the critical stress intensity factor, which is similar to effective fracture toughness used in the effective crack model by Karihaloo [9].

IOP Conf. Series: Materials Science and Engineering 1205 (2021) 012019 doi:10.1088/1757-899X/1205/1/012019

The initial cracking toughness stands for the capability to withstand an external load before the origin of crack propagation. The difference between these two parameters is the equivalent stress intensity factor caused by the cohesive forces on the fictitious crack called cohesive fracture toughness [12]. In this paper, the unstable fracture toughness and the cohesive fracture toughness were assessed first, and afterwards, the initial cracking toughness was calculated.

The cohesive softening function which characterizes the relationship between cohesive stress and effective crack opening displacement needs to be defined for calculation of the cohesive fracture toughness. In this paper, the nonlinear softening function according to Reinhardt [13] was chosen. The tensile strength as an input parameter of softening function was determined based on the identification process using the artificial neural network (ANN) [14]. The trained ANN is used to transfer the input data – a measured response in the form of an *F–d* diagram obtained from the fracture test – to the desired material parameters. Besides the tensile strength, the modulus of elasticity and fracture energy were also determined by the identification process. The details about the procedure of parameters identification can be found in [15].

The informative compressive strength was determined to complement material characteristics. The two parts of the specimen left after the fracture experiments were used for this purpose.

4. Results

The selected mechanical fracture characteristics of geopolymer mortars modified by graphite powder or carbon black as fine conductive fillers obtained by the above-mentioned fracture models are introduced in figures 1 and 2. The average value (determined based on 3 independent measurements) and standard deviation represented by error bars are presented.

The compressive strength of reference geopolymer mortar was about 50 MPa. The addition of carbon black did not have any effect on compressive strength in comparison with reference mortar. Contrarily, the addition of graphite powder did have a significant effect. While the addition of graphite in an amount of 5% caused an increase of compressive strength of about 24%, on the contrary, the addition of 10% of graphite powder caused a decrease of compressive strength of about 42%.

The modulus of elasticity of reference geopolymer mortar was about 6 GPa. The addition of both fine conductive fillers caused a decrease in modulus of elasticity by more than 30%. The modulus of elasticity determined by identification was about 0.5 GPa higher when the variability of the results is taken into the account for all investigated geopolymer mortars in comparison with the direct evaluation of the *F–d* diagrams, see table 3.

The effective fracture toughness of reference geopolymer mortar was about 0.3 $\text{MPa}\cdot\text{m}^{1/2}$. The addition of 1% of carbon black caused the decrease of fracture toughness by 15%. In the case of graphite, the decrease of fracture toughness was higher about 25 and 30% for filler amounts of 5 or 10%, respectively. Similar values of fracture toughness were obtained also from *F–CMOD* diagrams using the double-*K* fracture model, see figure 2.

The specific fracture energy was about 47 $\text{J}\cdot\text{m}^{-2}$. The addition of both fillers caused the decrease of fracture energy. This decrease was more significant in the case of the addition of graphite. The decrease of fracture energy is independent of filler amounts when the variability of the results is taken into account. Table 3 introduces the fracture energy obtained by identification. The differences in average values of fracture energy imply from their different physical sense – the values obtained by the identification process are primarily related to the material point, whereas the values obtained from the work of fracture method are related to the tested specimen and represent the average fracture energy.

Table 3 introduces also the tensile strength of investigated geopolymers mortars. The highest value was obtained for mortar with the addition of 0.5% of carbon black, which is in agreement with the maximum force obtained during the fracture experiments. These values were used together with fracture energy as input parameters of the nonlinear softening function used in the double-*K* fracture model.

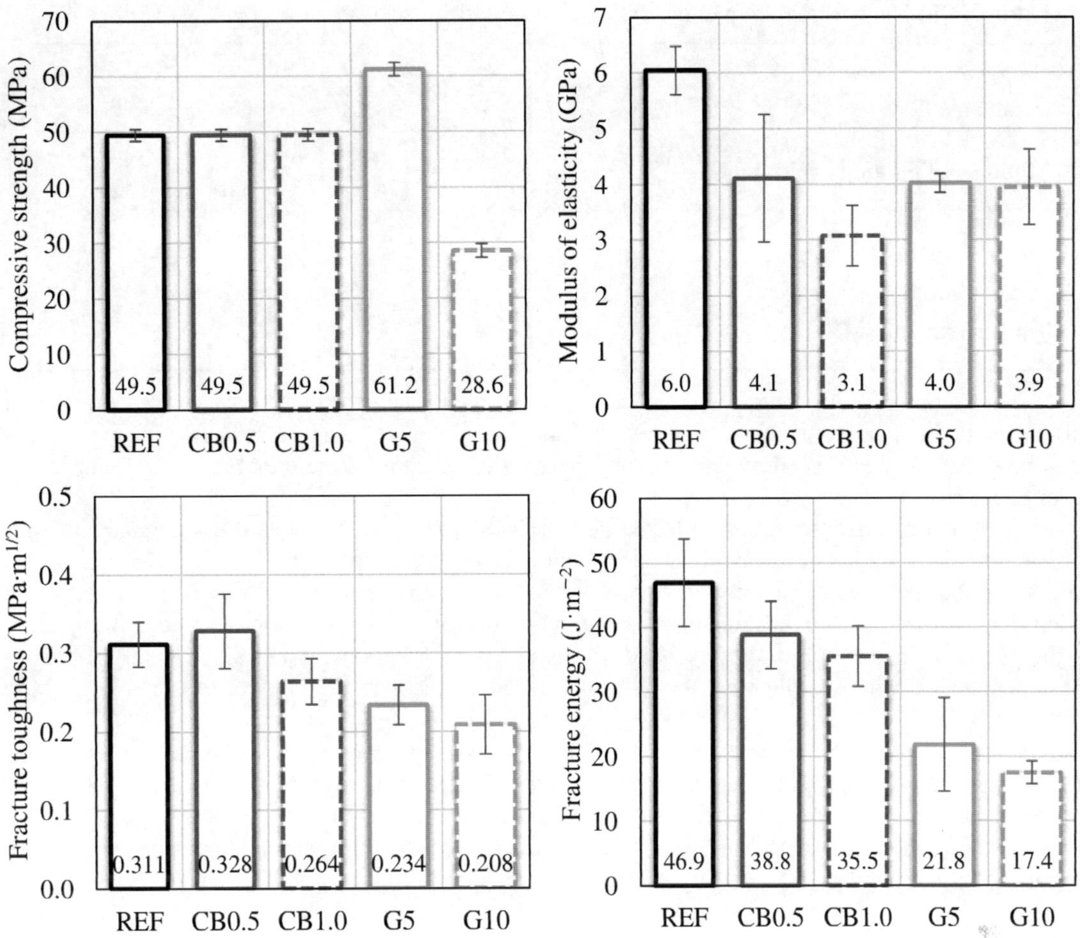

Figure 1. Compressive strength and mechanical fracture parameters obtained from $F-d$ diagrams.

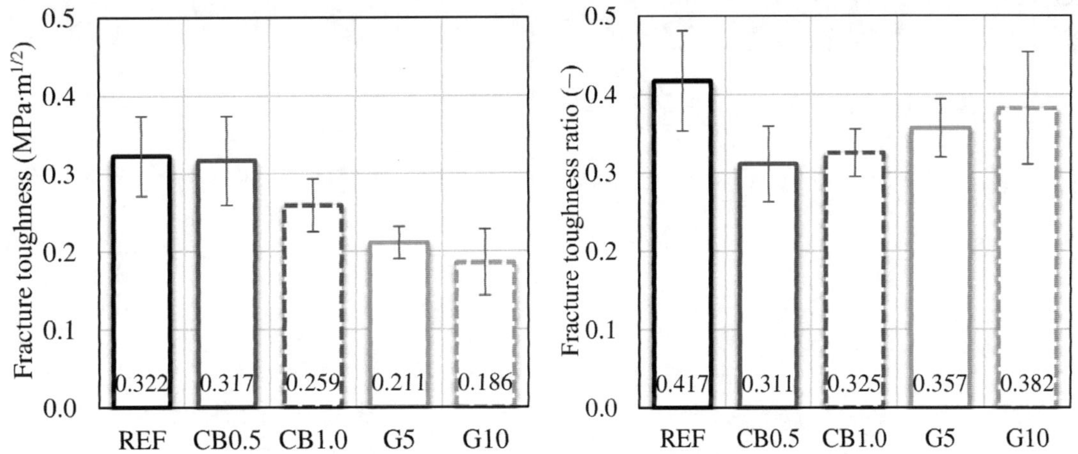

Figure 2. Fracture parameters obtained from $F-CMOD$ diagrams.

Table 3. Mechanical fracture parameters obtained by identification – mean values (CoV in %).

	Mortar				
Parameter	REF	CB0.5	CB1.0	G5	G10
Modulus of elasticity (GPa)	6.5 (11.1)	4.9 (31.5)	3.6 (16.1)	4.7 (5.0)	4.4 (21.7)
Fracture energy ($J \cdot m^{-2}$)	41.3 (32.7)	47.8 (12.7)	45.2 (7.5)	27.4 (37.7)	21.1 (8.7)
Tensile strength (MPa)	1.49 (1.0)	2.14 (8.6)	1.45 (20.6)	1.41 (22.4)	1.31 (1.1)

On the contrary to other fracture parameters, the resistance to stable crack propagation more decreased with the addition of carbon black than with the addition of graphite, see figure 2. The resistance to stable crack propagation is expressed by fracture toughness ratio, i.e. ratio of the initial cracking toughness to the unstable fracture toughness.

Microstructural characterization of tested geopolymer samples was provided by mercury intrusion porosimetry and scanning electron microscopy. Figure 3 shows a comparison of cumulative pore curves. Data show that the lowest porosity was observed for the reference sample. When carbon black was added the porosity only slightly increased but the amount of the admixture has practically no effect on the pore distribution. On the other hand, the addition of graphite powder significantly increased the total intruded volume which can be attributed to an increase in the number of pores larger than 1 μm.

The total porosity of the samples corresponds quite well to the trends in compressive strength of the specimens with carbon fillers compared to the reference geopolymer mortar. It is also in good correlation with the fracture toughness and fracture energy as these parameters decreased while the total porosity increased. The highest drop in the values of fracture parameters determined for the geopolymer mortars with graphite powder can be attributed to the morphology of graphite particles, which have a lamellar structure (figure 4). These particles can act as stress concentrators and allow easier crack propagation.

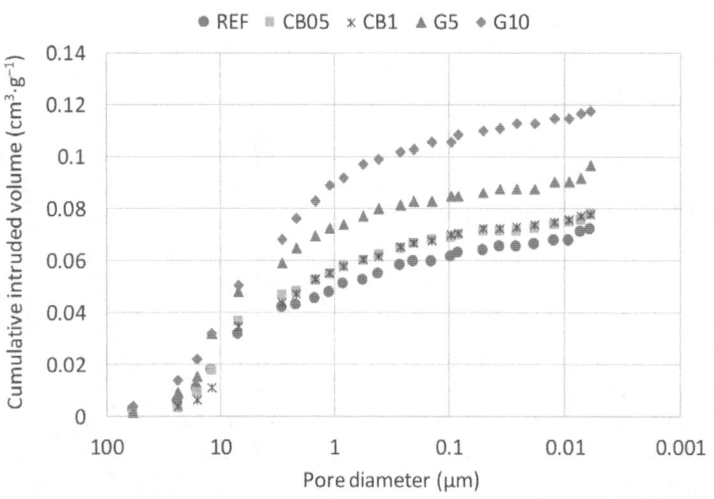

Figure 3. Pore distribution of tested composites provided as cumulative intruded volume.

Figure 4. Morphology of geopolymer matrix with lamellar graphite particles.

4. Conclusions

This study aimed to quantify the effect of the addition of carbon black and graphite powder on the mechanical fracture behaviour of fly ash geopolymer mortar. Despite both alternative and not commonly used fine fillers were primarily added to enhance the electrical properties of the investigated geopolymer mortar, their addition should not deteriorate other material properties depending on its prospective practical application. It is obvious from the presented results, that monitoring not only the compressive strength but also the other mechanical fracture parameters is necessary to obtain complex information about material behaviour, which is considerably affected by the addition of fine fillers. The addition of carbon fillers leads to the increase of porosity which can be attributed to an increase in the number of pores larger than 1 μm. It is in good agreement with the monitored fracture parameters which decreased while the total porosity increased.

Acknowledgement

This outcome has been achieved with the financial support of the Czech Science Foundation under project No. 19-11516S.

References

[1] Han B, Zhang L and Ou J 2017 Smart and multifunctional concrete toward sustainable infrastructures (Singapore: Springer)

[2] Han B, Yu X and Ou J 2014 Self-Sensing Concrete in Smart Structures (London: Butterworth Heinemann)

[3] Chung D D L 2012 Carbon materials for structural self-sensing, electromagnetic shielding and thermal interfacing *Carbon* **50**(9) pp 3342–3353

[4] Donnet J B, Bansal R C and Wang M J ed 1993 Carbon black: science and technology (New York: CRC Press)

[5] Provis J L and Van Deventer J S J ed 2009 Geopolymers: Structure, processing, properties and industrial applications (Cambridge: Woodhead Publishing Limited)

[6] Lamuta C, Candamano S, Crea F and Pagnotta L 2016 Direct piezoelectric effect in geopolymeric mortars *Materials & Design* **107** pp 57–64

[7] Hanjitsuwan S, Chindaprasirt P and Pimraksa K 2011 Electrical conductivity and dielectric property of fly ash geopolymer pastes *International Journal of Minerals, Metallurgy, and Materials* **18**(1) pp 94–99

[8] Frantik P and Masek J 2015 GTDiPS software (*Online* http://gtdips.kitnarf.cz/)
[9] Karihaloo B L 1995 Fracture Mechanics and Structural Concrete (New York: Longman Scientific & Technical)
[10] RILEM TC – 50 FMC 1985 Determination of the fracture energy of mortar and concrete by means of three-point bend test on notched beams *Materials and Structures* **18**(4) pp 287–290
[11] Kumar S and Barai SV 2011 Concrete Fracture Models and Applications (Berlin: Springer)
[12] Xu S and Reinhardt H W 1999 Determination of double-*K* criterion for crack propagation in quasi-brittle fracture, Part II: Analytical evaluating and practical measuring methods for three-point bending notched beams *International Journal of Fracture* **98** pp 151–177
[13] Reinhardt H W, Cornelissen H A W and Hordijk D A 1986 Tensile tests and failure analysis of concrete *Journal of Structural Engineering* **112** pp 2462–2477
[14] Novak D and Lehky D 2006 ANN Inverse Analysis Based on Stochastic Small-Sample Training Set Simulation *Engineering Application of Artificial Intelligence* **19** pp 731–740
[15] Lehky D, Lipowczan M, Simonova H and Kersner Z 2019 A neural network ensemble for the identification of mechanical fracture parameters of fine-grained brittle matrix composites *FraMCoS-X 10th International Conference on Fracture Mechanics of Concrete and Concrete Structures* ed G Pijaudier-Cabot, P Grasll and C La Borderie (France: IA-FraMCoS) pp 1–9

Fast material disintegration of cement concrete in the exterior

Z Nevosad[1], J Grosek[1], B Dohnalkova[1], T Zavrel[1] and O Machel[1]

[1] Department of Transport Infrastructure, Transport Research Centre – CDV, Lisenska 2657/33A, 63600 Brno, Czech Republic

E-mail: jiri.grosek@cdv.cz

Abstract. Over the recent years the amount of fast destruction defects of cement paste matrix has increased markedly for exposed concrete structures of highway and airport pavements. The accelerated destruction of silicate structure is unusual due to its irregular occurrence. It only occurs at 30 to 50% of newly built pavements (in continuous segments). The issue is applicable to loaded as well as unloaded pavements. Detailed diagnostics found that the probable cause is based on concurrently occurring expansive reactions in capillaries, cracks and pores in the cement paste matrix. In defect spots the presence of several expansive substances was proven, wherein the biggest question poses the unexpectedly high potassium concentration. Potassium is only contained at best in trace amounts in road concrete input components in the Czech Republic. Neither existing regulations nor standards sufficiently deal with this type of defects of cement concretes. If they mention the existence of this issue at all, they are just limited to recommendations. Since the modification of the existing commercially produced cements to pavements is still improbable, the solutions to this issue seem problematic.

1. Introduction

Quality of concrete structures or their concrete parts is evaluated based on the length of their service. Duration is rated among the decisive parameters for the evaluation of the quality of structures. In case it is significantly reduced, there are always good reasons to discover why. If they are found, the next step is searching and implementing suitable measures that reduce the probability of their recurrence in the future. The measures may be related to structure designing (static reasons), adaptation of concrete mixture composition properties (materials), changes in construction technology, type of maintenance, etc.

The prevailing opinion today believes that the material problems are not too crucial for concrete structures. Many issues can be dealt with by changed strength, a different reinforcement solution, changed shapes, or use of ingredients, admixtures and other components in the concrete mixture, etc. The set of these measures is not inexhaustible, it has its limits. Their research is continuously in progress, which is positive for development. Nevertheless, the environment impacts have such destructive effects on the existing concretes of the most exposed structures, that the above-mentioned scope of measures is clearly insufficient. Regarding particularly younger concrete surfaces of airports and highways, fast destruction of the silicate structure of concrete occurs at approximately half of the cases (line segments). With some exceptions the problem concerns the whole volume of the concrete pavement layer (or potentially concrete layers).

The new problem with materials is not manifested by the initial measurable reduced concrete compressive strength [1, 2]. It is indicated by a specific formation of cracks on the concrete surface. It is necessary to have some experience since the cracks in concrete also occur from other reasons. At first a network of fine cracks with the width lower than 0.1 mm, which are difficult to trace by a naked eye, appears on the surface. Over time these cracks interconnect and many of them further expand. They are already clearly visible from the widths of 0.3 mm. Subsequently, fine cracks of these widths (later even bigger) appear on the concrete surface. The cracks are disorderly arranged in irregular raster of usually 10 x 10 cm. The remaining initial cracks inside the surfaces of this raster do not expand in this phase. Subsequently, intensive surface dusting, and then chipping off of small pieces of concrete occurs. A specific feature is that the defect occurs irregularly and in variable intensity. Based on the performed diagnostics, the estimated occurrence of the defect ranges between 30 and 50% of volumes of all newly built concrete pavements in the Czech Republic. The defect occurrence is comparable on road pavements in operation (overrun) as well as on unoperated (unloaded by traffic) road pavements.

2. Experimental part

Airports and highways are specific monolithic concrete structures and there is an accelerated disintegration of silicate structure. A monolith, continuously built by finishers, often in thousands of m^3 of concrete, is usually additionally cut into so-called contraction units after the initial hardening of concrete. This is the way to reduce/eliminate length/height (smaller) concrete deformations due to thermal expansion, drying and saturating, contact with subbase, operational loading, etc. Cutting is performed based on the real hardening of concrete within approx. 8 to 48 hours after concrete installation. It is a very risky operation, which is nowadays, on the basis of empiric experience, managed in such way that generally neither visible cracks appear, nor any other concrete damage occur.

Over the last decades surface networks of fine cracks have appeared on some newly built concrete pavements at the age from approx. 3 months. Over time they interconnect into continuous omnidirectional cracks. Irregular networks with areas around 100 to 150 mm or longitudinal cracks running through many slabs are formed. Their length may even exceed 100 meters. It is strange that these defects only occur at new concretes. Comparable damage of older concretes, i.e., over 20 years of age, is either non-existent or very marginal.

Figure 1. Examples of visible crack networks on concrete pavements due to expansive reactions.

We were able to take/obtain cores from all concrete highways in the Czech Republic and from several airports as well [3, 5]. All cores were used for taking samples for the diagnostics of point chemical analysis in a microscope. Products of multiple types of expansion, in different concentrations, on the surface, in the middle and at the bottom of the cores, were found at all damaged new concretes. Expansive gels and crystals were only found in pores, cracks and capillaries. No obvious damage of aggregates that would prove its presence in expansive reactions was found by this diagnostic. The fastest destruction of the silicate structure is caused, probably in the same degree, by sulphate and alkali expansion. Other types of expansive reactions cause slower disintegration in the whole volume of the silicate structure.

3. Results

In particular, it was potassium that stood out from the performed chemical analysis of expansive components. It was present even in extreme concentrations (even over 30%) in products of expansive minerals (gels). The analysis of concrete paste matrix was performed with scanning electron microscopy and energy dispersive spectroscopy (SEM/EDS). The use of a Tescan Vega LSU microscope was the principal method of examination. EDS was used for point and line analyses. The area distribution of selected elements was determined. Figure 2 and table 1 show an assessment example of more than 1000 performed tests, while samples were taken from upper and bottom of the concrete slabs.

We were surprised by the fact that potassium was not found in concrete aggregates, or at best in trace amounts. This fact and the presence of otherwise undamaged aggregates brought us to a conclusion that aggregate expansion in road concretes in the Czech Republic has not been a major problem so far.

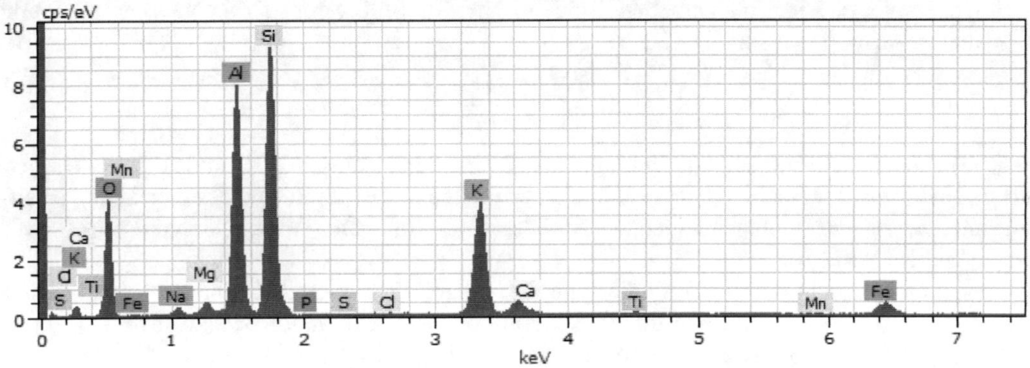

Figure 2. Ratios of main cations content – Analysis No. 1, example.

Table 1. Drill core of concrete pavement – chemical composition of concrete, example.

Analysis No.	1	2	3
SiO_2	48.82	59.7	48.85
TiO_2	0.0	0.0	0.24
Al_2O_3	31.39	4.3	34.28
FeO	1.93	2.0	2.05
MnO	0.0	0.0	0.0
MgO	1.82	0.7	2.48
CaO	3.52	25.9	0.96
Na_2O	0.80	2.0	1.27
K_2O	11.33	5.0	9.71
SO_3	0.15	0.0	0.0
Cl	0.24	0.4	0.16
Sum	100.00	100.00	100.00

The presence of potassium, which is a potential cofactor of alkali expansion, exclusively in pores, cracks and capillaries leads to a conclusion that expansive substances are transported by water/moisture in the structure of concrete. This movement is based on the thermal-moisture gradient. In the exterior the vector of humidity movement in time is very variable. It may be completely opposite within one day. It depends on current climatic conditions, on current amount of free water/moisture in concrete, on thermal-moisture conditions of the road subbase, or on all layers of the road structure, including the subgrade, etc.

The assumption of the decisive effect of thermal moisture gradients corresponds with the diagnosed fact that if potassium is presented in the concrete structure in more significant amounts, its concentration is very variable in the content of concretes in different points or places. The origin and formation of expansive minerals and gels in cement stone structure occurs during the hydration of the cement binder. The presence of other potential binders, except cement, is generally improbable for concrete road pavements.

Therefore, hydration cannot be perceived just as a separated relationship of components and properties of the concrete mixture. It is necessary to take into account potential reagents coming from the external environment that may secondarily enter hydration reactions. Such view may be respected by, for example, physical mathematical models. There are more forms of notations, we work with the hydration equation [2, 4].

$$A_h = \int_0^{t_h} \left(\frac{RTm}{\mu} - \frac{S_{pc}^4 m^5}{t^2 V^2} \right) dt \qquad (1)$$

A_h (J) is energy consumed by hydration, R is universal gas constant ($JK^{-1}mol^{-1}$), T (K) temperature, m (kg) binder mass (cement) + mass of all external reagents, μ ($kgmol^{-1}$) mass of reacting (dissolved) molecule, S_{pc} (m^2kg^{-1}) cement specific surface, t (s) time and V (m^3) is volume of hydrating cement/cement stone.

The relationship (1) ranks among the types of integrals we are still unable to solve. In case it is applied, it is always necessary to examine the definitional scope of validity in advance.

The equation has no relation to reached/reachable mechanic parameters of concrete/mortars. The equation makes it clear that, based on reactions over time, the mass m gradually decreases, and the hydration of consumed energy A_h increases. The equation has a character of a time decreasing function. The hydration is defined for time interval t, as long as the immediate increment of value ΔA_h does not drop to value zero (in reality to a value close to zero for concretes). By using the equation, it is possible to determine theoretical time of hydration. When using common alitic Czech cements CEM I in the amount of 400 kg/m^3, at the simplified calculation of average mass of basic clinker minerals, the theoretical duration is slightly over 16 years according to the above-mentioned equation. The real time will be many times longer, the prolongation is given by the fact that expected optimum reaction conditions cannot be reached with real concretes.

The equation shows that the most vulnerable hydration period is its initial time. The reaction is intensive and my easily be affected by the entered reagents. This can be explained by the fact that the existing fast expansive reactions in concrete are intensive just for newly cast concrete road pavements as well as by the fact that those fast expansive reactions do not occur or are minimal for old concrete in newly installed lanes. In addition, the equation explains that if expansive reactions in newly installed concrete occur comparably for overrun and unloaded pavements, it is a material, not static, defect.

We cannot get the answer from the equation why expansive reactions with the current fast disintegration of the silicate structure were nearly non-existent before. At the same time the construction of pavement layers remains similar. Neither the requirements for concrete, nor binding standards have changed dramatically technology-wise, the production is performed with conceptually the same technologies. In contrast to the past, two-layer concrete installation (it used to be one-layer installation) is preferred, qualitative requirements for aggregates have increased, and, instead of previously used road cement SC 70, CEM I 42.5R SC has been used (CEM II has been used on a trial as well since 2019).

Although the qualitative differences according to time based effective standards between previous SC 70 and CEM I 42.5R SC are virtually minimal. The material durability of the previous and current road concretes is significantly different; it is worse nowadays. This holds despite the fact that the existing cements fully meet the standards, and the performance of concrete installation is closely checked by investors and contractors, and generally no more serious deviations from standards or regulations are found.

4. Discussions

Nowadays, cement concrete road pavements are an important part of the majority of main domestic airports and highways. They concern the existing structures and the new structures that are to be built in the future. Results from all concrete segments of highways in the Czech Republic and from several civil and military airports were processed. The destructive processes in the concrete structure are very variable. The most crucial for expansive reactions appears to be the position of structures. Even identical structure compositions and identical concrete under identical traffic loading have different dominant types of expansive reactions in different segments. The greatest differences are related to the amount of potassium, whose presence in cement stone capillaries ranged between 0 and more than 30%.

Regarding materials, i.e., concrete mixture components, the situation is given by available sources of raw materials. Nowadays, the only trouble-free component are aggregates. The selection criteria for road concretes are currently strict and we have not proven in the Czech Republic that expansive reactions apply to it.

Cements, ingredients, admixtures, and water are parts of the cement stone. The expansive reactions of these types, if they are contained by concrete, spread abundantly. If the aggressive substances are bound to modifications of gels or crystals in capillaries, cracks, etc., cavities, they are already involved in the hydration system. In other words, it is very difficult or impossible to remove them. The expansion than cannot be stopped by no other way than by permanent reduction of the present free water. Such measure must be dealt with very carefully. It is a double-edge solution since potential long-term lack of water in concrete has negative effect on further hydration process.

The design of road pavements is not derived from compressive strengths, but from flexural strengths [4, 6]. Regarding material durability of concrete, mechanical properties of concrete are not decisive. The long-term experience from road pavement concretes, where compressive strengths over approx. 30 MPa apply, any higher strengths generally fail to have an impact on concrete material durability.

In case there was an interest to make significant progress regarding components, the dominant potential is entirely outside the criteria of the effective standard for concrete. The existing cements meet all the parameters, however, there is still an error somewhere. An interest in this issue would also arise in the research and development of cement. Apparently, this demand from the side of road engineering is made in many other countries outside Europe. If we have the right information, even the most advanced countries worldwide have failed to succeed with this demand to cement producers. If it is possible to use other cements than CEM I equally, which has been possible for some time in many countries except the Czech Republic, it is undoubtedly a step in the right direction. However, if we are to deal with the above-mentioned expansion defects in full, with the reference to the mentioned relationship, bigger intervention, i.e., big step, may only be the solution. Today we do not know how far it is.

5. Conclusions

We are in the situation that it is desirable to deal with the current state. We can see a number of potential trends and issues, in terms of structures and materials. Regarding the structure issues of repairs of the existing roads, the following are concerned:

- Define and diagnose the presence of aggressive substances.
- Prove conditions of the real transfer of aggressive substances into concrete, not only those from the air, but particularly those from road subbase layers and subgrade.
- Find and verify practical solutions for the road layers and subsequent treatment, which prevents or at least slows down the transfer of aggressive substances into concrete.

- Find a way to effectively drain the joint in between the remained subbase and the new concrete pavement.
- Find protection to a new concrete pavement.
- Find a technologically safe way to dispose of/recycle the concrete affected by the destruction of silicate structure due to expansive reactions.

Regarding the structure issue for new constructions

- Find suitable diagnostics for research of potential contamination of the subgrade.
- Design more effective drainage of the subgrade with outlets off the road.
- Find and verify practical solutions for prevention or at least a slowdown of the transfer of aggressive substances into subbase road layers.
- Consider potential effective drainage of the joint in between the subbase and concrete road pavement.
- Find protection to new concrete pavements.

Acknowledgements
This article was produced with the financial support of the Ministry of Transport within the programme of long-term conceptual development of research institutions.

References

[1] Chamberlain D A 2015 Protecting Concrete Infrastructure with Dual Crystal technology *Seminar Protecting Your Concrete Infrastructure* (Des Moines) pp 6–10
[2] Nevosad Z 2015 Main Problems of Cement Concrete *Seminar Protecting Your Concrete Infrastructure* (Des Moines) pp 14–17
[3] Chupík V *et al* 2009 *Petrografický rozbor betonů* (Brno: C&T press) pp 8–15
[4] Grošek J *et al* 2018 Nové možnosti pro testování prostředků sekundární ochrany cementobetonových krytů vozovek *Silniční obzor* **9** pp 261–264
[5] Frýbort *et al* 2018 Variations in the composition and relations between alkali-silica gels and calcium silicate hydrates in highway concrete *Materials Characterization* **137** pp 91–108
[6] Hodakova *et al* 2017 Impact of Climate Characteristics on Cement Concrete Pavements *SGEM: 17th Inter. Multidisc. Scien. GeoConf.* (Albena) pp 467–472

Hydrothermal synthesis and characterisation of zeolites from metakaolin and water glass

M Krejci Kotlanova[1,2,3], Z Dolnicek[4], I Khongova[1] and A Zezulova[1]

[1] Research Institute for Building Materials, Hněvkovského 65, 617 00, Brno, Czech Republic

[2] Department of Geological Sciences, Masaryk University, Kotlářská 267/2, 611 37 Brno, Czech Republic

[3] BIC Brno spol. s r.o., Technology Innovation Transfer Chamber, Purkyňova 648/125 612 00, Brno, Czech Republic

[4] Department of Mineralogy and Petrology, National Museum, Cirkusová 1740, 193 00, Prague 9, Czech Republic

Email: krejci@vustah.cz

Abstract. Zeolites were synthetised from two types of metakaolin mixed with water glass and in some samples with water. The mixtures were autoclaved for 24 hours at 130°C. Autoclaved samples were ground to analytical fineness and tested by X-ray diffraction and differential thermal analysis. Polished samples embedded in epoxy resin were studied using of electron microprobe including BSE imaging and WDX analysis of chemical composition of zeolites. Samples with the highest zeolite content were studied by SEM. The obtained results showed that the dominant synthetised zeolite in autoclaved samples is mineral chemically close to chabazite, less phillipsite. Zeolite Na-P1, gmelinite and boggsite were also identified by XRD analyses. The total content of zeolites is up to 29 wt. % in the sample prepared from a mixture containing metakaolin MK1/sodium water glass/water in weight ratio 1.5/2.5/1. The Si/Al of synthetised zeolites varies between 2.1 and 3.2.

1. Introduction

Natural zeolites are aluminosilicates with a porous structure, classified as tectosilicates. Tectosilicates have a spatial bond of SiO_4 tetrahedra. Most tectosilicates have several other ions in their crystal structure, such as Ca, K and Na. These elements can only enter the structure of the mineral if less valent ion replaces part of the Si^{4+} ions. In most cases it is Al^{3+} [1].

Zeolites are characterised by a three-dimensional bond between SiO_4 and AlO_4 tetrahedra defined by sharing corner oxygens [2]. Unlike other tectosilicates, zeolites have larger cavities in their crystal structure connected to each other by channels. Also, the number of pores in the structure of zeolites is significant. The size of these pores varies depending on the Si:Al ratio in the zeolite structure [3, 4]. Zeolites can retain large amounts of water and other substances. Water from the zeolite structure (so-called zeolite water) can be removed by slow heating [5].

To date, about 67 natural zeolites have been discovered, according to the International Zeolite Association. These minerals are mainly bound to volcanic rocks, in which they form fillings of cavities or amygdales. It can also be found in pegmatites, sedimentary rocks and hydrothermal veins [6].

IOP Conf. Series: Materials Science and Engineering 1205 (2021) 012021 doi:10.1088/1757-899X/1205/1/012021

The most common zeolites are heulandite, clinoptilolite, chabazite, stilbite, phillipsite, natrolite and analcime [7, 8].

More than 250 zeolites were prepared by hydrothermal synthesis [9]. From a structural point of view, zeolite groups (zeotypes) can be distinguished, which are identified by a three-letter code derived from a typical zeolite belonging to that group [10].

Synthetic zeolites can be prepared from a lot of chemical compounds containing Al and Si, and a lot of scientists deal with this problematic [f.e. 11–15].

Many authors tried to use natural and waste raw materials in the production of synthetic zeolites. For example, blast furnace slag [16–17], waste glass [18], fly ash [19–20], waste kaolin [21–22], natural kaolin or bauxite [23–24], aluminium waste [25], diatomite [26–27], rice husk ash [28], ash from corn cobs [29], volcanic glass [30], aluminium cans [31] or ash from bamboo leaves [32] can be used for the synthesis of zeolites.

Thanks to retaining water and other substances, zeolites are used in a wide range of applications. They are used for example as sorbents [33–34], molecular sieves [35–36], membranes [37], ion exchangers [38], catalysts [39], additive to fertilizers [40], water softeners [41], in papermaking industry [42] and animal nutrition [43].

Several researchers in the cement industry tried to use natural and synthetic zeolites as a supplementary cementitious material [44–48].

Joshaghani [45] deals with the use of zeolite as a cement replacement in concrete production. The author replaced Portland cement with zeolite in the amount of 15, 25 and 35 wt. %. Sixteen mixtures with different proportions of zeolite, with different sizes of aggregate particles and different w/c (water to cement ratio) were prepared. It was found that concrete permeability decreased when cement was replaced by up to 25 wt. % of zeolite. When the zeolite amount in the mixture is 35 wt. %, then tensile and compressive strengths of concrete decrease within 90 days. The used w/c of 0.37 seems to be the best in terms of reaching the highest strengths of the resulting concrete.

Sičáková et al [46] studied the long-term properties of composites with the addition of natural zeolite. Cement was doped with zeolite in an amount of 8, 13 and 25 wt. %. In one case, microsilica was added to the mixture. Concretes based on zeolite-doped cement show higher bulk density in 3 years of testing than concrete with ordinary Portland cement, on the contrary, 28-day strengths were lower. Higher zeolite content (25 wt. %) in the mixture reduced the compressive strength of the resulting concrete.

In this work, the initial results of the synthesis of zeolites from two different metakaolines in a mixture with sodium water glass are presented. The aim of the work is to verify the suitability of using two types of metakaolins, which differ in their composition and origin of the raw material. Metakaolin Mefisto L05 is made of claystone and Mefisto K05 is made of kaolin. Another goal of the work was to study in more detail the chemistry of synthesized zeolites using electrone microprobe. Work on this topic should lead to the production of a material with a high zeolite content, which could be used as a supplementary cementitious material.

2. Materials and methods

The following raw materials were selected for the hydrothermal synthesis of zeolites: metakaolines Mefisto K05 (in the text as MK1) and Mefisto L05 (in the text as MK2), which are kaolinitic clays calcined at 750°C, sodium water glass and water.

The chemical composition of metakaolines was studied on the X-ray spectrometer PANanalytical Axios PW4400/30 with Rh lamp, 40 kV/40 mA. The SiO_2 content of the water glass was determined by the gravimetric method, Na_2O by the atomic mass spectroscopy and H_2O gravimetrically by loss of drying and ignition according to ČSN 72 0102 [49]. The chemical composition of raw materials is given in tables 1 and 2.

Table 1. Chemical composition, loss of drying and silicate module of sodium water glass; values in wt. %.

	Sodium water glass
Na_2O	12.28
SiO_2	21.74
Loss of drying	41.95
Silicate module	1.65

Table 2. Chemical composition and loss of ignition of metakaolines; values in wt. %.

	MK1	MK2
SO_3	0.05	0.09
P_2O_5	0.07	0.17
TiO_2	1.56	1.23
SiO_2	53.32	53.21
Al_2O_3	37.09	42.94
Fe_2O_3	3.34	0.92
MgO	0.88	0.28
CaO	2.64	0.20
Na_2O	0.00	0.00
K_2O	1.23	0.78
Loss of ignition	2.10	1.59

Samples dried at 60°C and ground below 0.09 mm were analysed on the Bruker D8 Advance with Cu anode ($\lambda K\alpha$=1.54184 Å) with variable divergent aperture screens at Φ-Φ reflective Bragg-Brentano parafocusation geometry. Quantitative phase analysis was performed using the Rietveld method. In the case of amorphous phase determination, a standard (finely ground CaF_2) of 20 wt. % was mixed with the sample. The results of X-ray diffraction of metakaolines are given in table 3.

Table 3. XRD phase analysis of metakaolines (values in %).

Phase	MK1	MK2
Mullite	7.7	1.0
Quartz	4.7	6.3
Anatase	1.2	1.9
Kaolinite	4.2	3.6
Illite/muscovite	4.9	4.1
Microcline	6.8	-
Amorphous phase	70.5	83.1

Differential thermal analysis (DTA) was done on the dried and ground samples. The measurement was performed on device Perseus STA 449 (Netzsch) in the temperature range 35–1000°C with heating rate 10°C/min in dynamic atmosphere of synthetic air (N_2/O_2 80/20).

Samples of autoclaved mixtures ground below 0.09 mm were embedded in epoxy resin, ground on SiC papers with several grits and polished with diamond paste on polishing machine Struers Labo-Pol 30. Polished samples were coated with a thin layer (30 nm) of carbon in vacuo. Analyses of zeolites were carried out on electron microprobe Cameca SX-100. The following phases were used as standards: sanidine (Si, Al, K), albite (Na), apatite (P), Rb-Ge glass (Rb), wollastonite (Ca), celestite (S, Sr), halite (Cl), barite (Ba), hematite (Fe), LiF (F), BN (N), rhodonite (Mn), ZnO (Zn), chalcopyrite (Cu), vanadinite (Pb). In point analyses, an acceleration voltage of 15 kV, beam current of 4 nA and beam diameter of ~1µm were used. Peak counting times (CT) were usually 20 s (for N 120 s), CT for each background were half of that of peak. Raw counts were converted to wt. % using standard PAP procedure. Contents of the elements mentioned above, which are not included in the tables, were consistently below the detection limits usually ranging between 300 and 1000 ppm, in case of Pb, Sr, N and F between 2000 and 3000 ppm. Photographs of the samples were taken in BSE.

Scanning electron microscopy (SEM) analysis was performed on a TESCAN MIRA3 device operating at 10 kV. Powder sample was placed under a sample holder where scans were acquired at different points within the sample with different magnifications.

2.1. Mixtures preparation

The required amount of metakaolin was weighed. Subsequently, sodium water glass was added in and the mixture was stirred in a mixer at medium speed for 5 minutes. Then, water (for MS1, SK1) was added to the mixture, followed by stirring for another 10 minutes. Mixtures MS2 and SK2 were stirred for 15 minutes. The resulting mixtures were casted into a 20 × 20 × 100 mm silicone mould and subsequently cured at room temperature for 48 hours (for mixtures marked with the letter "a"). After that, the samples were demoulded and autoclaved for 24 hours at 130°C. The mixtures marked "b" were autoclaved in the form of a gel/liquid mixture immediately after stirring. The mixture proportions are given in table 4.

Table 4. Mixture proportions, values in grams.

Sample	MK1	MK2	Sodium water glass	Water
MS1a		177.8	302.2	120
MS2a		222	377.8	
MS1b		177.8	302.2	120
MS2b		222	302.2	
SK1b	177.8		302.2	120
SK2b	222		377.8	

3. Results

3.1. XRD analysis

Autoclaved samples consist mainly of the amorphous component, which is most represented in the sample MS1a, the least in the samples SK (table 5). Except for the amorphous component, the following phases were identified by XRD analyses: quartz [SiO_2], muscovite [$KAl_2(AlSi_3O_{10})(OH)_2$], anatase [$TiO_2$], trona [$Na_3H(CO_3)_2 \cdot 2H_2O$], mullite [$3Al_2O_3 2SiO_2$], kaolinite [$Al_2O_3$ $2SiO_2 \cdot 2H_2O$], thermonatrite $Na_2CO_3 \cdot H_2O$] and microcline [$KAlSi_3O_8$]. The results of Rietveld analysis are summarised in table 5. Zeolites in the samples are represented by chabazite [$(Ca,K_2,Na_2)_2[Al_2Si_4O_{12}]_2 \cdot 12H_2O$], phillipsite [$Na_2[Al_2Si_3O_{10}] \cdot 2H_2O$], boggsite [$Ca_8Na_3(Si,Al)_{96}O_{192} \cdot 70H_2O$] and gmelinite [$Na_4(Si_8Al_4)O$ $_{24} \cdot 11H_2O$]. Also, Na-P1 zeolite [$Na_6Al_6Si_{10}O_{32} \cdot 12H_2O$] was identified, but it wasn´t possible to determine its percentage in the sample, whereas this phase is missing in the program evaluating the quantity of individual components in the sample. Highest zeolite contents (up to 29%) were found in SK samples (table 5).

3.2. DTA analysis

Autoclaved mixtures ground below 0.09 mm were analysed by differential thermal analysis and the measurement was performed in temperature range 35–1000°C with heating rate 10°C/min. The results of DTA analysis are shown in figure 1 and table 6. The dehydration temperature of zeolites is characterised by an endotherm in the range of 139 to 285°C (table 6, figure 1) and most closely corresponds to the published curves for chabazite [50]. Other zeolites mainly occur in small amounts in the samples and probably therefore no visible changes were recorded for these zeolites in the DTA, TG and DTG curves.

Table 5. XRD phase analysis of autoclaved samples.

Sample	MS1a	MS1b	MS2a	MS2b	SK1b	SK2b
Phase				%		
Quartz	5.3	3.1	4.4	4.2	0.7	0.9
Muscovite/illite	5.4	5.6	5.7	8.0	13.9	14.3
Phillipsite	2.0	3.8	6.6	1.9	6.9	8.7
Gmelinite	1.2	1.9	-	1.7	-	-
Mullite	0.9	0.7	0.7	1.1	-	-
Trona	3.9	5.3	3.1	1.9	-	-
Thermonatrite	0.5	-	-	-	-	-
Chabazite	2.7	7.0	2.8	6.9	16.5	17.3
Kaolinite	0.2	1.8	-	1.3	-	-
Zeolite Na-P1	?	?	?	?	-	-
Anatase	1.8	1.3	1.8	1.8	0.4	0.7
Boggsite	-	-	-	-	2.8	3.0
Microcline 8	-	-	-	-	11.9	10.2
Amorphous phase	76.1	69.5	74.9	71.2	46.9	44.9
Sum of zeolites	5.9	12.7	9.4	10.5	26.2	29.0

Not very significant endotherm in the range of 400–600°C visible on DTG curves for all samples and in the case of SK samples also on the DTA curve probably indicates the loss of crystal water from the zeolite structure. The exotherm in the range between 859 and 874°C, which is most evident in the SK2 mixture is apparently related to the destruction of the framework of the minerals and probably the transformation of zeolite into basic members of the plagioclase group. However, we have not verified this by XRD or WDX analysis. The total loss of mass of the samples heated to a temperature of 1000°C is in the range from 11.4 to 15.8%, whereas the highest loss of mass was analysed in samples SK1 and SK2, which contain the highest content of zeolites.

Table 6. Thermoanalytical data of autoclaved samples.

Sample	Temperature of dehydration (°C)	Total loss of mass up to 1000 °C (%)	Temperature of transformation, destruction of framework of zeolite (°C)
MS1a	139-201	14.15	
MS1b	144-195	13.75	
MS2a	150-194	13.02	
MS2b	140-182	11.43	
SK1b	145-200	15.08	
SK2b	152-285; 499	15.84	874

3.3. WDX analysis and SEM

Autoclaved samples were analysed by electron microprobe to identify the chemical composition of synthetised zeolites. It has been found that the synthesized zeolites by their chemistry do not correspond to the published data for individual minerals, but chemically they are in some cases close to them. The main monovalent cations that enter the structure of the zeolite is Na, significantly less K, Ca and NH_4. These constituents are isomorphically substituted in the crystal structure of the mineral. In addition, the zeolites also contain trace amounts of Fe, Mg and F (tables 7–8). Contents of other elements (S, P, Zn, Cu, Sr, Ba, Pb, Rb) are mostly negligible (below 0.02 apfu) (tables 7 and 8). It was found that the most abundant zeolite composition in the samples is close to chabazite-(Na), less phillipsite, which also corresponds to the results of XRD analyses (see table 5). The representative WDX analysis of chabazite-(Na) are shown in the table 7 and phillipsite in table 8. In the MS samples, a number of particles of unreacted components are visible in the backscattered electron images (figure 2 a–d), while in the SK samples the spherical aggregates of zeolites already significantly dominate (figure 2 e, f). Samples with the highest content of zeolite (SK1 and SK2) were studied using a scanning electron microscope. Zeolite aggregates have been found to be up to 30 micrometers in size. Two types of zeolites with different structures were found in the SK1 sample (figure 3 a, b).

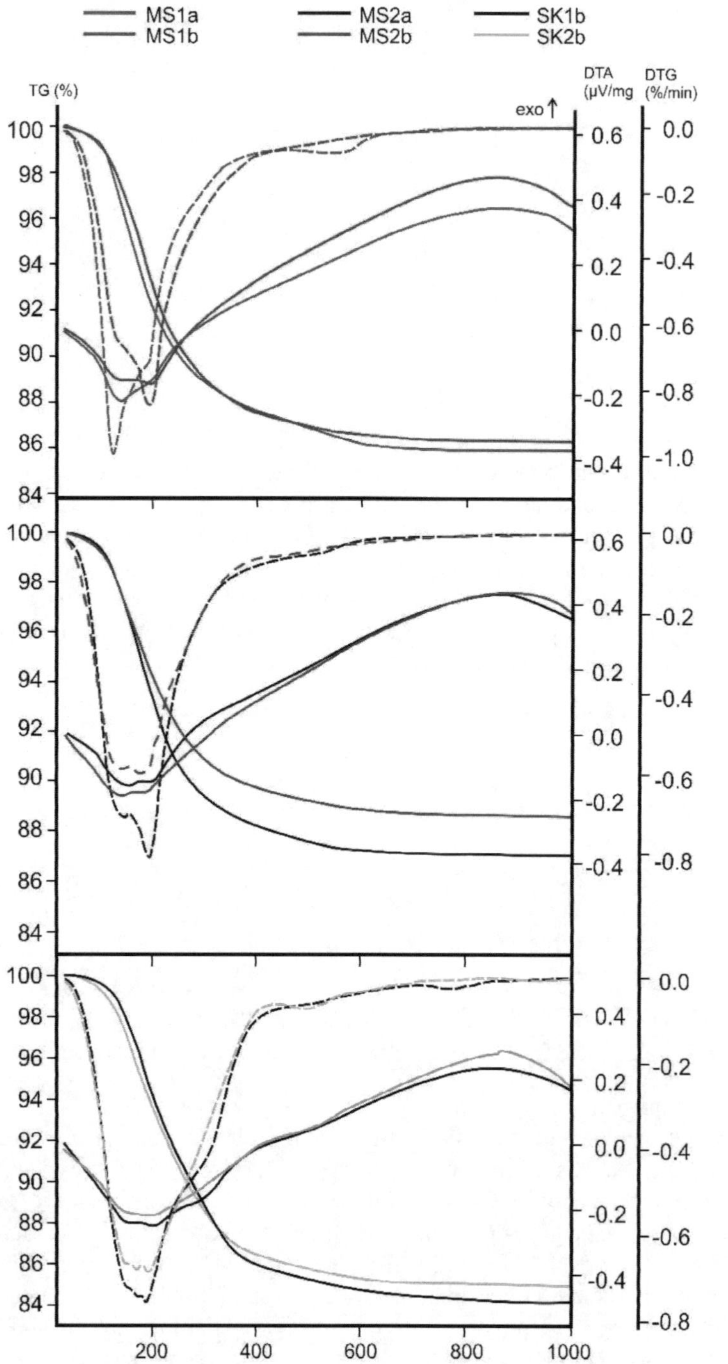

Figure 1. TG, DTG and DTA curves of autoclaved samples.

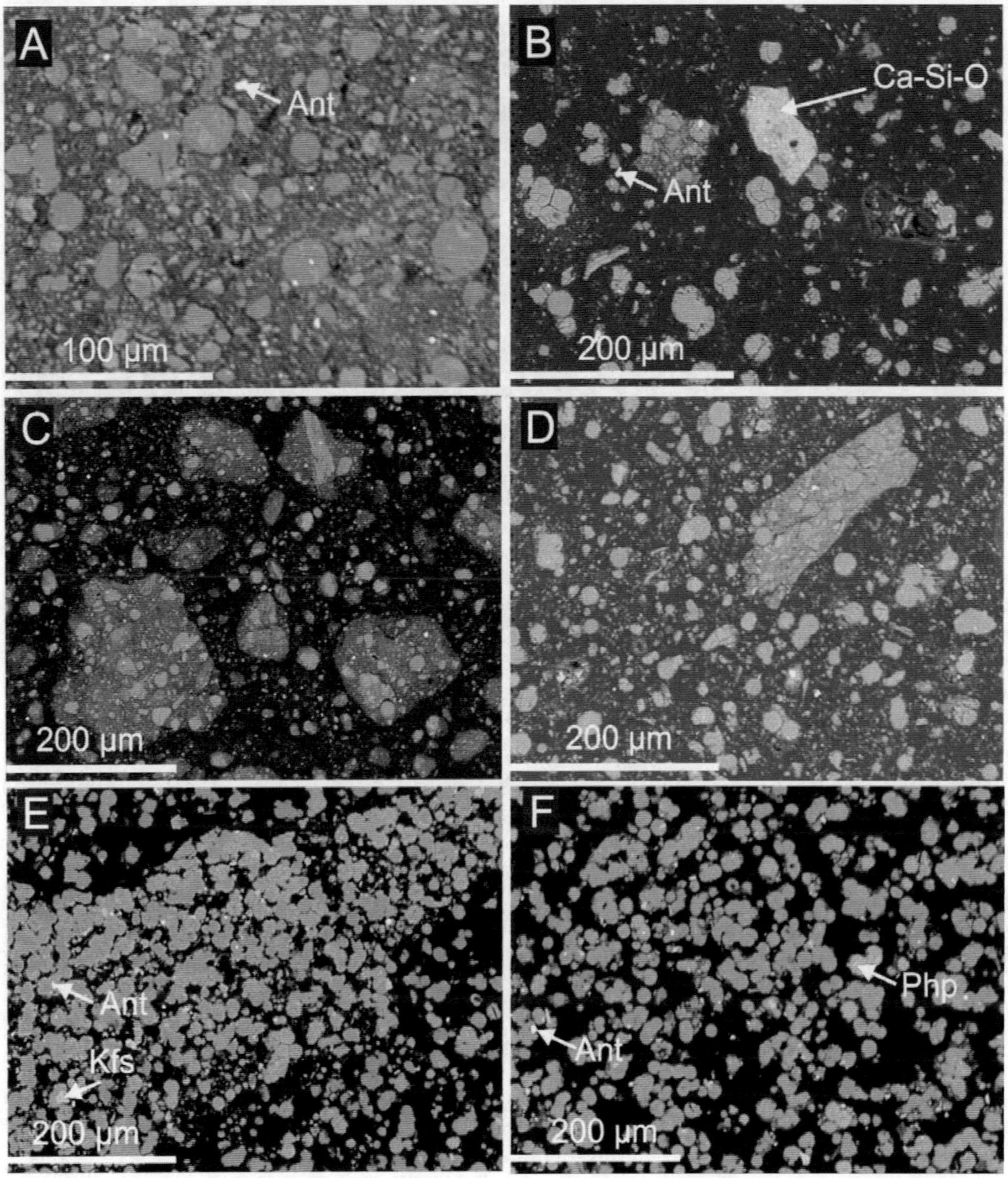

Figure 2. BSE images of autoclaved samples: a) MS1a, b) MS1b, c) MS2a, d) MS2b, e) SK1b, f) SK2b; spherical aggregates represent zeolites; ant – anatase, Kfs – potassium feldspar, php – phillipsite.

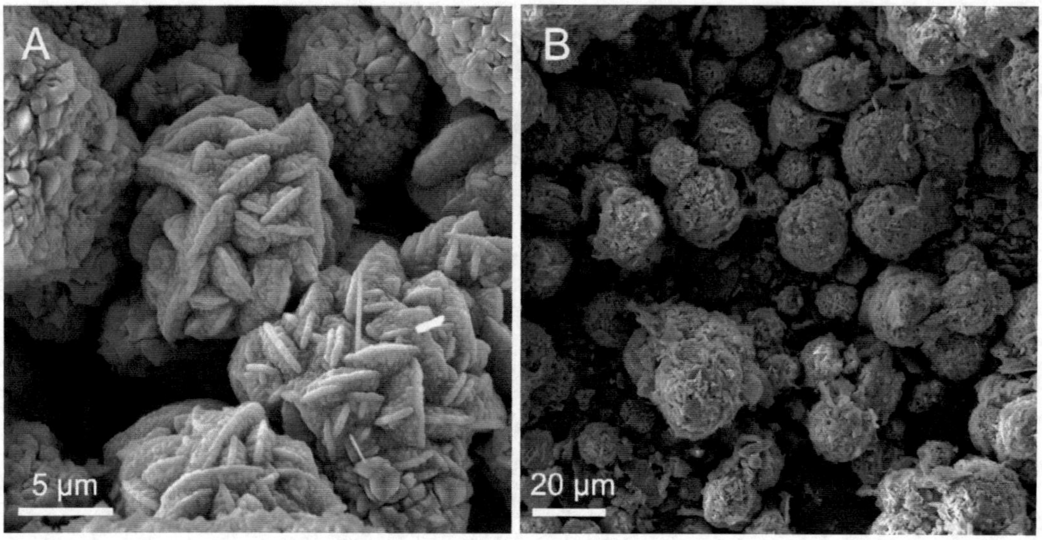

Figure 3. SEM images of zeolites: a) sample SK1b, b) sample SK2b.

4. Discussion

Crystallisation of zeolites takes place quickly in samples with MK1, zeolite content in these samples is up to 29%. This can be due to the fact that metakaolin MK1 contains a larger amount of K-feldspars and mullite, from which a large amount of zeolites can be formed by hydrothermal synthesis [51, 52]. Formation of thermonatrite and trona in MS samples probably relates to the reaction of unreacted Na with airborne CO_2 and water (in the case of trona). Higher amounts of zeolites were formed in samples autoclaved in the form of gel/liquid immediately after stirring the mixture. In the case of samples which were firstly cured at room temperature for 48 hours and subsequently autoclaved, the contents of formed zeolites are slightly lower (see table 5).

The used metakaolines contain only a small amount of Ca and K and water glass has a high Na content, it was probable that predominantly Na will enter the zeolite structure and thus Na-rich members of chabazite, phillipsite and gmelinite were formed. In the ternary diagram of Na_2O-K_2O-CaO we can see that sodium is the dominant element in the structure of most zeolites (figure 4). Only, in 6 point analysis, K-rich members of zeolites were identified (figure 4).

In some analysis, minute inclusion of a foreign phases, such as quartz or feldspar, were detected to be enclosed in zeolites. Thus, it is probable that some of the analyses correspond to a mixture of zeolite and unreacted components originated from raw materials and the contents of Si, Al and alkalis are significantly affected in these cases.

Compared to other research, in our work metakaolin with calcination temperature 750°C was mixed with water glass, in contrast, in Novembre et al [53] metakaolin calcined at 650°C was mixed with a solution of NaOH and SiO_2 in various proportions. The mixture has been cured at a temperature of 68°C and ambient pressure. Na-A zeolite began to form after about 8 hours, hydroxysodalite after 190 h and Na-P zeolite after 382 hours.

In work of Pereira et al [54], zeolite-A was synthesised from two types of calcined kaoline, white and red. The calcination temperature of kaoline was 600°C and the endurance was 12 hours. The used metakaolins had slightly higher contents of K_2O and TiO_2 compared to the metakaolines used in our research. The metakaolin was then mixed with NaOH in a NaOH/metakaolin ratio of 8:1. The mixture was heated at 80°C for 24 hours. The synthesised zeolites had the following chemical composition 43.52-44.35 wt. % SiO_2, 33.82-34.39 wt. % Al_2O_3, 19.82-20.59 wt. % Na_2O and traces of Ti, Fe, Mg, Ca, Mn and K. In the case of our research, zeolites with a more variable chemical composition are formed (see tables 7 and 8). As verified in our study, it´s not necessary to add NaOH as a reagent to the mixture. Water glass can also perform this function, but the growth rate of the zeolite crystals isn´t probably so fast.

Table 7. WDX analyses of zeolites with a chemical composition close to chabazite-(Na), apfu values are recalculated on the 24 atoms of O per formula unit; n.a. – not analysed, b.d.l. – below detection limit.

	MS1a		MS2a				SK1b				MS1b
SO_3	0.03	0.13	0.07	0.10	b.d.l.	0.03	0.02	0.03	0.03	0.06	0.10
P_2O_5	0.05	0.04	0.05	0.18	b.d.l.	b.d.l.	0.03	0.06	b.d.l.	0.03	0.02
SiO_2	42.01	45.63	55.01	48.28	55.33	50.43	56.05	54.77	55.56	52.84	50.21
Al_2O_3	17.61	20.67	21.89	22.96	22.31	22.67	22.56	21.61	20.82	22.19	18.92
MgO	0.07	0.10	b.d.l.	0.13	b.d.l.	b.d.l.	0.26	0.10	0.12	0.12	0.09
CaO	0.20	0.20	0.10	0.19	0.42	0.33	0.91	0.32	1.52	1.39	0.53
SrO	0.10	0.13	b.d.l.	0.15	0.09	0.20	b.d.l.	0.04	0.03	b.d.l.	0.04
BaO	b.d.l.	b.d.l.	0.06	0.09	b.d.l.	0.03	0.04	b.d.l.	b.d.l.	0.05	b.d.l.
FeO	0.35	0.41	0.41	0.56	0.24	0.23	0.31	0.30	0.30	0.35	0.10
MnO	0.01	0.04	b.d.l.	0.07	b.d.l.	0.03	b.d.l.	b.d.l.	b.d.l.	b.d.l.	0.06
Na_2O	7.27	7.30	11.10	9.93	8.14	9.89	6.84	9.82	7.96	9.20	10.63
K_2O	0.69	0.54	0.82	0.91	1.60	1.27	3.58	3.14	3.35	3.04	4.52
$(NH_4)_2O$	0.83	1.94	0.56	1.06	b.d.l.	b.d.l.	n.a.	n.a.	n.a.	n.a.	n.a.
Cl	0.17	0.12	0.05	0.09	0.03	b.d.l.	0.04	0.02	0.04	0.05	0.02
F	0.23	0.18	0.02	b.d.l.	b.d.l.	b.d.l.	0.21	b.d.l.	b.d.l.	b.d.l.	b.d.l.
SiO_2/Al_2O_3	2.39	2.21	2.51	2.10	2.48	2.22	2.48	2.53	2.67	2.38	2.65
S^{6+}	0.004	0.016	0.008	0.012	b.d.l.	0.004	0.001	0.003	0.003	0.007	0.012
P^{5+}	0.008	0.006	0.006	0.024	b.d.l.	b.d.l.	0.004	0.008	b.d.l.	0.004	0.003
Si^{4+}	8.032	7.838	8.174	7.694	8.152	7.822	8.079	8.201	8.332	8.022	8.130
Al^{3+}	3.968	4.184	3.834	4.312	3.874	4.144	3.892	3.813	3.680	3.970	3.610
Mg^{2+}	0.020	0.026	b.d.l.	0.030	b.d.l.	b.d.l.	0.056	0.022	0.027	0.027	0.022
Ca^{2+}	0.040	0.036	0.016	0.032	0.066	0.054	0.143	0.051	0.244	0.226	0.092
Sr^{2+}	0.012	0.012	b.d.l.	0.014	0.008	0.018	b.d.l.	0.003	0.003	b.d.l.	0.004
Ba^{2+}	b.d.l.	b.d.l.	0.004	0.006	b.d.l.	0.002	0.002	b.d.l	b.d.l.	0.003	b.d.l.
Fe^{2+}	0.056	0.058	0.050	0.074	0.030	0.030	0.037	0.038	0.038	0.044	0.014
Mn^{2+}	0.002	0.006	b.d.l.	0.010	b.d.l.	0.004	b.d.l.	b.d.l.	b.d.l.	b.d.l.	0.008
Na^{+}	2.694	2.430	3.172	3.068	2.324	2.974	1.958	2.851	2.314	2.708	3.337
K^{+}	0.168	0.118	0.156	0.184	0.300	0.252	0.666	0.600	0.641	0.589	0.934
NH^{4+}	0.183	0.425	0.112	0.246	b.d.l.	0.000	n.a.	n.a.	n.a.	n.a.	n.a.
Cl^{-}	0.056	0.034	0.012	0.024	0.008	0.002	0.010	0.005	0.010	0.013	0.005
F^{-}	0.140	0.098	0.010	b.d.l.	b.d.l.	b.d.l.	0.100	b.d.l.	b.d.l.	b.d.l.	b.d.l.
Si/Al	2.02	1.87	2.13	1.78	2.10	1.89	2.07	2.15	2.26	2.02	2.25
Si+Al	12.00	11.99	12.01	12.01	12.03	11.97	11.97	12.01	12.01	11.99	11.74

It is known from several studies that the Si/Al content in zeolite, which is used as a supplementary material, significantly affects the properties of the resulting cement. Further research shows that the zeolites with a higher Si/Al ratio have higher pozzolanic activity [55]. In our study, zeolites have a SiO_2/Al_2O_3 varying between 2.1 and 3.2 (see tables 7 and 8). These materials could have relatively high pozzolanic activity. Verification of these properties will be the subject of further research.

Table 8. WDX analyses of zeolites with a chemical composition close to phillipsite, apfu values are recalculated on the 32 atoms of O per formula unit; b.d.l. – below detection limit.

	MS2b			MS1b			SK1b	SK2b		
SO_3	b.d.l.	b.d.l.	b.d.l.	0.11	b.d.l.	b.d.l.	b.d.l.	0.07	b.d.l.	b.d.l.
P_2O_5	b.d.l.	b.d.l.	b.d.l.	b.d.l.	b.d.l.	0.08	0.08	b.d.l.	0.13	b.d.l.
SiO_2	50.60	51.42	49.71	57.74	57.14	54.10	55.64	52.45	54.01	52.89
Al_2O_3	20.24	20.82	21.54	22.25	21.84	16.96	22.12	22.93	19.96	20.42
MgO	0.11	0.15	b.d.l.	b.d.l.	0.05	0.07	0.39	0.13	0.29	0.08
CaO	0.51	0.89	0.22	0.51	0.42	0.49	0.60	0.59	0.13	b.d.l.
SrO	b.d.l.	b.d.l.	0.09	b.d.l.	0.16	b.d.l.	0.16	0.06	0.07	0.21
BaO	b.d.l.	b.d.l.	b.d.l.	0.05	b.d.l.	b.d.l.	0.07	0.09	b.d.l.	b.d.l.
FeO	0.26	0.17	0.07	0.58	0.22	0.13	0.32	0.27	0.34	0.24
Na_2O	7.67	6.91	9.94	2.29	2.39	3.70	5.05	3.66	4.78	6.16
Rb_2O	b.d.l.	b.d.l.	b.d.l.	b.d.l.	b.d.l.	b.d.l.	b.d.l.	0.10	b.d.l.	b.d.l.
K_2O	7.72	10.31	11.13	3.53	3.48	4.42	8.88	5.37	8.25	6.11
$(NH_4)_2O$	b.d.l.	b.d.l.	b.d.l.	0.21	b.d.l.	b.d.l.	b.d.l.	b.d.l.	b.d.l.	0.29
F	b.d.l.	b.d.l.	0.07	0.06	0.16	0.08	0.13	b.d.l.	0.05	0.21
SiO_2/Al_2O_3	2.50	2.47	2.31	2.60	2.62	3.19	2.52	2.29	2.71	2.59
S^{6+}	b.d.l.	b.d.l.	b.d.l.	0.016	b.d.l.	b.d.l.	b.d.l.	0.011	b.d.l.	b.d.l.
P^{5+}	b.d.l.	b.d.l.	b.d.l.	0.000	0.006	0.014	0.014	b.d.l.	0.023	b.d.l.
Si^{4+}	10.768	10.664	10.270	11.464	11.514	11.866	10.910	10.907	11.173	11.042
Al^{3+}	5.076	5.088	5.244	5.206	5.186	4.384	5.112	5.620	4.866	5.024
Mg^{2+}	0.034	0.046	b.d.l.	b.d.l.	0.016	0.022	0.114	0.040	0.089	0.025
Ca^{2+}	0.116	0.188	0.048	0.108	0.090	0.116	0.126	0.131	0.029	0.004
Sr^{2+}	0.000	0.000	0.010	b.d.l.	0.018	b.d.l.	0.018	0.007	0.008	0.025
Ba^{2+}	b.d.l.	b.d.l.	b.d.l.	0.004	b.d.l.	b.d.l.	0.006	0.007	b.d.l.	b.d.l.
Fe^{2+}	0.046	0.030	0.012	0.096	0.038	0.024	0.052	0.047	0.059	0.042
Na^+	3.164	2.778	3.982	0.882	0.934	1.574	1.920	1.476	1.917	2.493
Rb^+	b.d.l.	b.d.l.	b.d.l.	b.d.l.	b.d.l.	b.d.l.	b.d.l.	0.013	b.d.l.	b.d.l.
K^+	2.096	2.728	2.934	0.894	0.894	1.236	2.222	1.424	2.177	1.627
NH^{4+}	b.d.l.	b.d.l.	b.d.l.	0.096	b.d.l.	b.d.l.	b.d.l.	b.d.l.	b.d.l.	0.109
F	b.d.l.	b.d.l.	0.046	0.038	0.102	0.028	0.080	b.d.l.	0.033	0.139
Si/Al	2.12	2.12	1.96	2.20	2.22	2.71	2.13	1.94	2.30	2.20
Si+Al	15.84	15.75	15.51	16.67	16.70	16.25	16.11	16.53	16.04	16.07

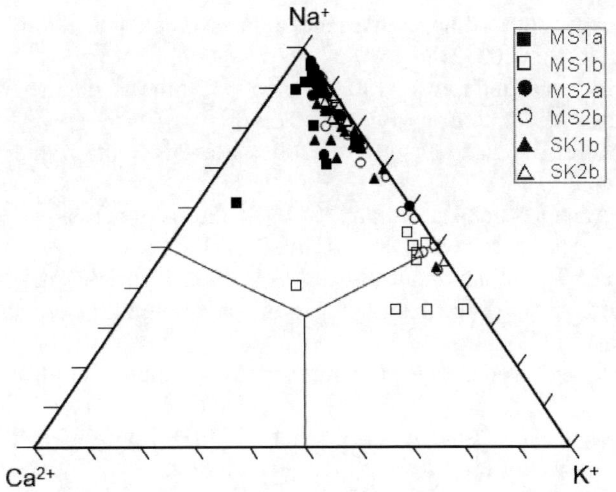

Figure 4. Ternary Ca^{2+}-K^{+}-Na^{+} plot for zeolites from studied samples.

5. Conclusion

The results presented in this publication represent initial data for the synthesis of zeolites from Czech metakaolins and water glass. Metakaolin Mefisto K05 proved to be a more suitable raw material in a mixture with water glass in the synthesis of zeolites. This is probably because the original raw material contains a higher amount of feldspar, from which zeolites are formed. A mixture with a zeolite content of up to 30% was prepared by hydrothermal synthesis and studied by XRD, DTA and WDX analysis. The synthesised zeolites correspond to their chemical composition close to Na-rich members of chabazite and phillipsite and the Si/Al varies between 2.1 and 3.2. Work on this topic continue and lead to developing of a high-grade zeolite material that could be used as a substitute for Portland cement in the concrete industry.

Acknowledgments

This work was created with the financial support of the Ministry of Industry and Trade within the institutional financing of long-term conceptual development of a research organisation Research Institute of Building Materials 2018–2022, EXPRO 2019 project of the Czech Science Foundation (No. 19-29124X). Electron-microprobe analytical work was supported by the Ministry of Culture of the Czech Republic (long-term project DKRVO 2019-2023/1.II.c to National Museum, 00023272).

References

[1] Duffin Ch 2006 Silicate minerals: An Overview (BMS Occasional Paper) p 29
[2] Coombs D S, Alberti A, Armbruster T *et al* 1997 Recommended nomenclature for zeolite minerals: Report of the subcomittee on zeolites of the international mineralogical association, commision on new minerals and mineral names *The Canadian Mineralogist* **35** pp 1571–1606
[3] Munthali M V, Elsheikh M A, Johan E and Matsue N 2014 Proton Adsorption selectivity of Zeolites in Aqueous Media: Effect od Si/Al Ratio of Zeolites *Molecules* **19** pp 20468–20481
[4] da Silva L S, Araki C A, Marcucci S M P, da Silva dos S T and Arroyo P A 2019 Desilication of ZSM-5 and ZSM-12 Zeolites with different Crystal Size: Effect on Acidity and Mesoporous Initiation *Materials Research* **22** 2
[5] Xu R, Pang W, Yu J, Huo Q and Chen J 2007 Chemistry of zeolites and related porous materials: Synthesis and structure
[6] Pauliš P 2014 Nejzajímavější mineralogická naleziště Čech – zeolity a doprovodná mineralizace (KUTTNA) p 280
[7] Marantos I, Christidis G E and Ulmanu M 2011 Zeolite formation and deposits *Natural Zeolites Handbook* pp 19–36

[8] Bish D L and Boak J 2001 Clinoptiolite-Heulandite Nomenclature *Reviews in Mineralogy and Geochemistry* **45** 1 pp 207–216
[9] Mgbemere H E, Ekpe I C and Lawal G I 2017 Zeolite synthesis, characterization and application areas: A review *Int. Res. J. Environmental Sci.* **6** 10 pp 45–59
[10] Baerlocher Ch, Meier W M and Olson D H 2001 Atlas of zeolite framework types (5th revised edition)
[11] Abrishamkar M, Azizi S N and Kazemian H 2010 Synthesis of borosilicate MFI type zeolite using different aging techniques *Chem. Met. Alloys* **3** pp 12–17
[12] Bu X, Feng P, Gier T E, Zhao D and Stucky G D 1998 Hydrothermal synthesis and structural characterization of zeolite-like based on galium and aluminium germanates *J. Am, Chem. Soc.* **120** pp 13389–13397
[13] Dubanská V 1992 Hydrothermal preparation of vanadium zeolites. Ceramics-Silikáty **36** pp 31–38
[14] Hemmann F, Hackebeil J, Lisner A and Mertens F 2018 Rapid, Non-Toxic Synthesis of Beta Zeolites *ChemRxiv* **6** pp 1–4
[15] Özkan V, Aydin Z and Ozkan A 2018 Comparison of hydrothermal method and ultrasonic method in zeolite synthesis *NESciences* **3** 3 pp 282–291
[16] Guo H, Tang L, Yan B, Wan K and Li P 2017 NaA zeolite derived from blast furnace slag: its application for ammonium removal *Water Sci. Technol.* **76** 5–6 pp 1140–1149
[17] Wajima T 2014 Synthesis of Zeolite from blast furnace slag using alkali fusion with addition od EDTA *Advanced Materials Research* 1044–1045 pp 124–127
[18] Majdinasab A and Yuan Q 2019 Microwave synthesis of zeolites from waste glass cullet using landfill leachate as a novel alternative solvent *Materials Chemistry and Physics* **223** pp 613–622
[19] Musyoka N M 2009 Hydrothermal synthesis and optimisation of zeolite Na-P1 from South African coal fly ash (Msc. Thesis, University of Nairobi) p 196
[20] Zou J, Guo Ch, Wei C, Li F and Jiang Y 2016 Synthesis of pure Na-X and Na-P zeolite from acid-Extracting residues of CFB fly ash by a single-step hydrothermal method *Materials transactions* **57** 5 pp 726–731
[21] Temuujin J, Okada K and Mackenzie J D 2002 Zeolite formation by hydrothermal treatment of waste solution from selectively leached kaolinite *Materials Letters* **52** pp 91–95
[22] da Silva Filho S H, Bieseki L, da Silva A R, Maia A B B, San Gil R A S and Pergher S B C 2015 Synthesis of Zeolite A empolying Amazon kaolin waste *Cerâmica* **61** 360 pp 409–413
[23] Gougazeh M and Buhl J-Ch 2014 Syntesis and characterization of zeolite A by hydrothermal transformation of natural Jordanian kaolin. Journal of the Association of Arab Universities for Basic and Applied Sciences **15** 1 pp 36–42
[24] Kwakye-Awuah B, Von-Kiti E, Buamah R, Nikrumah I and Williams C 2014 Effect of Crystalization Time on the hydrothermal synthesis of zeolites from Kaolin and bauxite *International Journal of Scientific and Engineering Research* **5** 2 pp 734–741
[25] Sánchez Hernández R 2018 Complete transformation of aluminium waste into zeolite and its use in the removal of pollutants from aqueous solution (PhD. Thesis, Consejo Superior de investigaciones Cientificas, Centro Nacional de Investigaciones Metalúrgicas) p 254
[26] Garcia G, Cardenas E, Cabrera S, Hedlund J and Mouzon J 2016 Synthesis of zeolite Y from diatomite as silica source *Microporous and Mesoporous Materials* **219** pp 29–37
[27] Sanhueza V, Kelm U, Cid R and López-Escobar L 2004 Synthesis of ZSM-5 from diatomite: A case of zeolite synthesis from a natural material *Journal of Chemical Technology and Biotechnology* **79** 7 pp 686–690
[28] Santos E, Sousa A B, Leite R Ch N, Laborde H, Menezes R R and Rodrigues M 2014 Preparation of Zeolite MCM-22 Using the Rice Husk Ash as Silica Source *Materials Science Forum* **805** pp 646–650
[29] Pangan N, Gallardo S, Gaspillo P, Kurniawan W and Hinode H 2015 Hydrothermal synthesis and

characterization of zeolite NAA from corncob ash *Proceedings of research World International Conference, Osaka, Japan* pp 22–26

[30] Khodabandech S and Davis M E 1996 Synthesis of a heulandite-type zeolite by hydrothermal conversion of zeolite P1 *Chem. Commun.* pp 1205–1206

[31] Ahmedzeki N S, Joodee A M, Abbas M N and Jaed Y M 2018 Waste resource utilization for zeolite a synthesis *Journal of Chemical Technology and Mettalurgy* **33** 2 pp 11–16

[32] Setiadji S, Nurazizah S I, Sundari C, Darmalaksana W, Nurbaeti D F, Noviati I, Azizah T B N, Adburrahman D and Ivansyah A L 2018 Synthesis of zeolite ZSM-11 using bamboo leaf as silica source *IOP Conference Series Materials Science and Engineering* **434** 1 p 6

[33] Klamrassamee T, Pavasant P and Laosiripojana N 2010 Synthesis of Zeolite from Coal Fly Ash: Its Application as Water Sorbent. *Engineering Journal* **14** 1 pp 37–44

[34] Orjioke M N, Uchechukwu O, Igwe Ch N and Ajah N U 2016 Synthesis and characterization of zeolite and ots application in adsorption of nickel from Aqueous solution *Journal of pharmaceutical, chemical and biological sciences* **3** 4 pp 592–600

[35] Flanigen E 1980 Molecular sieve zeolite technology: The first twenty-five years *Pure and applied chemistry* **52** 9 pp 2191–2211

[36] Flanigen E 2001 Zeolites and molecular sieves: An historical perspective *Studies in Surface Science and Catalysis* **137** pp 11–35

[37] Feng C, Khulbe K C, Matsuura T, Farnood R and Ismail A F 2015 Recent progress in Zeolite/Zeotype membranes *Journal of membrane Science and Research* **1** pp 49–72

[38] Tsitsishvili V, Nanuli D, Urotadze S, Alelishvili M, Mirdzveli N and Nijaradze M 2017 Ion Exchange properties of Georgian natural zeolites *Chemistry Journal of Moldova* 12

[39] Izci E and Citak A 2007 Dielectric behavior of the catalyst zeolite NaY *Turkish Journal of Chemistry* **31** 5 pp 523–530

[40] Cairo P C, de Armas J M, Artiles P T, Martin B D, Carrazana R J and Lopez O R 2017 Effects of zeolite and organic fertilizerz on soil quality and yield of sugarcane *Australian Journal of Crop Science* **11** 6 pp 733–738

[41] Ghadamnan E, Nabavi S R and Abbasi M 2019 Nano LTE Zeolite in water softening process: synthesis, characterization, kinetic studies and process optimization by response surface methodology (RSM) *J. Water Environ. Nanotechnol.* **4** 2 pp 119–138

[42] Narayanan S, Batchelor W and Webley P A 2011 Advances of zeolites in papermaking to create value added products-A revieew

[43] Schneider A F, de Almeida D S, Yuri F M, Zimmermann O F, Gerber M W and Gewehr C E 2016 Natural zeolites in diet or litter of broilers *Nutrition and Metabolism* pp 257–263

[44] Emam E and Yehia S 2017 Performance of concrete containing zeolite as a supplementary cementitious material *International research journal of enginnering and technology* **4** 12 pp 1619–1625

[45] Joshaghani A 2016 The Effects of Zeolite as Supplementary Cement Material on Pervious Concrete *2016 International Concrete Sustainability Conference* (Washington DC) p 11

[46] Sičáková A, Špak M, Kozlovská M and Kováč M 2017 Long-Term Properties of Cement-Based Composites Incorporating Natural Zeolite as a Feature of Progressive Building Material *Advances in Materials Science and Engineering* p 8

[47] de la Cruz J C, del Campo J M and Colorado D 2014 A much better concrete with zeolite additions-state of the art review *2nd International Conference on Emerging Trends in Engineering and Technology* (London) pp 153–159

[48] Tanijaya J and Hardjito D 2008 Experimental Study on the Use of Natural zeolites as Partial Replacement for Cement in Concrete in Yeong-Bin Yang et al. (ed) *The Eleventh East Asia Pacific Conference on Structural Engineering and Construction* (EASEC-11)

[49] ČSN 72 0102 Základní postup rozboru silikátů - Stanovení ztráty sušením

[50] Foldvári M 2011 Handbook of thermogravimetric system of minerals and its use in geological practice *Occasional Papers of the Geological Institute of Hungary* **213** 106

IOP Conf. Series: Materials Science and Engineering 1205 (2021) 012021 doi:10.1088/1757-899X/1205/1/012021

[51] Su S, Ma H and Chuan X 2016 Hydrothermal synthesis of zeolite A from K-feldspar and its crystallization mechanism *Advanced Powder Technology* **27** 1 pp 139–144
[52] Yuan J, Yang J, Ma H and Chang Q 2017 Preparation of Zeolite F as Slow Release Fertilizers from K-Feldspar Powder *Chem. Select* **2** 10722
[53] Novembre D, di Sabatino B, Gimeno D and Pace C 2011 Synthesis and characterization of Na-X, Na-A and Na-P zeolites and hydroxysodalite from metakaolinite *Clay Minerals* **46** pp 339–354
[54] Pereira P M, Ferreira B F, Oliveira N P et al 2018 "Synthesis of Zeolite A from Metakaolin and Its Application in the Adsorption of Cationic Dyes" *Appl. Sci.* **8** 4 p 608
[55] Liguori B, Aprea P, Gennaro B, Iucolano F, Colella A and Caputo D 2019 Pozzolanic Activity of Zeolites: The Role of Si/Al Ratio. *Materials* **12** 4231

IOP Conf. Series: Materials Science and Engineering 1205 (2021) 012022 doi:10.1088/1757-899X/1205/1/012022

Testing of biocidal properties of thermal insulation system during material life cycle

I Kukletova[1] **and I Chromkova**[1]

[1] Research Institute for Building Materials, Hnevkovskeho 65, 617 00 Brno

Email: kukletova@vustah.cz

Abstract. With the growing number of realized thermal insulation systems, fast growing number of buildings with facades attacked by microorganisms occurs. Such surface attack can occur even in the first years after thermal insulation completion, and it can be a serious problem in a very short time. Problem is both in realization and maintaining of thermal insulation, and directly in the used material. The paper presents results of laboratory experiments on resistance to biological attack, carried out for five commercial materials available on the Czech market. Before the experiments, materials were subjected to accelerated ageing for 4 and 10 years. Resistance testing comprises regulated application of algae and mould strains on samples and incubation in a defined environment. Materials resistant after ageing simulation of the longer time period were found as well as materials susceptible to attack already after completion of a facade set.

1. Introduction

Currently there is a growing number of buildings, where the facade surface is attacked by microorganism biofilm growth. This phenomenon directly connects to the trend of thermal insulation of buildings. Biofilms on facades are taken mainly as an aesthetic problem, however, it is primarily a technological and healthy risk [1, 2, 3].

Microorganisms get on the facade surface from the air. Under suitable conditions, the microorganisms reproduce creating visible coloured films (biofilms). Biofilm composition depends on climatic conditions and a lot of other factors, among them humidity is the most important. As nutrients serve microparticles of organic matter brought with dust or contained in plaster [1]. Possibilities and degree of building deterioration by microorganisms depend mainly on the chosen material, performance of problematic details, building location, insolation, type of surrounding vegetation, pH of adjacent soil and, last but not least, on operation and way of using the building [2]. The principal condition for the growth and reproduction of microorganisms on building facades is humidity. Mainly at thermally insulated buildings, there is moisture occurrence in upper layers of thermal insulation. The reason is water condensation on the building surface, which is not heated up by the heat escaping from the building. On the north- and north-west building sides also water from rainfalls contributed to the moisture content in the given material. The sun in these places does not dry the surfaces, as well as no biocidal effects of UV radiation can have an impact [2, 4].

There are many facade materials on the market, however, not all of them are resistant to biological components. Methods for testing of facade system resistance to biological components are not embodied in the Czech standards. Based on current regulations and experiments from other countries, screening methods were developed to evaluate the material resistance to algae and moulds. These methods were verified on real samples. Methods include mainly macroscopic observation of algae and mould growth on building materials in defined conditions.

In practice, it is found out that facade materials lose their properties with time, mainly by weather conditions. The retaining time of biocidal properties differs according to the composition and producer of the given material. Therefore, as a suitable tool, a simulation of accelerated ageing precedes complex testing of resistance to biological components. Simulation of accelerated ageing was developed in the Research Institute for Building Materials as a part of a complex method for durability evaluation of fibre-cement composite elements. It includes cyclic changing of dominant factors of the summer and winter seasons in the Czech Republic. Weather mode is patterned on long-term average annual data from 1961 to 2010. A correlation of results from accelerated ageing and long-term storage of samples in real conditions was achieved on a long-term basis.

For the study, five commercial materials available on the market in the Czech Republic were selected, the materials were tested nameless, and exposed to accelerated ageing for 4 and 10 years.

2. Methodology

For sample preparation there were purchased five different sets (kits) of ETICS, from which samples of facade systems were prepared in this composition: plaster, undercoat, glass fabric, reinforcing layer and insulation material – polystyrene. For testing purposes, the samples were cut to dimensions 5 × 5 × 2.3 cm. The samples were tested after 14 days of curing and after accelerated ageing for 4 and 10 years.

Method of accelerated ageing was developed for the equipment (figure 1), where the effect of seasons of the year on tested samples was simulated. One model year takes one week, and it includes summer conditions – heavy rainfall, 10 times. After each rainfall, there is a break followed by radiant heating by using IR lamps. The radiant heat is supplemented with xenon discharge tubes, which create UV radiation. Effect of radiant heat is applied 10 times. Radiant heat output is set to ca. 70°C on the sample surface. The effect of radiant heat and UV radiation is determined by the total average duration in the summer season. This time is divided into 10 parts to change with rainfalls always with a break for a couple of minutes. The model summer season is complemented with the winter season model when the tested material is exposed to freezing at -20°C followed by thawing in water of 20°C. The winter season is defined as one freezing/thawing cycle per year.

Figure 1. Equipment for accelerated ageing.

Subsequently, the samples were tested to resistan to green algae and moulds. Methodology for testing resistance to green algae is based on the standard ČSN EN 15458 (672032) Paints and varnishes – Laboratory method for testing the efficacy of film preservatives in coating against algae [5]. For inoculation of samples, a suspension of two algal types was used: *Nostoc commune*, *Stichococcus bacillaris*. The algal cultures were purchased from the collection of autotrophic organisms CCALA Třeboň (492 *Stichococcus bacillaris* Naegeli, 118 *Nostoc commune* Vaucher). The collection strains were cultivated, and inoculation suspension according to the standard with approximate density 10^6 CFU/ml was prepared. For the purpose of algae cultivation as well as the testing BBM (Bold-Basal/Bristol Medium) is used as the medium [6]. The tested samples were laid on the agar surface in individual Petri dishes. The samples and the agar surface were covered with a thin layer of mixed algal culture with a density of 10^6 CFU/ml; 10 ml of the culture was used to one Petri dish, the culture was slowly inoculated on the sample, and subsequently, it runs down on agar. The samples were tested in three parallel assessments. Incubation occurred under these conditions: 23±2°C, 1000±200 lx, light regime 16 hrs light/8 hrs dark. The whole test took 35 days. During the test and after the test, the algae growth was visually evaluated on the samples, both at a macroscopic level and with using stereomicroscope. Material resistance degree is described by percentual estimation of the sample covering and by the scale: 0 no growth, 1 slight growth, 2 medium growth, 3 intense growth.

Methodology of testing the resistance to moulds is based on the standard ČSN EN 15457 (672031) Paints and varnishes - Laboratory method for testing the efficacy of film preservatives in a coating against fungi [7], EN 60068-2-10 (345791) Environmental t-sting - Part 2-10: Tests - Test J and guidance: Mould growth [8] and ČSN 72 4310 (724310) Testing of mould proofness of building products and materials [9]. The samples are laid on the surface of Malt agar in individual Petri dishes. The test took place in three parallel assessments. Inoculation mould culture consisted of 6 collection strain types (from CCM Czech Collection of Microorganisms): *Aspergillus niger* CCM 8155, *Penicilium brevicompactum* CCM 8040, *Paecilomyces variotii* CCM F-398, *Alternaria alternata* CCM F-397, *Trichoderma viride* CCM F-728, *Cladosporium cladosporioides* CCM F-348). Suspension is prepared by washing off spores from a sporulated culture with mineral solution (3 g $NaNO_3$, 1 g KH_2PO_4, 0.5 g $MgSO_4$. 7 H_2O, 0.5 g KCl, 0.01 g $FeSO_4$ .7H_2O to 1000 ml distilled water, a sterilization 30 min at 0.1 MPa). The spore concentration in the solution should be $1–2 \times 10^6$ spores/ml. Inoculation took place under sterile conditions. The samples are covered by a thin layer of the mixed mould culture. 1 ml suspension was used on the sample surface. Incubation took place under these conditions: temperature 24±2°C and relative humidity 96±2%. Petri dishes with samples are laid in a sealed cultivation vessel, where the high relative humidity is kept by adding saturated solution Na_2SO_4. The whole test took 28 days. During the incubation, the cultivation vessel is opened once a week in a biohazard box, air is changed and the state of each sample is recorded. During the test and after the test the mould growth is visually assessed on the samples and surrounding nutrient soil. Mould spores in positive examinations should germinate to 6 days from the start of incubation. The resistance is assessed as follows: fungicide material – mould is not generated, inhibition zone is on the subjacent agar, fungistatic material – mould is not generated on the sample, colonies are not visible even by the microscope, growing on soil, non-resistant material – report 1–5 according to the scale (1 – slight growth, visible only by microscope, scattered colonies, 2 – slow growth – numerous small colonies, up to 25% of the surface, 3 – intensive growth, up to 50% of the surface, 4 – very intensive growth, up to 75% of the surface, 5 – total growth, 100% of the surface).

3. Results and discussion

Test results are stated in Table 1 and 2 and supplemented by photo documentation – figure 2 and 3. From the collected data, it can be concluded that some materials keep relatively good resistance even after accelerated ageing for 10 years, on the other hand, other materials lose their biocidal properties already in the first four years of exposure to climatic conditions. At one material, the problem with resistance to mould attack was recorded already at time = 0.

IOP Conf. Series: Materials Science and Engineering 1205 (2021) 012022 doi:10.1088/1757-899X/1205/1/012022

Table 1. Test results of resistance to green algae.

Test indication	**Growth degree of algal cells on samples – only plaster surface (35 days of incubation)**			**Average covering of observed sample surface by algal cells after test finish (35 days of incubation) /%**		
	Accelerated ageing			Accelerated ageing		
Sample indication	0 years	4 years	10 years	0 years	4 years	10 years
1	0	0	1	0	0	5
2	0	1	2	0	2	20
3	0	0	1	0	0	1
4	0	3	3	0	70	70
5	0	1	2	0	1	30

Table 2. Test results of resistance to moulds.

Test indication	**Growth degree of moulds on samples – only plaster surface (28 days of incubation)**			**Average covering of observed sample surface by mould growth after test finish (28 days of incubation) /%**		
	Accelerated ageing			Accelerated ageing		
Sample indication	0 years	4 years	10 years	0 years	4 years	10 years
1	2	2	3	10	20	30
2	1	3	4	2	30	60
3	1	1	3	1	2	30
4	1	4	5	2	75	90
5	1	3	2	1	25	20

Figure 2 shows photos of the sample 2 after finishing individual tests. This sample proves average results; however, it can be seen that in the test with moulds after accelerated ageing for four years already, a noticeable mould attack occurs. In figure 3, we can see a relatively large growth of algae and mould at the sample 4 already after four years of accelerated ageing, this sample proved biocidal properties at time = 0.

Figure 2. Photodocumentation of test results of the sample 2 – number of years of accelerated ageing is given under individual photos.

Figure 3. Extreme growths at the sample 4 after accelerated ageing for four years – exposure to algae (left), exposure to moulds (right).

Testing of facade systems across the Czech market confirmed real-life experience that a lot of materials do not meet announced properties or they lose them after a very short time. There were found materials resistant even after accelerated ageing for a longer time as well as materials prone to attack already after realization of a facade set.

Within a choice of suitable material during realization of thermal insulation, it is very difficult to predict expiration of biocidal properties of facade systems without material testing, therefore, the European commission for standardization prepares the standard „Assessment of the resistance of insulation products against mould development" (WI = 00088439) that should be issued in the next year. This standard is based on experts′ knowledge and on already standardized methods for other materials. It includes mainly methods ASTM C1338 [10], ASTM G21 - 96 [11] and other methods arising from them.

Verdier et al. [12] present a summary of individual methods of testing resistance to moulds with a comparison of selected experimental parameters – microbial strains, humidity, inoculation technique, incubation conditions, incubation duration. In conclusion, he states that although these tests ensure sustainable and accelerated microbial growth and provide a view of material resistance, in reality, they differ from a real condition of building material exposure to the environment. Within interpretation, it is really necessary to take it into consideration, yet the test results provide us with important data on material behaviour and, last but not least, it is possible to compare the resistance of different materials. In literature [13], testing of resistance to moulds in combination with accelerated ageing is described in the scope of using biocomposites from agricultural materials. Accelerated ageing, in this case, represents exposure to unfavourable conditions 30°C and 90% RH for three months. Besides visual assessment, also digital image correlation (DIC) with the aim to accelerate evaluation and increase objectivity rate is employed. Regarding the fact that plant materials are the topic, natural contamination of materials with moulds is used.

Parracha et al. [14, 15] describe tests of various ETICS systems with a purpose to compare mould growth on the surface and other properties, mainly moisture transport properties (capillary water absorption, water vapour permeability, water absorption under low pressure, and drying kinetics), thermal conductivity, surface properties (colour, gloss, and roughness). The results show that deeper knowledge of correlation among ETICS properties can effectively contribute to the effective assessment and long-term durability of these systems.

In the first study [14], Parracha deals with the correlation of hygrometric properties – capillary absorption and drying capacity with testing of mould growth at 7 ETICS sets. One group of samples was inoculated by *Aspergillus niger, Penicillium funiculosum* and *Aureobasidium pullulans* and the second one was retained with natural contamination. Incubation took place under these conditions: temperature 22±1°C and 70±5% RH. The samples showed no growth for four weeks, after that, a growth occurred, mainly at samples with insulation layer from cork. The fact that at no ETICS a growth occurred on the surface after four weeks of testing the author explained by a presence of biocide in the surface layer.

The newest study [15] describes testing of 12 various ETICS sets to above mentioned technological properties and tendency to moulds. The samples are inoculated only by *Aspergillus niger* and *Penicillium funiculosum.* The incubation took place under these conditions: temperature 22±1°C and 70±5% RH. The results of resistance to moulds are similar to the first paper. In both studies [14, 15], new produced systems are examined that had not been exposed to any type of ageing or weather effects, therefore, it is necessary to complete it in the following research. The author stated that although the ETICS samples showed no or small growth of moulds at the end of test, according to further observations, it turned out that some of them are more sensitive to attack. It supports the necessity of regular maintenance of ETICS, including testing of a biological attack.

Kvande et al. [16] describe the testing of ETICS in Norway with regard to system defects. A part of laboratory experiments is also accelerated ageing of materials, both exposure to natural weather conditions and testing by means of laboratory equipment. In the equipment, it is possible to revolve the samples among four climatic zones. Exposure is always 1 hr in each climatic zone in the given order. These factors affect the samples: UV and IR light of various intensity and wavelength, heating up to 63°C, freezing at -20°C and thawing at 20°C, water sprinkling 15 $dm^3/(m^2h)$. The purpose of this accelerated ageing is to compare the properties of various materials. Tests, therefore, do not provide exact data on lifetime expectancy expressed with a number of years. Detailed research of this method and other methods of accelerated ageing is done by Jelle [17], the method is always chosen by the material and climate. The described methods correspond with the method used for our purposes.

4. Conclusions

The aim of the study was to evaluate five commercial facade sets across the Czech market with respect to resistance to biological attack. The materials were tested after their preparation and they were subjected to accelerated ageing for 4 and 10 years before the testing. Tests of resistance were carried out by means of controlled application of algal and mould strains on the samples and incubation in a defined environment. Some materials were resistant to accelerated ageing for a longer period, some of the materials prone to attack already after a facade set preparation. The results confirmed experience from real life that a lot of materials do not meet the declared properties, or they lose these properties after a very short time. According to our methodology, accelerated ageing of materials appears to be a suitable supplementary method for testing the resistance of such materials.

Acknowledgements
This paper was elaborated with the institutional support for long-term development of research organizations by the Ministry of Industry and Trade of the Czech Republic.

References

[1] Wasserbauer R 2001 *Biological deterioration of buildings* (in Czech) 1. ed (Praha: ABF, Nakladatelství ARCH) p 257
[2] Büchli R and Raschle P 2004 *Algae and fungi on façade* (in Czech) 2. ed (Ostrava: Nakladatelství MISE, s.r.o.) p 108
[3] Flannigan B, Samson R A and Miller J D 2001 *Microorganisms in home and indoorworkenvironments: diversity, health impacts, investigation and control* (BocaRaton [Fla.]: CRC Press) p 529
[4] Adan O C G and Samson R A 2011 *Fundamentals of mold growth in indoor environments and strategies for healthy living* (Wageningen: Wageningen Academic Publishers) p 523
[5] ČSN EN 15458 (672032)2015 Paints and varnishes – Laboratory method for testing the efficacy of film preservatives in coating against algae(Praha, Úřad pro technickou normalizaci, metrologii a státní zkušebnictví) p 16
[6] Kukletova I and Chromkova I 2019 Proposal of methodology for testing the resistance of building materials against mold infestation *IOP Conf. Series: Materials Science and Engineering* **583** 012029
[7] ČSN EN 15457 (672031) Paints and varnishes - Laboratory method for testing the efficacy of film preservatives in a coating against fungi
[8] ČSN EN 60068-2-10 (345791) 2008 Environmental Impact Testing - Part 2–10: Tests - Test J and Instructions: Mold Growth (Praha,Český normalizační institut) p 36
[9] ČSN 72 4310 (724310) Testing of mould proofness of building products and materials (in Czech)
[10] ASTM C1338 2014 Standard test method for determining fungi resistance of insulation materials and facings (ASTM International) p 3
[11] ASTM G21 - 96 2015 Standard practice for determining resistance of synthetic polymeric materials to fungi (ASTM International) p 6
[12] Verdier T, Coutand M, Bertron A and Roques CH 2014 A review of indoor microbial growth across building materials and sampling and analysis methods. *Building and Environment* **80** pp 136–149
[13] Viel M, Collet F, Lecieux Y, François M, Colson V, Lanos C, Hussain A and Lawrence M 2018 Resistance to mold development assessment of bio-based building materials *Composites Part B* doi: https://doi.org/10.1016/j.compositesb.2018.09.063
[14] Parracha J, Cortay A, Borsoi G, Veiga R and Nunes L 2020 Evaluation of ETICS Characteristics that Affect Surface Mould Development *Current Topics and Trends on Durability of Building Materials and Components* Serrat C, Casas J R and Gibert V (Eds) URL https://www.scipedia.com/public/Parracha_et_al_2020a
[15] Parracha J L, Borsoi G, Flores-Colen I, Veiga R, Nunes L, Dionísio A, Gomes M G and Faria P 2021 Performance parameters of ETICS: Correlating water resistance, bio-susceptibility and surface properties, *Construction and Building Materials* **272** 121956
[16] Kvande T, Bakken N, Bergheim E and Thue J V 2018 Durability of ETICS with Rendering in Norway – Experimental and Field Investigations, *Buildings* 2018 **8(7):93** https://doi.org/10.3390/buildings8070093
[17] Jelle B P 2012 Accelerated climate ageing of building materials, components and structures in the laboratory *J Mater Sci* **47** pp 6475–6496 https://doi.org/10.1007/s10853-012-6349-7

Fibre-reinforced cementitious composite: parameter identification using Ohno shear beam test

D Lehký[1], R Pukl[2], D Novák[1] and M Lipowczan[1]

[1] Institute of Structural Mechanics, Faculty of Civil Engineering, Brno University of Technology, Brno, Veveří 95, 602 00 Brno, Czech Republic
[2] Červenka Consulting, Na Hřebenkách 55, Praha 5, 150 00 Prague, Czech Republic

Email: novak.d@fce.vutbr.cz

Abstract. Computational-experimental methodology based on artificial neural networks used to identify the material parameters of fibre-reinforced cementitious composite is presented and applied for Ohno shear beam test. The aim is to provide techniques for an advanced assessment of the mechanical fracture properties of these materials, and the subsequent numerical simulation of components/structures made from them. The paper describes the development of computational and material models utilized for efficient material parameter determination with regards to a studied composite. The data is used in inverse analysis based on artificial neural networks together with sensitivity analysis which plays an important role in the process. Developed software tool FRCID-S is also briefly described.

1. Introduction

A key aspect when performing nonlinear fracture mechanics modelling of structures made of fibre-reinforced composite (FRC) is the knowledge about the tensile softening model and its parameters, as well as the corresponding fracture energy which dissipates during the cracking process, as was emphasized e.g. in [1, 2]. The aim of the performed research is to deepen existing knowledge about the behaviour of the studied FRC material especially in relation to its resistance to crack propagation. There are several recent applications of artificial neural networks (ANN) for FRC or general engineered cementitious materials. For example, in [3], selected properties of engineered cementitious composites were predicted using an ANN which had been trained using data collected from the literature. Another practical example can be found in [4], where an ANN was employed for the prediction of tensile and compressive strengths at 28 days of hardening. The utilization of the ANN in this paper is quite different. Here, selected material parameters of FRC are identified based on inverse analysis and the virtual simulation of a fracture test. The ANN is used here as a surrogate model describing the inverse relationship between structural response parameters and material parameters.

The research consists of several parts: First, a suitable constitutive law for FRC cementitious composites was developed within ATENA nonlinear fracture mechanics software [5]. The computational model was verified using experimental data of Ohno shear beam test [6]. Second, sensitivity analysis was performed. It showed the importance of individual material parameters related to the response of specimens tested under laboratory conditions in specific shear test configuration. With the help of the sensitivity analysis results, software for the identification of material parameters was developed. This program, which is called FRCID-S, implements the inverse analysis method based on an artificial neural network (ANN) [7, 8] in combination with efficient statistical simulation.

2. Shear test

Identification of material parameters is performed by the inverse analysis method with the use of laboratory test results. Here, the focus is on the shear test of the so-called Ohno beam with dimensions and load as shown in figure 1. The inner part of the Ohno beam (marked in grey in the figure) is subjected to pure shear stress (the bending moment is zero in the middle). Flexural reinforcement is placed in the outer parts of the beam to prevent bending failure. Average shear stress is calculated as:

$$\tau_{av} = \frac{2}{3}\tau_{max} = \frac{2}{3}\frac{F}{2A} = \frac{F}{3A}, \tag{1}$$

where F is the magnitude of the force load and A is the cross-sectional area of the beam at the point of the largest shear stress (the inner part of the beam). The shear strain corresponds to the oblique direction indicated by the green arrow in figure 1. Strain was determined in the laboratory using the Optotrak system, see [6] for details.

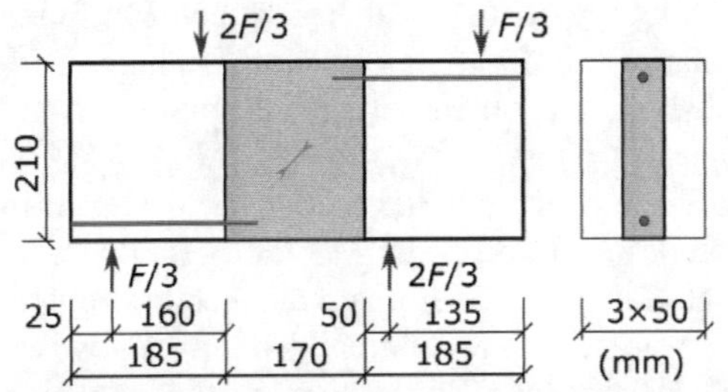

Figure 1. A schematic view of Ohno beam subjected to shear test.

When creating and setting the computational model of the Ohno beam tested in shear, the result of the laboratory test in the form of a shear stress vs. shear strain diagram was used, see figure 2. A DRECC composite containing 7% by volume of Dramix steel fibres 6 mm long and 0.15 mm in diameter was tested. The ratio of water to cement in the mix was 0.27. It also contained the following additives – silica fume (ratio to cement weight was 0.20), coupling agent (0.0025) and super-plasticizer (0.03). Specimen was moist cured for 24 hours before demoulding, and subsequently cured in water for four weeks. The age at testing was 6 weeks. DRECC composite shows relatively high tensile and compressive strength and average ductility. The value of the strength at the formation of the first crack was 4.9 MPa, the corresponding relative deformation 0.17%, the value of the strength at maximum load was 9.89 MPa, the maximum deformation was 0.7%. Further details on the material used, its properties and details of laboratory testing can be found in [6].

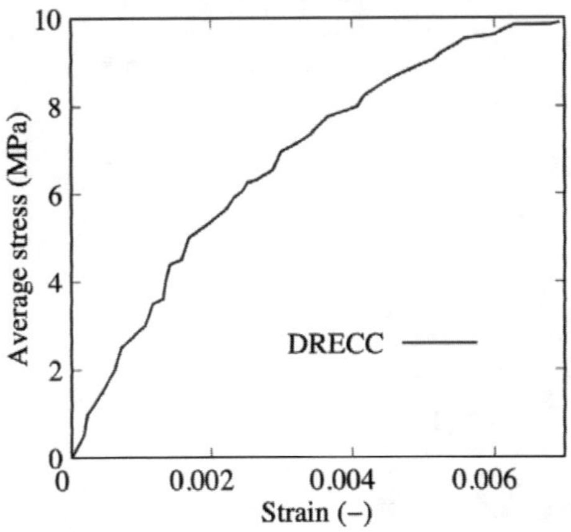

Figure 2. Experimental diagram of average shear stress vs. shear strain obtained from the Ohno beam test.

3. Computational model

The nonlinear numerical simulation of the shear test (figure 1) is performed using the finite element method (FEM) based program ATENA [5]. The constitutive relationship at the material point plays the most important role in nonlinear calculation and determines to what extent the computer model captures the actual behaviour of the structure. Since fibre-reinforced concrete is a complex heterogeneous material with a strongly nonlinear response, the nonlinear material model "3DNonlinearCementitious2User" is used to realistically calculate the response of this composite, capturing all important aspects of its behaviour in tension, compression, and shear.

The concrete tensile damage model is based on nonlinear fracture mechanics combined with the crack band model and the concept of smeared cracks. The main material parameters here are the tensile strength and the shape of the softening function (curve characterizing the crack opening versus the residual tensile stress). The real discrete crack is replaced in the model by a band of localized strains. The relative strain corresponding to the crack width is related to the finite element size. The softening function in the material law for the smeared crack model must therefore be determined individually for each finite element so that the prescribed relationship for the crack width is maintained. Only such a model based on an energy formulation ensures the objectivity of the solution and independence from the finite element network [5].

The proposed function to capture the tensile softening phase of the studied composite is shown in figure 3a. Its shape tries to capture the behaviour of the composite according to the modelled test. The shape of the function is described by other three parameters T_1–T_3 in addition to the tensile strength, as can be seen in figure 3a. The response of the material to shear loading and failure is described by two functions, which characterize the reduction of the shear stiffness (see figure 3b) and the reduction of the shear strength of the material (see figure 3c) due to cracks. Bilinear functions were chosen as the basic model with the parameters S_1, S_2, R_1, R_2 defining the breakpoint of the shear softening function and shear strength reduction function (figure 3b,c).

The developed FEM model of the Ohno shear beam is shown in figure 4. The forces are applied in accordance with the real test at the load locations. The nonlinear solution is initially implemented using a modified Newton-Raphson iterative method, which is later changed to the Arc-length method in order to more accurately determine the ultimate capacity of the model.

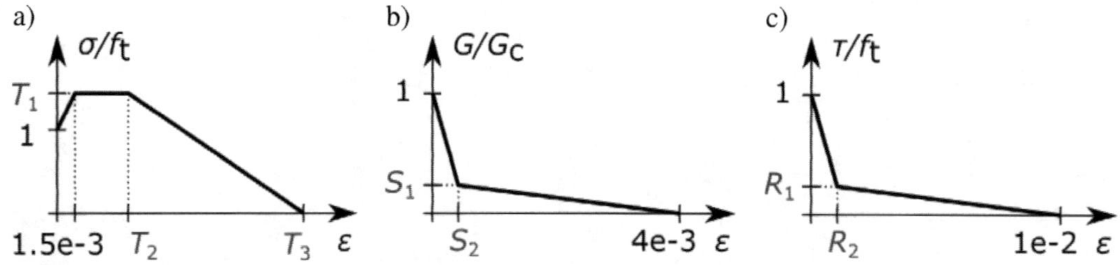

Figure 3. Proposed functions for the studied composite: a) tensile softening, b) shear stiffness reduction, c) shear strength reduction.

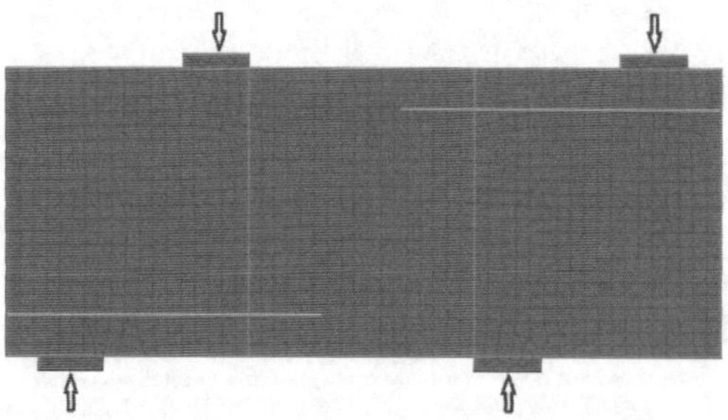

Figure 4. FEM computational model of Ohno shear beam.

4. Sensitivity analysis

A parametric study was performed to understand the influence of the individual input parameters of the material model on the response of the beam in shear. The aim was to identify the so-called dominant parameters, to identify parameters that have no effect on the response in the test and thus exclude them from the inverse analysis, to determine the realistic ranges of the individual parameters and to propose a set of suitable response parameters to be used as input to the inverse analysis and to allow unambiguous identification of the material parameters of interest. The results of the sensitivity analysis yielded the following conclusions relevant to the identification presented below:

- Tensile strength is one of the dominant parameters. An increase in its value leads to an increase in the overall resistance of the specimen. As the value increases, the initial linear phase of the diagram also increases.
- The parameter T_1 related to the activation of fibres in the cement matrix shows a similar effect as the tensile strength, however, the length of the linear part of the diagram does not change with increasing value – the point where the change in stiffness due to microcracks occurs is maintained.
- Parameters T_2 and T_3 do not affect the overall load capacity of the specimen but have some influence on the value of the maximum relative deformation achieved. This effect is more pronounced for T_2, while for T_3 a reasonable maximum limit is soon reached.
- Of the shear parameters S_1, S_2, R_1 and R_2, the only parameter with some degree of influence on the response shape is the S_1 parameter related to shear stiffness reduction. With a change in its value, a slight change in the stiffness of the final part of the diagram occurs when shear failure of the specimen occurs.
- The modulus of elasticity E has an expected effect on the stiffness of the whole specimen and thus on the initial slope of the whole response diagram.

For the purpose of creating a stochastic model for the inverse analysis and subsequent identification, the results show that due to the relatively unambiguous linear dependence of the elastic modulus E on the value of the initial stiffness of the test specimen and its easy determination independently of the other parameters, it can be excluded from the identification by neural networks and thus reduce the complexity of the inverse problem. Parameters that have no effect on the resulting shear response can be excluded from the identification and considered as default values in the calculation. Four parameters of the material model remain to be identified – tensile strength f_t, tensile parameters T_1, T_2 and shear parameter S_1.

Based on the interpretation of the results in terms of the selection of a set of appropriate response parameters for the inverse analysis, it appears appropriate to use the following input parameters for the neural network: force value at shear strain of 0.002, force value at shear strain of 0.005, maximum force value from the entire test, and shear strain value corresponding to the maximum force.

5. Methodology and software implementation

An inverse analysis method based on artificial neural network (ANN) is used to identify the parameters of the studied composite. The theoretical details and identification procedure is beyond the scope of this paper and can be found in papers [7, 8, 9]. The basic cornerstone of the method is an artificial neural network that transforms the measured data obtained from fracture experiments into the desired material parameters. The parameters of the network that determine its behaviour, such as synapses and biases, must first be set in a phase called network training. Network training is performed with the help of a training set, which is a set of known ordered input-output pairs. The individual elements of the training set are obtained using a virtual stochastic simulation of the test using the ATENA model described above. The first step is to create a stochastic model for the identified material parameters, the set of which was generated from the above sensitivity analysis. The parameters are assigned a probability distribution and the means (initial rough parameter estimates), and standard deviations are defined. A suitable distribution is a uniform distribution, which allows the boundaries of each parameter to be defined and leads to a uniform distribution of realizations over the range of their values. From the defined variables, random realizations of all parameters are generated using the Latin hypercube sampling simulation method. Here, 30 simulations were used. Note that the importance of training sample preparation was previously emphasized and tested by Tong and Liu [10], including the LHS scheme. In spite of the fact that these authors concluded that number-theoretic methods appear to be the most efficient, the LHS scheme also provided very good results. Moreover, our focus on LHS is also determined by the general applicability of this small-sample simulation technique for practical statistical, sensitivity and reliability analyses in many fields of engineering.

A set of numerical FEM analyses is then performed with the generated realizations to provide a random response of the test specimen. With a suitably chosen stochastic model, the obtained responses well represent the range of experiments performed and contribute to good convergence in network training and subsequent identification. The random response in the form of average shear stress vs. shear strain diagrams are obtained. Appropriate response parameters are determined using sensitivity analysis, in this case the shear stresses at shear strains of 0.002 and 0.005, the shear stresses corresponding to the maximum load capacity and the shear strain corresponding to the maximum load capacity.

The random realizations of the material parameters together with the random response parameters of the structure are used as the training set for the neural network whose parameters are optimized. Once the network is properly trained to respond to the input information (random response) with the corresponding output (material parameters), it can be simulated with the experimental response as its input, see ANN in figure 5. The output of the simulation is the set of identified material parameters. With the obtained parameters, a validation numerical simulation can be performed, which should lead to the best possible agreement with the experimental measurements.

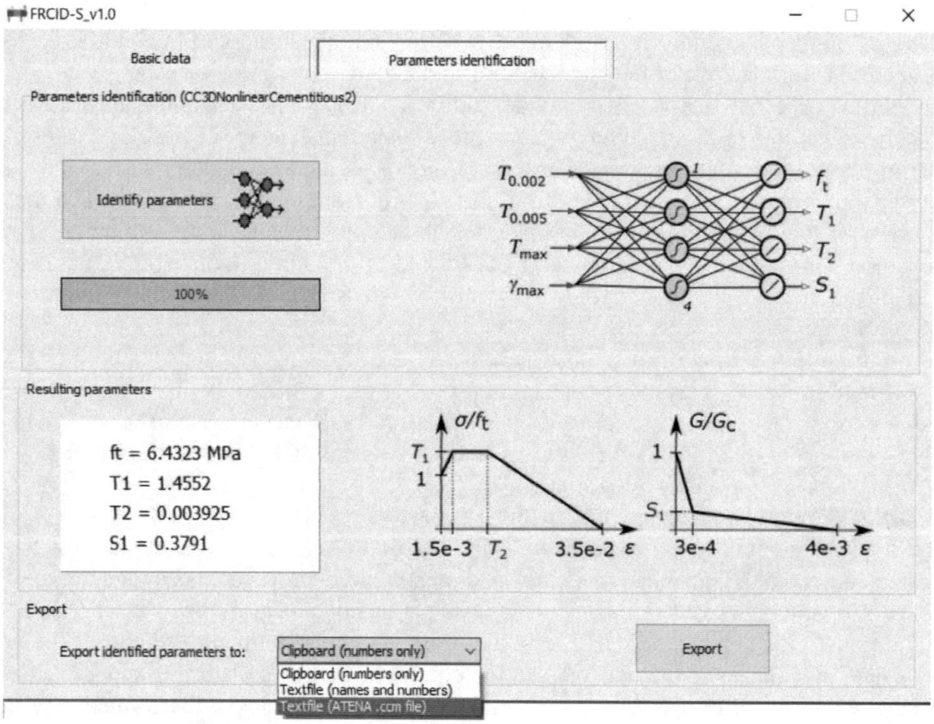

Figure 5. FRCID-S software – parameter identification and export.

The above-described method of parameter identification, which combines nonlinear simulations with the training of an ANN, is relatively time consuming and of high complexity. Therefore, the whole procedure was implemented in FraMePID-3PB software [8] and successfully used for the material parameter identification of plain concrete. The software has now been modified and updated for fibre-reinforced concrete tested in shear, including the tensile softening function and two shear functions, as described in the section on the computational model. A screenshot of the first version of FRCID-S software is shown in figure 5.

6. Verification and application

Initial verification of the software and the implemented ANN was performed using a set of four randomly simulated shear responses of specimens made of the studied cement composite. Given the knowledge of the input set of material parameters, it was possible to directly compare their values with those obtained by identification (this is not possible for purely experimental data). A graphical comparison for the four tested specimens can be seen in figure 6, while a numerical comparison is given in table 1. Furthermore, a classical comparison was made in the form of the resulting responses of the specimens, i.e. shear stress–strain diagrams, see figure 7. The conclusions of the sensitivity analysis, i.e. the significantly greater dominance of the tensile strength f_t and parameter T_1 of the tensile softening model compared to the other two identified parameters T_2 and S_1, were confirmed.

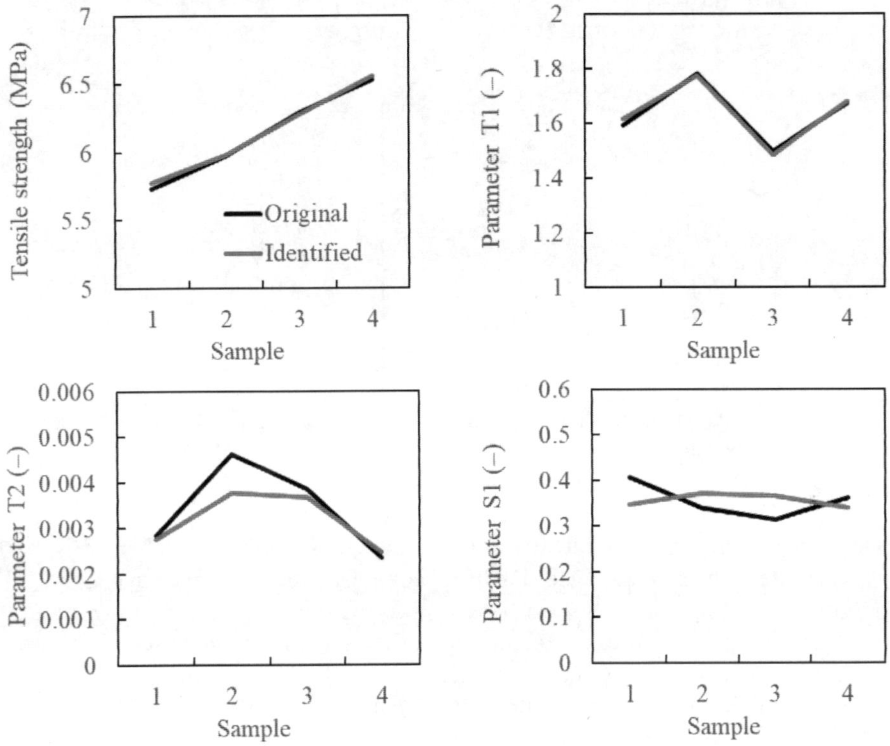

Figure 6. Comparison of original and identified responses for 4 test samples.

Table 1. Set of original and identified (in brackets) values of material model parameters for 4 test specimens.

Specimen	f_t (MPa)	T_1 (–)	T_2 (–)	S_1 (–)
1	5.728 (5.775)	1.592 (1.616)	2.83×10^{-3} (2.76×10^{-3})	4.07×10^{-1} (3.46×10^{-1})
2	5.971 (5.981)	1.780 (1.774)	4.62×10^{-3} (3.76×10^{-3})	3.39×10^{-1} (3.71×10^{-1})
3	6.292 (6.275)	1.496 (1.484)	3.84×10^{-3} (3.67×10^{-3})	3.14×10^{-1} (3.65×10^{-1})
4	6.531 (6.557)	1.671 (1.680)	2.34×10^{-3} (2.47×10^{-3})	3.61×10^{-1} (3.38×10^{-1})

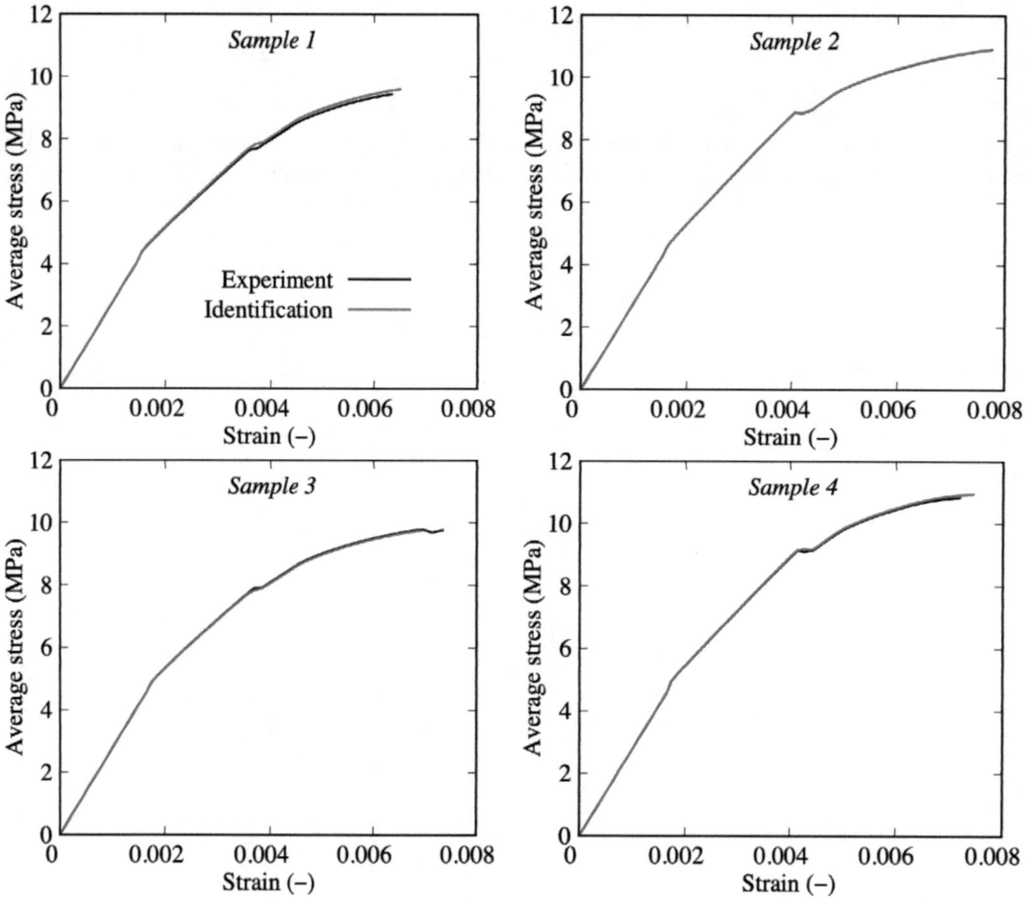

Figure 7. Comparison of original and identified responses for 4 test samples.

The proposed material parameter identification method was applied together with the FRCID-S software to a selected specimen of the DRECC composite described in section 2. The result of the laboratory test in the form of shear stress–strain diagram is shown in figure 2. Compared to the simulated data in figure 7, the experimental curve shows a greater instability in the measurements, as is typical in real tests of fibre-reinforced concrete.

The input data in the form of the shear stress–strain diagram was loaded to the FRCID-S, processed and passed to the ANN for simulation, resulting in a set of corresponding material parameters, which are listed in table 2.

Table 2. A set of identified values of material model parameters for a sample of DRECC composite.

Specimen	f_t (MPa)	T_1 (–)	T_2 (–)	S_1 (–)
DRECC	6.432	1.455	3.93×10^{-3}	3.79×10^{-1}

Verification of the obtained results was then carried out using numerical simulation of the test with the identified parameters. A comparison of the original experimental response with the response obtained by numerical calculation is shown in figure 8. It is evident from the comparison that the identified parameters define the tested composite very well and the resulting simulated response fits the experimental test well. Thus, the proposed ANN is a good surrogate model for the inverse problem of identifying the material parameters of the studied cement composite with Dramix steel fibres.

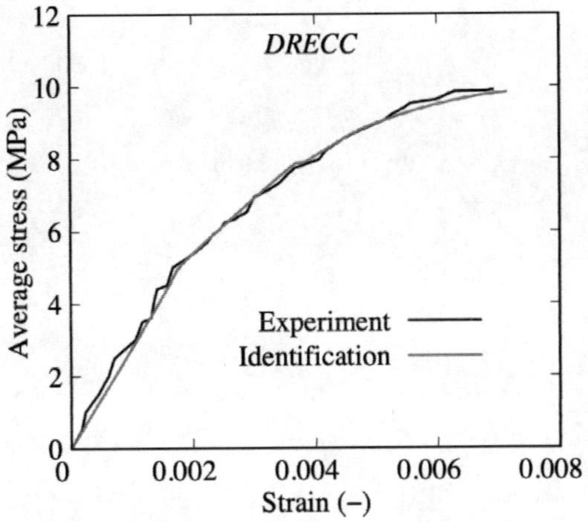

Figure 8. Comparison of experimental and identified responses for the tested DRECC composite.

7. Conclusions

The paper describes a methodology and software tool which can be routinely used for the indirect determination of the mechanical fracture parameters of FRC. Experimental results of shear test beam were used for development of the user-friendly software consisting of a predefined and trained artificial neural network for the fast identification of material parameters. The whole concept is based on a combination of the statistical simulation method, finite element modelling and ANN.

Based on sensitivity analysis four parameters of the material model of FRC were identified as key parameters for inverse analysis – tensile strength f_t, tensile parameters T_1, T_2 and shear parameter S_1. The results of identification and the subsequent performance of identified parameters on tested FRC samples proved the ANN's efficiency and ability to identify material parameter values leading to the accurate simulation of the response of the studied composite with Dramix steel fibres.

Acknowledgment

This work has been supported by project "MUFRAS" No. 19-09491S, awarded by the Czech Science Foundation of the Czech Republic.

References

[1] Bažant Z P and Planas J 1998 *Fracture and size effect in concrete and other quasibrittle materials* (CRC Press, Boca Raton, Florida)
[2] Bažant Z P and Becq-Giraudon E 2002 *Cem. Concr. Res.* **32** 529
[3] Shi L, Lin S T K, Lu Y, Ye L and Zhang Y X 2018 *Constr. Build. Mater.* **174** 667
[4] Hossain K M A, Anwar M S and Samani S G 2018 *Neural Comput. Appl.* **29** 631
[5] Červenka V, Jendele L and Červenka J 2019 *ATENA program document. – Part 1: theory* (Cervenka Consulting, Prague, Czech Republic)
[6] Li V C, Mishra D K, Naaman A E, Wight J K, LaFave J M, Wu H C and Inada Y 1994 *Adv. Cem. Based Mater.* **1**(3) 142
[7] Novák D and Lehký D 2006 *Eng. Appl. Artif. Intell.* **19** 731
[8] Lehký D, Keršner Z and Novák D 2014 *Adv. Eng. Softw.* **72** 147
[9] Novák D, Lehký D and Pukl R 2020 *Adv. Mater. Lett.* **11**(3) 20031488
[10] Tong F and Liu X L 2004 *Proc. of the 4th Int. Conf. on Engineering Computation Technology* 131 (Civil-Comp Press, Stirling, Scotland)

AUTHOR INDEX